NONLINEAR SOLID MECHANICS FOR FINITE ELEMENT ANALYSIS: DYNAMICS

Designing engineering components that make optimal use of materials requires consideration of the nonlinear static and dynamic characteristics associated with both manufacturing and working environments. The modeling of these characteristics can only be done through numerical formulation and simulation, which requires an understanding of both the theoretical background and associated computer solution techniques.

By presenting nonlinear solid mechanics, dynamic conservation laws and principles, and the associated finite element techniques together, the authors provide in this second book a unified treatment of the dynamic simulation of nonlinear solids.

Alongside a number of worked examples and exercises are user instructions, program descriptions, and examples for two MATLAB computer implementations for which source codes are available online.

While this book is designed to complement postgraduate courses, it is also relevant to those in industry requiring an appreciation of the way their computer simulation programs work.

JAVIER BONET is a Professor of Engineering, currently Deputy Vice-chancellor for Research and Enterprise at the University of Greenwich and formerly Head of the College of Engineering at Swansea University. He has extensive experience of teaching topics in structural mechanics and dynamics, including large strain nonlinear solid mechanics, to undergraduate and graduate engineering students.

ANTONIO J. GIL is a Professor in the Zienkiewicz Centre for Computational Engineering, Swansea University. He has numerous publications in various areas of computational mechanics, with specific experience in the field of large strain nonlinear mechanics.

RICHARD D. WOOD is an Honorary Research Fellow in the Zienkiewicz Centre for Computational Engineering at Swansea University. He has over 20 years' experience of teaching the course Nonlinear Solid Mechanics for Finite Element Analysis at Swansea University, which he originally developed at the University of Arizona.

NONLINEAR SOLID MECHANICS FOR FINITE ELEMENT ANALYSIS: DYNAMICS

Javier Bonet
University of Greenwich

Antonio J. Gil
Swansea University

Richard D. Wood
Swansea University

CAMBRIDGE
UNIVERSITY PRESS

CAMBRIDGE
UNIVERSITY PRESS

University Printing House, Cambridge CB2 8BS, United Kingdom

One Liberty Plaza, 20th Floor, New York, NY 10006, USA

477 Williamstown Road, Port Melbourne, VIC 3207, Australia

314–321, 3rd Floor, Plot 3, Splendor Forum, Jasola District Centre, New Delhi – 110025, India

79 Anson Road, #06–04/06, Singapore 079906

Cambridge University Press is part of the University of Cambridge.

It furthers the University's mission by disseminating knowledge in the pursuit of education, learning, and research at the highest international levels of excellence.

www.cambridge.org
Information on this title: www.cambridge.org/9781107115620
DOI: 10.1017/9781316336083

First published 2021

Printed in the United Kingdom by TJ Books Limited, Padstow Cornwall

A catalogue record for this publication is available from the British Library.

Library of Congress Cataloging-in-Publication Data
Names: Bonet, Javier, 1961– author. | Gil, Antonio J. author. | Wood, Richard D., 1943– author.
Title: Nonlinear solid mechanics for finite element analysis : dynamics / Javier Bonet, University of Greenwich, Antonio J. Gil, Swansea University, Richard D. Wood, Swansea University.
Description: New York : Cambridge University Press, 2020. | Includes bibliographical references and index.
Identifiers: LCCN 2020020233 (print) | LCCN 2020020234 (ebook) | ISBN 9781107115620 (hardback) | ISBN 9781316336083 (ebook)
Subjects: LCSH: Continuum mechanics. | Finite element method. | Dynamics.
Classification: LCC QA808.2 .B657 2020 (print) | LCC QA808.2 (ebook) | DDC 531/.3–dc23
LC record available at https://lccn.loc.gov/2020020233
LC ebook record available at https://lccn.loc.gov/2020020234

ISBN 978-1-107-11562-0 Hardback

To Catherine, Clare, Doreen, and our children

A fragment from the poem
"An Essay on Criticism"
by Alexander Pope (1688–1744)

A little Learning is a dang'rous Thing;
Drink deep, or taste not the Pierian Spring:
There shallow Draughts intoxicate the Brain,
And drinking largely sobers us again.
Fir'd at first Sight with what the Muse imparts,
In fearless Youth we tempt the Heights of Arts,
While from the bounded Level of our Mind,
Short Views we take, nor see the lengths behind,
But more advanc'd, behold with strange Surprize
New, distant Scenes of endless Science rise!

CONTENTS

PREFACE

This book extends a previous text entitled *Nonlinear Solid Mechanics for Finite Element Analysis: Statics* (referred to in this book as NL-Statics) into the field of transient dynamic processes. As discussed in the NL-Statics volume:

> A fundamental aspect of engineering is the desire to design artifacts that exploit materials to a maximum in terms of performance under working conditions and efficiency of manufacture. Such an activity demands an increasing understanding of the behaviour of the artifact in its working environment together with an understanding of the mechanical processes occurring during manufacture.

Clearly such an assertion includes the time-dependent behavior of solids, be they structures or general solids subject to geometric and material nonlinear effects. The material developed in this book is important in order to understand the fundamental ideas behind, and correct use of, many software packages that are being widely employed in engineering for applications such as: impact and crash simulation by the automotive and aerospace industries; modeling metal forming and manufacturing processes; or, more recently, in virtual reality simulation being developed for computer games or even surgery simulators. For such an understanding, consideration must be given to the concepts of thermodynamics and thermoelastic constitutive modeling, which may become significant for large deformation, large strain processes undergoing transient motion. Recognizing its widespread adoption, particularly as a graduate training platform, this present dynamics text employs MATLAB®* for the implementation of the finite element analyses, the software being freely available at `www.flagshyp.com` or `www.cambridge.org/9781107115620`.

READERSHIP

In common with the first volume, NL-Statics, this text is most suited to a postgraduate level of study by those either in higher education or in industry who

* Mathworks, Inc.

have graduated with an engineering or applied mathematics degree. However, the material is equally applicable to first-degree students in the final year of an applied maths course or an engineering course containing some additional emphasis on maths and numerical analysis. A familiarity with elementary dynamics and vibration analysis is assumed, as is some exposure to the principles of the finite element method. It will also be assumed that the reader is already familiar with the nonlinear solid mechanics concepts described in the NL-Statics volume. However, a primary objective of the book is that it be reasonably self-contained with regards to the new material on solid dynamics and thermodynamics.

LAYOUT

Chapter 1 Introduction

Here the nature of nonlinear dynamic equilibrium is discussed in the context of very simple one- and two-degrees-of-freedom spring-mass examples. Leap frog and mid-point time integration schemes are introduced as examples of explicit and implicit procedures. As an illustration of physical invariants, conservation of energy is discussed. A very simple two-degrees-of-freedom nonlinear spring-mass model is used by way of an example in order to illustrate these issues. Employing the mid-point rule for this example enables the introduction of the concepts of the directional derivative and the tangent matrix. A short MATLAB code used to implement this example is also presented. In essence, this code is the prototype for the main finite element program discussed later in the book.

Chapter 2 Dynamic Analysis of Three-Dimensional Trusses

This chapter is largely independent of the remainder of the text and deals with the large strain elastic dynamic behavior of trusses. The chapter first develops the dynamic equilibrium equations, which lead to the common simplification of a lumped mass assumption. In preparation for future chapters, equilibrium is also presented in the context of a variational approach, leading to the introduction of concepts such as the Lagrangian, the action integral, and the derivation of equilibrium via Hamilton's principle of least action. This is equivalent to the common derivation of static equilibrium as the minimum of the total potential energy presented in the NL-Statics volume. Leap-frog and mid-point time integration schemes are re-derived via a discrete approximation of the least action variational principle. Although not derived in the same manner, the trapezoidal time integration scheme is introduced and is shown to be a particular case of the Newmark family of schemes. Global conservation laws dealing with linear and angular momentum and energy are discussed. This is followed by a comprehensive discussion of Hamiltonian formulations involving momentum variables, phase space, and the Hamiltonian map. The notion of simplecticity, the conservation of phase space volume, and the relation to simplectic time integrators is discussed in some detail, drawing on parallels with the deformation gradient described in NL-Statics. A number of illustrative examples are provided using the dynamic version of the FLagSHyP code.

Chapter 3 Dynamic Equilibrium of Deformable Solids

This chapter describes the dynamic equilibrium of a solid in the large strain context, the key difference between this and static equilibrium being the addition of the inertial forces. The resulting equations are presented in the reference and spatial configurations as well as in differential and integral form via the dynamic version of the principle of virtual work. This chapter also derives the global conservation properties of the dynamic equations, namely momentum and energy, from the principle of virtual work. It finishes by extending the variational concepts of the action integral and Hamilton's principle, presented earlier in the discrete context of trusses, to the continuum.

Chapter 4 Discretization

This chapter describes the classical finite element discretization of the nonlinear solid dynamic equations when these are expressed in the form of the principle of virtual work. The resulting concept of mass matrix that emerges from discretizing the inertial term and the common practice of nodal mass lumping are described.

The reader is referred to the NL-Statics text for a full exposition of the consistent tangent matrix. The concept of the Lagrangian, previously discussed in Chapter 2 with respect to trusses, is now extended to the discretized continuum. Building on Chapter 2, variational time integrators are employed to re-derive the leap-frog and mid-point time integration schemes. Global conservation properties are again discussed within the context of the discretized equations of the solid.

Chapter 4 also introduces some commonly used time integration schemes and briefly describes the nonlinear solution process involved in their implementation. However, it must be emphasized that it is not the aim of the book to provide a comprehensive study of time or space discretization techniques. The aim is instead to explain the underlying continuum mechanics principles on which such discretizations are based. The chapter, therefore, does not provide an in-depth study of the effects of the discretization, or the errors associated with various time integration schemes or their stability properties.

Chapter 5 Conservation Laws in Solid Mechanics

This chapter formulates again the laws governing solid dynamics in the form of a system of first-order conservation laws, where both velocities and strains are problem variables. This approach differs from the formulation presented in Chapter 3 in that the equations in this chapter include only first-order derivatives in time, as opposed the second-order derivatives present in the acceleration terms used previously. The chapter begins by establishing a generic conservation law which is then particularized to the various conservation laws such as momentum, energy, or deformation gradient. These are considered in detail for the motion of a solid described in the total Lagrangian, updated Lagrangian, and Eulerian frameworks, including possible thermal effects. Of special interest is the introduction of a geometric conservation law relating to the time evolution of the deformation gradient.

Chapter 6 Thermodynamics

Nonlinear dynamic transient processes are often accompanied by changes in temperature and heat flow. This chapter aims to introduce the basic concepts in thermodynamics that are necessary to understand such thermomechanical processes. The approach followed is based on the concept of energy conjugacy; that is, entropy is introduced as the conjugate variable to temperature. The energy conservation law is re-expressed as the first law of thermodynamics and in terms of internal energy rate as well as entropy rate. The chapter also introduces the entropy conservation inequality. A number of simple thermo-elastic material models are described, which include the Mie–Grüneisen equation of state widely used for metals and the simple entropic elasticity model used for rubbers.

Chapter 7 Space and Time Discretization of Conservation Laws in Solid Dynamics

This chapter presents the finite element solution of finite deformation solid dynamic equilibrium equations presented in the form of a system of first-order conservation laws. This approach leads to a more accurate representation of strains and stresses as they become primary variables rather than obtained through derivatives of the geometry. Both isothermal and thermoelastic constitutive models will be explored. The topic is introduced using a simple one-dimensional linear elastic example. The well established Petrov–Galerkin stabilized methodology, widely used in computational fluid dynamics, is described for this purpose. This is combined with a very simple explicit Runge–Kutta time integration scheme, which is also widely used in the literature. A solution algorithm is presented which is implemented as a MATLAB program discussed in Chapter 9.

Chapter 8 Computer Implementation for Displacement-Based Dynamics

In this chapter information is presented on the nonlinear finite deformation finite element computer program FLagShyP (**F**inite Element **Large** **S**train **H**yperelastoplastic **P**rogram). This program, thoroughly introduced in NL-Statics, is extended in this chapter to the case of dynamic loading by accommodating the various time integration schemes discussed in Chapter 4. Usage and layout of the MATLAB program is discussed together with the function of the various key subroutines.

This free program is available from `www.flagshyp.com` or `www.cambridge.org/9781107115620`. Alternatively, it can be obtained by `e-mail` request to the authors `j.bonet@greenwich.ac.uk` or `a.j.gil@swansea.ac.uk`. The authors would like to acknowledge the assistance given by Dr. Rogelio Ortigosa in the development of this computer program.

Chapter 9 Computer Implementation for Conservation-Law-Based Explicit Fast Dynamics

This chapter presents a new computer implementation for the analysis of explicit fast dynamics using a first-order conservation-law-based formulation. Specifically, a new program entitled PG_DYNA_LAWS (**P**etrov–**G**alerkin Explicit **Dyna**mics for Conservation **Laws**) is introduced for the solution of finite deformation problems expressed in the form of a system of first-order conservation laws as described in Chapter 7.

This free program is available from `www.flagshyp.com` or `www.cambridge.org/9781107115620`. Alternatively, it can be obtained by `e-mail` request to the authors `j.bonet@greenwich.ac.uk` or `a.j.gil@swansea.ac.uk`. The authors would like to acknowledge the assistance given by Dr. Chun Hean Lee in the development of this computer program.

Appendix – Shocks

Dynamic problems sometimes involve solutions where by the variables experience sudden discontinuities across moving surfaces. These are known as shocks and are common in problems involving impact or the sudden application of loads, such as as a result of explosions. The appendix describes the conditions that the main dynamic variables need to satisfy across shock surfaces in order to meet the global conservation laws. In fluid dynamics these are known as Rankine–Hugoniot equations, and the appendix will derive similar conditions in the context of Lagrangian solid mechanics. The use of these conditions, together with experimental measurements of shock speeds in order to derive constitutive models, will also be explored.

Finally, the appendix will derive equations for the traction and velocity at contact points between solids impacting against each other.

Bibliography

A bibliography is provided that enables the reader to access the background to the more standard aspects of solid dynamics and thermodynamics, from Hamiltonian and conservation law standpoints. Also listed are texts and papers that have been of use in the preparation of this book or that cover similar material in greater depth.

CHAPTER ONE

INTRODUCTION

1.1 NONLINEAR SOLID DYNAMICS

The nonlinear analysis of the mechanical behavior of solid continua can be categorized in a number of ways. Solids may exhibit material or geometric nonlinearity. In the former the constitutive behavior, that is, the stress-strain relationship will be nonlinear and in the latter geometric changes, such as large rotations, affect the behavior. In many situations, such as metal forming, both occur simultaneously. A further category is whether the response of the solid to loading, be it forces or temperature, is dynamic or static, or in other words time dependent or not, and to be more precise whether inertial forces are relevant or can be ignored. In a previous text entitled *Nonlinear Solid Mechanics for Finite Element Analysis: Statics* the authors covered the fundamental nonlinear continuum mechanics necessary for the development of the equilibrium equations and their eventual solution using finite element discretization. The present text extends that development into the nonlinear dynamic realm and, whereas it is reasonably self-contained, it is useful to have an awareness of the material in the companion Statics text.

The dynamic response of solids may be linear or nonlinear. Linear response is generally associated with small deformation vibration behavior about an equilibrium position where geometrically nonlinear effects are normally insignificant, but not always, as in the case of a vibrating string in tension. Examples of dynamic behavior that can be considered in the linear regime are the vibration of buildings under moderate earthquakes which do not take the structure anywhere near its possible failure. Nonlinear dynamic behavior is characterized primarily by the presence of large rotation in addition to possible large strain. For example, satellites exhibit large rotation small strain behavior, whereas the collapse of, say, a building due to an extreme earthquake or a high speed impact situation is the result of both large rotation and large strain.

As the title implies, the solid continua are modeled in a discrete manner by an assemblage of three-dimensional finite elements for which formulations are developed to represent the geometric and material behavior due to the motion of the solid with respect to time as a result of various types of loading. In general the spatial representation of the body in terms of finite elements is similar between dynamic and static applications. The main difference emerges in relation to the introduction of inertial forces due to the mass and acceleration of the body and the presence of time as a key independent variable.

Time being the essential difference between static and dynamic behavior means that procedures need to be devised in order to progress the motion of the solid as time progresses. These procedures are generally known as *time-stepping schemes* whereby the motion, that is, primarily the velocity and thence the position, are discretized in time. This means that the motion between discrete time steps is approximated in some way. A number of such time-stepping schemes are presented in this text, from the simple leap-frog scheme to more complex and general Newmark schemes.

Fundamental to the description of dynamic behavior, linear or nonlinear, is Newton's Second Law of Motion, giving a dynamic equilibrium equation relating force, mass, and acceleration. Motion is progressed by determining the acceleration and then using the time-stepping scheme to find the velocity and advance the position. In mathematical terms this implies the solution of a second-order equation in time by an appropriate time-stepping scheme.

An alternative approach presented in this text involves the reformulation of the dynamic equilibrium of a solid into a system of first-order conservation laws for the physical and geometric variables describing the solid and its motion. This process leads to a mixed set of unknowns incorporating both velocities and strains. Similar sets of conservation laws are used extensively in computational fluid dynamics and the discretization of such laws via the finite element method is well understood and routinely applied. In this text both the traditional displacement-based approach, leading to a second-order system of equations in time, and the first-order set of conservation laws will be presented to solve solid dynamics problems. In the final chapters a number of examples will be solved using both these approaches to demonstrate their validity and general applicability.

The general aim of this book is to provide the reader with a good understanding of the necessary continuum mechanics concepts and theory required to successfully model by finite elements the time-dependent large deformation of solids, including possible thermal effects. It will cover important continuum concepts such as virtual work, potential and kinetic energy, the Lagrangian, and Hamilton's principle, as well as thermodynamics and thermoelasticity. It will also present time-discretization issues such as energy and momentum conserving schemes and simplectic integrators or advanced finite element discretization technologies such as Petrov–Galerkin methods for first-order conservation laws. Although the authors have made every effort to keep the text self-contained, the reader would benefit from some degree of familiarity with the contents of the companion statics volume: *Nonlinear Solid Mechanics for Finite Element Analysis: Statics*. This text will be referred to in the remainder of this book as the NL-Statics Volume.

The remainder of this chapter sets out to provide a gentle introduction to nonlinear dynamic behavior via simple one- or two-degrees-of-freedom examples. These examples are used to introduce simple time integration schemes such as the leap-frog method or the mid-point rule and discuss important issues associated with these schemes such as stability, or to compare implicit versus explicit methodologies.

1.2 ONE-DEGREE-OF-FREEDOM NONLINEAR DYNAMIC BEHAVIOR

In this section the torsion spring supported single rigid column discussed in Section 1.2.2 of NL-Statics is re-examined to demonstrate a simple example of nonlinear dynamic behavior. In order to introduce inertial forces, the column supports a mass which is restrained by a linear elastic torsion spring and viscous torsion damping at the bottom hinge, see Figure 1.1. The governing equations do not admit an analytical solution and consequently a simple numerical "leap-frog" integration in time will be introduced in order to obtain the behavior of the column with respect to time.

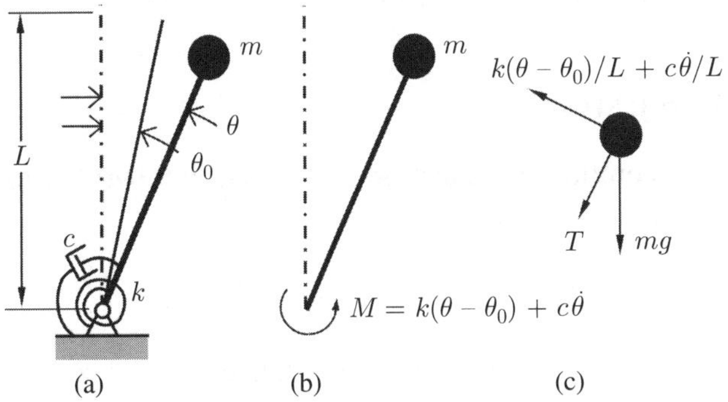

FIGURE 1.1 Simple column.

1.2.1 Equation of Motion

The column shown in Figure 1.1, where the angular motion of the mass m, from an initial angle of θ_0, is constrained by a torsion spring having stiffness k and, in parallel, torsional viscous damping having a coefficient c. The effect of introducing viscosity into the system will be demonstrated below. The torsion spring produces a moment $k(\theta-\theta_0)$ which in turn is equivalent to a tangential force on the mass equal to $k(\theta-\theta_0)/L$. The viscous damping produces a similar force but now equal to $c\dot{\theta}/L$, where $\dot{\theta}$ is the angular velocity. Elementary kinematics gives the tangential and radial acceleration of the mass m in terms of $\dot{\theta}$ and the tangential acceleration $\ddot{\theta}$ as $L\ddot{\theta}$ and $L\dot{\theta}^2$, respectively.

Employing Newton's Second Law of Motion the dynamic equilibrium equations in the tangential and radial directions are

$$mL\frac{d^2\theta}{dt^2} = mg\sin\theta - \frac{c}{L}\frac{d\theta}{dt} - \frac{k}{L}(\theta-\theta_0), \tag{1.1a}$$

$$mL\left(\frac{d\theta}{dt}\right)^2 = T + mg\cos\theta. \tag{1.1b}$$

Observe that the first of these equations is sufficient to determine the motion of the column via the evaluation of $\theta(t)$, whereas the second equation enables the calculation of the tension, T, in the column once the angle $\theta(t)$ and angular velocity $\dot{\theta}(t)$ have been obtained. The geometric nonlinearity in, for example, the first of the above equations is enshrined in the tangential gravitational force term $mg\sin\theta$.

As mentioned above, unlike the static column example considered in NL-Statics Section 1.2.2, the above equations do not readily admit an analytical solution and have to be solved using a numerical time integration scheme which enables the angular velocity $\dot{\theta}(t)$ and hence the angular position $\theta(t)$ to be calculated after the acceleration $\ddot{\theta}(t)$ has been found from Equation (1.1a). Numerous time integration schemes exist and a number of these will be introduced in this book, but for now a simple robust scheme known as leap-frog time integration will be employed.

1.2.2 Leap-Frog Time Integration

The column dynamic tangential equilibrium Equation (1.1a) can be rewritten to give the tangential acceleration as

$$\mathbf{a} = \frac{g}{L}\sin(\mathbf{x}) - \frac{c}{mL^2}\mathbf{v} - \frac{k}{mL^2}(\mathbf{x}-\mathbf{X}); \quad \mathbf{a} = \frac{d^2\theta}{dt^2}; \quad \mathbf{v} = \frac{d\theta}{dt}; \quad \mathbf{x} = \theta, \tag{1.2}$$

where the notation for acceleration $\mathbf{a}$, velocity $\mathbf{v}$, and coordinate position $\mathbf{x}$ provide for a general description of the time-stepping scheme* that will be valid for multiple-degree-of-freedom problems.

* For the column problem $\mathbf{x} = \theta$, $\mathbf{v} = \dot{\theta}$, and $\mathbf{X} = \theta_0$.

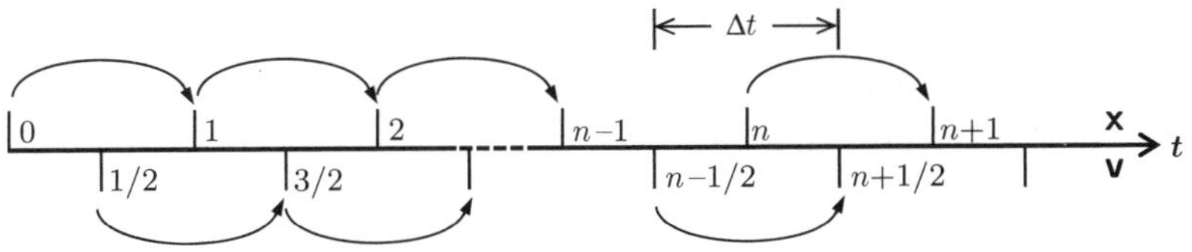

FIGURE 1.2 Leap-frog time integration.

The time during which the motion takes place is now divided into a number of equal time steps Δt, where a typical time is labeled n, see Figure 1.2. Using the above acceleration $\mathbf{a}$, assuming $\mathbf{v}$ and $\mathbf{x}$ to be known, an approximate integration over a time step Δt can be employed to calculate the velocity at the half-time-step $t_{n+1/2}$ in terms of the previous half time step velocity and the time step $\Delta t = t_{n+1/2} - t_{n-1/2}$ as

$$\mathbf{v}_{n+1/2} = \mathbf{v}_{n-1/2} + \int_{t_{n-1/2}}^{t_{n+1/2}} \mathbf{a}\, dt \approx \mathbf{v}_{n-1/2} + \mathbf{a}_n \Delta t. \tag{1.3}$$

Similarly the updated coordinates $\mathbf{x}_{n+1}$ are found using the velocity at the previous half time step above to yield

$$\mathbf{x}_{n+1} = \mathbf{x}_n + \int_{t_n}^{t_{n+1}} \mathbf{v}\, dt \approx \mathbf{x}_n + \mathbf{v}_{n+1/2} \Delta t. \tag{1.4}$$

This process, illustrated in Figure 1.2, is an example of a staggered time-stepping scheme.

Notice in Figure 1.2 that at the first time step $n = 0$ the velocity at $\mathbf{v}_{1/2}$ is required to start the process. Given an initial velocity $\mathbf{v}_0$ and an initial acceleration $\mathbf{a}_0$ calculated from Equation (1.2), $\mathbf{v}_{1/2}$ can be found using the approximation

$$\mathbf{v}_{1/2} = \mathbf{v}_0 + \mathbf{a}_0 \frac{\Delta t}{2}. \tag{1.5}$$

Observe that, given the staggered nature of the scheme, unless the viscosity vanishes, that is $c = 0$, the velocity in Equation (1.2) is not generally known at time n and needs to be taken at $t_{n-1/2}$ in order to allow a direct or explicit evaluation of the acceleration, $\mathbf{a}_n$, at time n as

$$\mathbf{a}_n = \frac{g}{L} \sin(\mathbf{x}_n) - \frac{c}{mL^2} \mathbf{v}_{n-1/2} - \frac{k}{mL^2} (\mathbf{x}_n - \mathbf{X}). \tag{1.6}$$

The exception to this rule occurs at $n = 0$, when $\mathbf{v}_0$ is actually known from the initial conditions, and hence $\mathbf{a}_0$ can be evaluated as[†]

$$\mathbf{a}_0 = \frac{g}{L} \sin(\mathbf{x}_0) - \frac{c}{mL^2} \mathbf{v}_0 - \frac{k}{mL^2} (\mathbf{x}_0 - \mathbf{X}). \tag{1.7}$$

[†] $\mathbf{X}_0$ need not be equal to ; see Exercise 5 of Chapter 2.

It is clear that the replacement of $\mathbf{v}_n$ by $\mathbf{v}_{n-1/2}$ in Equation (1.6) introduces an error in the calculation of $\mathbf{a}_n$. This can be avoided by using the correct $\mathbf{v}_n$ expressed as $\mathbf{v}_n = \mathbf{v}_{n-1/2} + (\Delta t/2)\mathbf{a}_n$ to give an expression for $\mathbf{a}_n$ as

$$\mathbf{a}_n = \frac{g}{L}\sin(\mathbf{x}_n) - \frac{c}{mL^2}\left(\mathbf{v}_{n-1/2} + \frac{\Delta t}{2}\mathbf{a}_n\right) - \frac{k}{mL^2}(\mathbf{x}_n - \mathbf{X}). \tag{1.8}$$

Consequently, $\mathbf{a}_n$ is now implicit in its own evaluation. For a one-degree-of-freedom problem this is not a problem as it would suffice to move all terms containing $\mathbf{a}_n$ to the left of the equation and divide the remaining right side by the accumulated coefficient of $\mathbf{a}_n$, to yield

$$\mathbf{a}_n = \Big(\frac{g}{L}\sin(\mathbf{x}_n) - \frac{c}{mL^2}\mathbf{v}_{n-1/2} - \frac{k}{mL^2}(\mathbf{x}_n - \mathbf{X})\Big)/\big(1 + \frac{c}{mL^2}\frac{\Delta t}{2}\big). \tag{1.9}$$

Whereas Equation (1.6) is an explicit equation, Equation (1.8) is an implicit equation requiring a solution for $\mathbf{a}_n$. In the single-degree-of-freedom case, solving Equation (1.9) is trivial, but for realistic simulations containing large numbers of degrees of freedom k, c, and m are matrices, which leads to a system of equations requiring solution. Generally, time-stepping schemes in which the variables can be updated without solving systems of equations are known as *explicit*, whereas schemes that require the solution of a system of equations are known as *implicit*, and if this is required at every time step such schemes are computationally costly.

In comparison to an implicit scheme, the greater efficiency of an explicit scheme would seem preferable; however, it will be seen later that they suffer from severe limitations with respect to time-step size before producing grossly inaccurate solutions. These time-step limitations are known as stability restrictions and will be discussed in Section 1.3.3 below. Implicit time-stepping schemes are usually constructed in such a manner that avoids stability restrictions and consequently allow much larger time steps to be used, albeit at greater cost per step. The leap-frog algorithm is shown in Box 1.1. This explicit algorithm is of general applicability and as a consequence does not include the correction implied by Equation (1.9).

BOX 1.1: Leap-frog algorithm

- INPUT geometry, material properties, and solution parameters
- INITIALIZE $\mathbf{x}_0$, $\mathbf{v}_0$
- SET $\mathbf{a}_0$ (1.7) using initial values
- SET $\mathbf{v}_{1/2}$ (1.5)
- DO WHILE $t < tmax$ (time steps)
  - SET $\mathbf{x} = \mathbf{x} + \mathbf{v}\,\Delta t$ (1.4)
  - FIND $\mathbf{a}$ (1.6)

(continued)

Box 1.1: *(cont.)*

- SET $\mathbf{v} = \mathbf{v} + \mathbf{a}\,\Delta t$ (1.3)
- ENDDO

1.2.3 Column Examples

The following examples employ the leap-frog time integration discussed in Section 1.2.2. In Figure 1.3 the viscous coefficient is $c = 0$ and in Figure 1.4 the viscous coefficient is $c = 650$.‡ Both examples have an initial angle of $\theta_0 = 45°$, $L = 10$, $m = 100$, $g = 9.81$, $k = 1000$, $\Delta t = 0.0001$, and the initial tangential velocity is $\mathbf{v}_0 = 0$. Allowing the time to extend to $t = 200$ reveals that the damped solution converges to the static solution of $\theta \approx 167.42°$.

1.3 TWO-DEGREES-OF-FREEDOM EXAMPLE

The spring-mass system shown in Figure 1.5 will be used to introduce the effects of geometric nonlinearity in a dynamic situation. The internal force in the spring is $\mathbf{T}$ and the spring has a stiffness k. At time $t = 0$ the orientation of the spring is

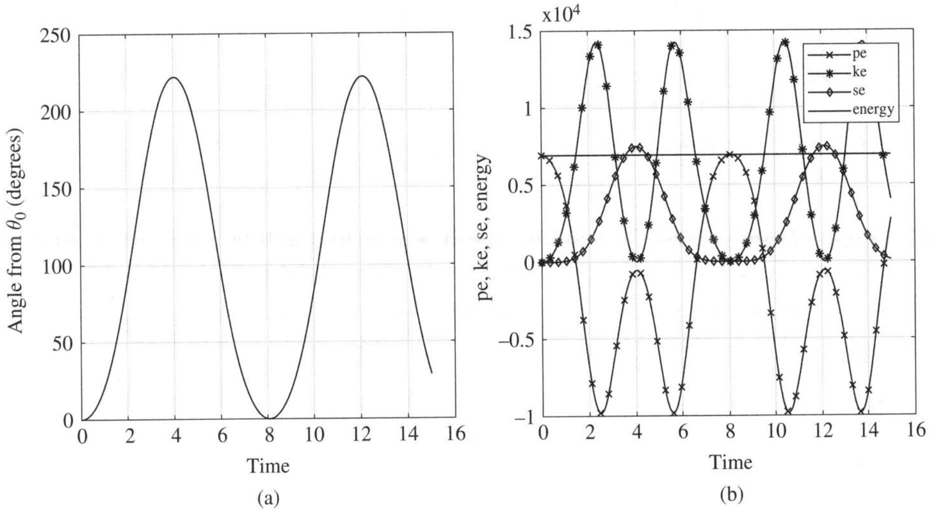

FIGURE 1.3 Column leap-frog with $\theta_0 = 45°, c = 0$: (a) Angle from θ_0–time; (b) Energy–time (pe = potential energy, se = elastic (strain) energy, ke = kinetic energy, total energy).

‡ For this case the correction for a_n given by the last term in Equation (1.9) involving the viscosity c is negligible and is not employed in the solution.

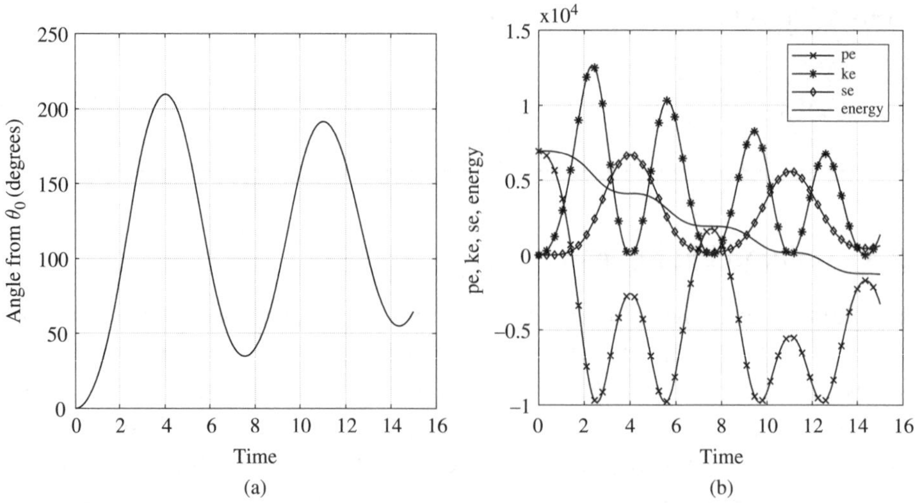

FIGURE 1.4 Column leap-frog with $\theta_0 = 45°, c = 650$: (a) Angle from θ_0–time; (b) Energy–time.

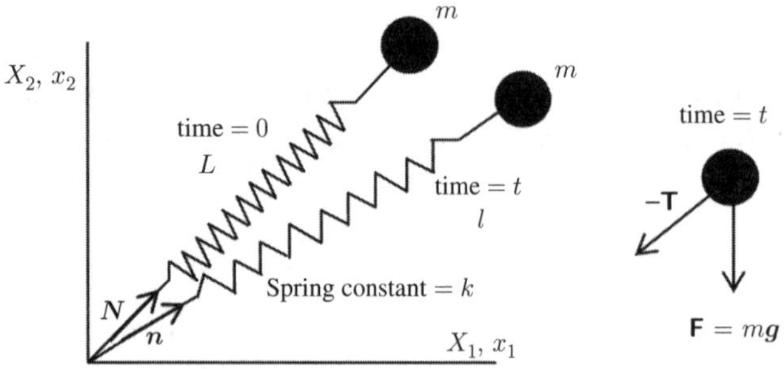

FIGURE 1.5 Two-degrees-of-freedom example.

given by the unit normal $\boldsymbol{N}$ and the length is L, while at time t the orientation is given by $\boldsymbol{n}(t)$ and the length is $l(t)$. The unit vectors $\boldsymbol{N}$ and $\boldsymbol{n}(t)$ at time $t = 0$ and t respectively are determined by the corresponding mass coordinates $\mathbf{X}$ and $\mathbf{x}(t)$ as

$$\mathbf{X} = \begin{bmatrix} X_1 \\ X_2 \end{bmatrix}; \quad \boldsymbol{N} = \frac{1}{L}\begin{bmatrix} X_1 \\ X_2 \end{bmatrix}, \tag{1.10a,b}$$

$$\mathbf{x}(t) = \begin{bmatrix} x_1 \\ x_2 \end{bmatrix}; \quad \boldsymbol{n}(t) = \frac{1}{l}\begin{bmatrix} x_1 \\ x_2 \end{bmatrix}; \quad l = \sqrt{(\mathbf{x}\cdot\mathbf{x})}. \tag{1.10c,d,e}$$

Time derivatives of the coordinate $\mathbf{x}(t)$ give the velocity and acceleration as

$$\mathbf{v}(t) = \begin{bmatrix} v_1 \\ v_2 \end{bmatrix}; \quad \mathbf{v}(t) = \frac{d\mathbf{x}}{dt}; \quad \mathbf{a}(t) = \begin{bmatrix} a_1 \\ a_2 \end{bmatrix}; \quad \mathbf{a}(t) = \frac{d^2\mathbf{x}}{dt^2}. \tag{1.11a,b,c,d}$$

For simplicity the explicit dependency of quantities upon time will be removed, for example, $\mathbf{x}(t)$ becomes $\mathbf{x}$, etc. The internal spring force $\mathbf{T}$ and the vertical gravitational force $\mathbf{F}$ are

$$\mathbf{T}(\mathbf{x}) = T\boldsymbol{n}; \quad T = k(l - L); \quad \mathbf{F} = \begin{bmatrix} 0 \\ -mg \end{bmatrix}. \tag{1.12a,b}$$

Obviously, on Earth at least, $g = 9.81\,\mathrm{m/s^2}$.

1.3.1 Equations of Motion

Employing Newton's Second Law of Motion applied to the mass m at time t, the equations of motion are written in terms of the acceleration $\mathbf{a}$, internal force $\mathbf{T}$, and external force $\mathbf{F}$ as

$$\begin{bmatrix} m & 0 \\ 0 & m \end{bmatrix} \begin{bmatrix} a_1 \\ a_2 \end{bmatrix} = \begin{bmatrix} 0 \\ -mg \end{bmatrix} - \frac{T}{l} \begin{bmatrix} x_1 \\ x_2 \end{bmatrix} \quad \text{or simply } \mathbf{Ma} = \mathbf{F} - \mathbf{T}(\mathbf{x}). \tag{1.13a,b}$$

Equation (1.12a,b) reveals that the internal force $\mathbf{T}$ at time t is a function of the length l and unit normal $\boldsymbol{n}$, both being functions of the current position $\mathbf{x}$ of the mass. Consequently, the equations of motion (1.13a,b) are geometrically nonlinear. Such equations do not admit an analytical solution and have to be solved numerically in the following section using the leap-frog scheme.

EXAMPLE 1.1: Energy conservation

The two-degrees-of-freedom spring-mass system described in this section is a convenient example with which to illustrate the conservation of physical quantities such as energy. The equilibrium Equation (1.13a,b) for the mass at time t and position $\mathbf{x}$ can be rewritten as

$$m\mathbf{a} + k(l - L)\boldsymbol{n} = m\boldsymbol{g},$$

where

$$\mathbf{a} = \frac{d\mathbf{v}}{dt}; \; \boldsymbol{n} = \frac{\mathbf{x}}{l}; \; l = \|\mathbf{x}\|; \; \boldsymbol{g} = -g\begin{bmatrix} 0 \\ 1 \end{bmatrix}.$$

(continued)

Example 1.1: *(cont.)*

Multiplying the equilibrium equation by $\mathbf{v}$ and rearranging gives

$$m\mathbf{a} \cdot \mathbf{v} + k(l - L)\boldsymbol{n} \cdot \mathbf{v} + mg\, v_2 = 0,$$

noting that

$$\mathbf{a} \cdot \mathbf{v} = \frac{1}{2}\frac{d}{dt}(\mathbf{v} \cdot \mathbf{v}) = \frac{1}{2}\frac{dv^2}{dt};\ v = \|\mathbf{v}\|;\ v_2 = \frac{dx_2}{dt},$$

and

$$\boldsymbol{n} \cdot \mathbf{v} = \frac{\mathbf{x}}{l} \cdot \frac{d\mathbf{x}}{dt} = \frac{1}{2}\frac{1}{l}\frac{d}{dt}(\mathbf{x} \cdot \mathbf{x}) = \frac{1}{2}\frac{1}{l}\frac{dl^2}{dt} = \frac{dl}{dt}$$

gives, after some simple algebra,

$$\frac{d}{dt}\left[\frac{1}{2}mv^2 + \frac{1}{2}k(l - L)^2 + mgx_2\right] = 0.$$

The term in the square bracket is the total energy of the mass m comprising the kinetic, elastic, and potential components, thus demonstrating the conservation of total energy.

1.3.2 Leap-Frog Examples

Using the two-degrees-of-freedom formulation, the column problem given in Section 1.2.3 can be rerun with a high linear spring constant of $k = 10^5$ to approximate a rigid column. This is equivalent to the column case shown in Figure 1.1 with a torsion spring constant of value zero. Figure 1.6 shows the results for the two-degrees-of-freedom simulation. When comparing with Figure 1.3, note that due to the presence of the torsion spring in Figure 1.3 the maximum angle is, as expected, less than that given in Figure 1.6 where the torsion spring stiffness is necessarily zero.

The next example is essentially a simple linear spring-mass problem with one degree of freedom. In order to achieve this, the mass is constrained to move along a fixed axis, as shown in Figure 1.7 (see also Exercise 2). This example will show that for a stable implementation of the leap-frog solution the value of the time step must be such that $\Delta t < 2/\omega$ where $\omega = (k/m)^{\frac{1}{2}}$. For a mass of unity and a spring stiffness $k = 1000$, Figure 1.8 shows the leap-frog solutions for time step $\Delta t = 0.01/\omega$ (solid plot) and $\Delta t = 2.01/\omega$ (dashed plot). For $\Delta t < 2/\omega$ the solution is identical to that given analytically in Exercise 2, having a maximum amplitude of $(1 - \cos(\omega t)) = 2$, whereas for $\Delta t > 2/\omega$ the numerical solution is clearly unstable. More interesting examples will be reserved for Section 1.3.6.

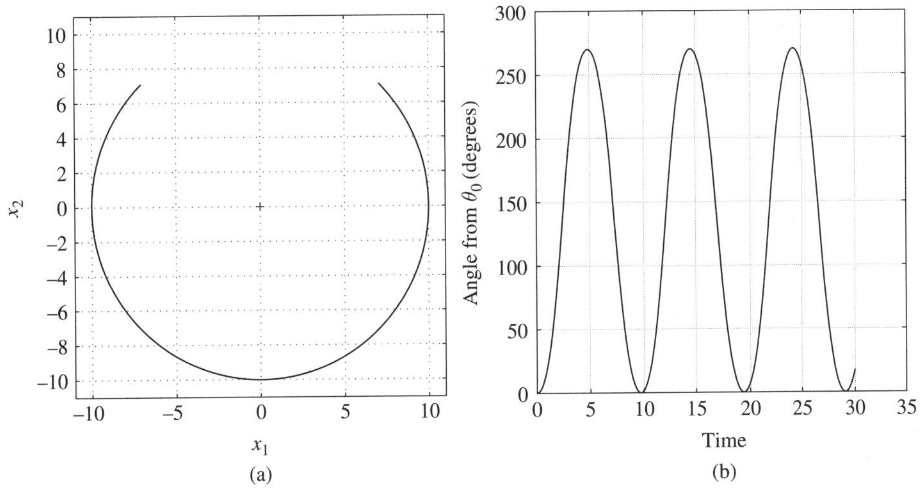

FIGURE 1.6 Column with torsion spring constant $k = 0$ as a 2 dof problem, linear spring constant $k = 10^5$, $\theta_0 = 45°$. (a) Motion; (b) Angle from θ_0–time.

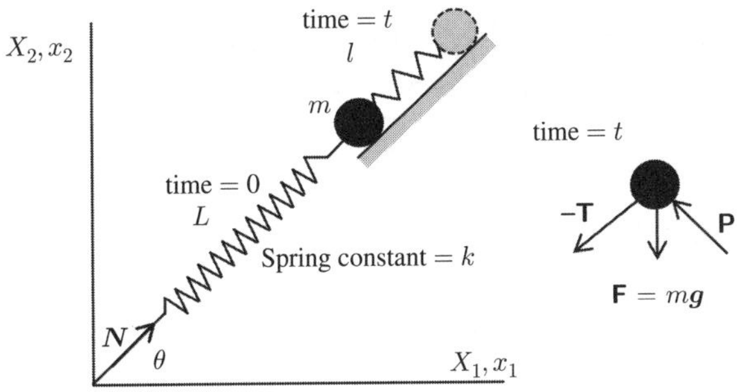

FIGURE 1.7 Constrained spring-mass system.

1.3.3 Time Integration – Stability

The preceding example demonstrated the problem of the numerical stability of the time integration scheme. This is generally complex to analyze for nonlinear systems with multiple degrees of freedom but can readily be predicted in the case of the single-degree-of-freedom system discussed above. In this case, when $\theta = 0$ (see Figure 1.7), gravity is not effective, and thus, in the absence of the forcing term, the dynamic equilibrium equation is simply

$$m\frac{d^2x}{dt^2} + k\,x = 0; \quad x = x_1 - X_1, \tag{1.14a,b}$$

which has the analytical solution

$$x = \frac{v_0}{\omega}\sin(\omega t) + x_0\cos(\omega t); \quad \omega^2 = \frac{k}{m}, \tag{1.15a,b}$$

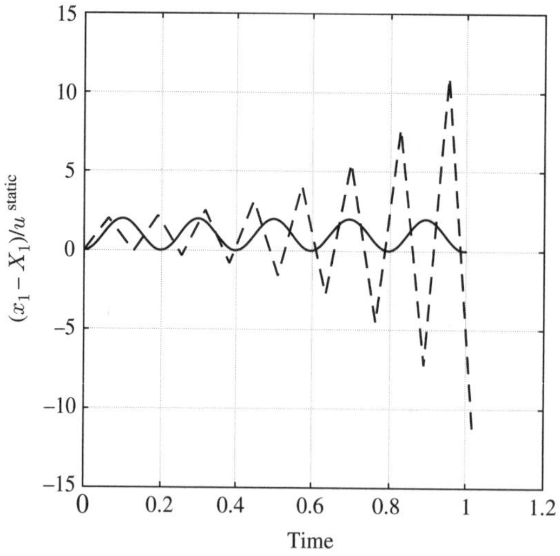

FIGURE 1.8 Constrained spring mass, $\theta_0 = 45°$ numerical stability.

where x_0 and v_0 are the initial displacement and velocity respectively. Integration of Equation (1.14a,b) using the leap-frog method gives

$$v_{n+1/2} = v_{n-1/2} - \Delta t \frac{k}{m} x_n, \tag{1.16}$$

and

$$\begin{aligned} x_{n+1} &= x_n + \Delta t v_{n+1/2} \\ &= x_n + \Delta t \left(v_{n-1/2} - \Delta t \frac{k}{m} x_n \right). \end{aligned} \tag{1.17}$$

The intermediate velocity $v_{n-1/2}$ in the above equation can be eliminated by noting that $v_{n-1/2} = (x_n - x_{n-1})/\Delta t$ to give, after some simple algebra, a difference equation relating x_{n-1}, x_n, and x_{n+1} as

$$x_{n+1} - (2 - \omega^2 \Delta t^2) x_n + x_{n-1} = 0; \quad \omega^2 = \frac{k}{m}. \tag{1.18}$$

This is the discrete version of the continuous differential Equation (1.14a,b) that results from the use of the leap-frog integration scheme. The values of x_n for $n = 1, \ldots, N$ can be predicted analytically by assuming a pattern of the type

$$x_{n+1} = A x_n, \; x_n = A x_{n-1}, \ldots, \tag{1.19a,b}$$

where A is the amplification factor between contiguous steps. Substituting this Equation $(1.19\text{a,b})_a$ for x_{n+1} into Equation (1.18) and using Equation $(1.19\text{a,b})_b$ to express x_{n-1} in terms of x_n gives, after dividing by x_n, an equation for A as

$$A^2 - (2 - \omega^2 \Delta t^2) A + 1 = 0, \tag{1.20}$$

which has solutions

$$A = \left(1 - \frac{\omega^2 \Delta t^2}{2}\right) \pm \sqrt{\left(1 - \frac{\omega^2 \Delta t^2}{2}\right)^2 - 1}. \tag{1.21}$$

Note that for values of A such that $|A| > 1$ the solution x_n will grow exponentially, very much unlike the analytical solution which has an oscillating nature. This is known as an unstable time integration process and must be avoided by adequate choice of Δt. A simple inspection of Equation (1.21) shows that for small enough values of Δt the discriminant is negative and the solution is therefore complex, that is,

$$A = \left(1 - \frac{\omega^2 \Delta t^2}{2}\right) \pm i\sqrt{1 - \left(1 - \frac{\omega^2 \Delta t^2}{2}\right)^2}. \tag{1.22}$$

In such cases it is easy to show that $|A| = 1$ and the solution for x_n exhibits a similar oscillating behavior as the exact solution. The boundary between $|A| = 1$ and $|A| > 1$ is defined by the expression

$$1 - \left(1 - \frac{\omega^2 \Delta t^2}{2}\right)^2 \geq 0, \tag{1.23}$$

which leads to the constraints

$$-1 \leq \left(1 - \frac{\omega^2 \Delta t^2}{2}\right) \leq 1. \tag{1.24}$$

Simple algebra now gives the stability limit for Δt as

$$0 \leq \Delta t \leq \frac{2}{\omega}. \tag{1.25}$$

Values of Δt between these boundaries ensure that the numerical solution to Equation $(1.14\text{a,b})_a$ is stable and "well behaved" in that it exhibits the oscillatory nature of the analytical solution. However, it is worth emphasizing that the critical value $\Delta t_{\text{crit}} = 2/\omega$ represents a rather large proportion of the natural period of vibration $T = 2\pi/\omega$ of the system. In particular,

$$\Delta t_{\text{crit}} = \frac{2}{\omega} = \frac{2}{2\pi/T} = \frac{T}{\pi}. \tag{1.26}$$

Usually, in order for the numerical integration to follow accurately a sinusoidal function in time, Δt should be in the region $T/10$ to $T/20$. Hence the critical time-step limit for stability is well beyond the value that would be necessary to use to achieve reasonable accuracy. This would seem to indicate that stability is not generally a problem. Recall, however, that the analysis carried out above only refers to one-degree-of-freedom problems. Real situations have multiple degrees of freedom and consequently assessing an adequate time step is much more complicated.

For example, in the two-degree-of freedom-problem described in Section 1.3, for high values of the spring stiffness k the motion will resemble the oscillation of a mass pendulum with a much larger period of vibration than that given a spring-mass system where the stiffness k is such that the spring can extend. In fact, even in this simple example that ratio between the period of the (high k) pendulum oscillation and the spring-mass system can be several orders of magnitude. Unfortunately the stability restriction will be constrained by the smaller of the two periods as

$$\Delta t_{\text{crit}} \leq \min\left\{\frac{T_{\text{spring}}}{\pi}; \frac{T_{\text{pendulum}}}{\pi}\right\} = \frac{T_{\text{spring}}}{\pi}. \tag{1.27}$$

For a stiff spring with constant k this value can be many orders of magnitude smaller than the time step required to follow accurately the pendulum motion, which would be of the order of $T_{\text{pendulum}}/20$. Consequently, stability imposes very small time steps in relation to the actual motion being integrated. Problems of this nature, where the maximum and minimum natural vibration periods (or frequencies) are many orders of magnitude apart, are known as "stiff" systems (in the sense that the spring is very stiff in relation to the pendulum). However, this is not always the case; for instance, in the two-degree-of-freedom problem above, if the spring constant k is very low so that the ratio between the natural period of the spring and pendulum is small then the system would not be stiff. Using time-stepping schemes with stability restrictions of the type described in Equation (1.25) can result in the need to carry out a very large number of time steps in order to advance the solution over the time span required. It is therefore essential that each time step can be evaluated as quickly and efficiently as possible, ideally without the solution of a linear or nonlinear system of equations.

Given that the time-step constraints for stiff problems can be computationally onerous, the following section will present an example of an "implicit" time-stepping scheme that avoids such restrictions but requires the solution of a nonlinear system of equations at each time step.

1.3.4 Mid-Point Rule

In anticipation of the use of a Newton–Raphson solution technique, the equilibrium Equation (1.13a,b)$_b$ is recast in terms of a residual force $\mathbf{R}(\mathbf{x}, \mathbf{a})$ as

$$\mathbf{R}(\mathbf{x}, \mathbf{a}) = \mathbf{M}(\mathbf{a}) + \mathbf{T}(\mathbf{x}) - \mathbf{F} = 0. \tag{1.28}$$

The mid-point rule scheme is an alternative time integration technique which, despite appearing similar to the leap-frog scheme, emerges as an implicit scheme. It is defined by the approximate integration of velocities and positions at time step $n+1$ as

$$\mathbf{v}_{n+1} = \mathbf{v}_n + \Delta t\, \mathbf{a}_{n+1/2}; \quad \mathbf{x}_{n+1} = \mathbf{x}_n + \Delta t\, \mathbf{v}_{n+1/2}, \tag{1.29a,b}$$

where $\mathbf{a}_{n+1/2}$ is given from the equilibrium Equation (1.13a,b)$_b$ or (1.28) by

$$\mathbf{M}\mathbf{a}_{n+1/2} = \mathbf{F} - \mathbf{T}(\mathbf{x}_{n+1/2}), \tag{1.30a,b}$$

and, by definition, $\mathbf{x}_{n+1/2}$ and $\mathbf{v}_{n+1/2}$ are

$$\mathbf{x}_{n+1/2} = \frac{1}{2}(\mathbf{x}_n + \mathbf{x}_{n+1}); \quad \mathbf{v}_{n+1/2} = \frac{1}{2}(\mathbf{v}_n + \mathbf{v}_{n+1}). \tag{1.31a,b}$$

Inspection of Equation (1.30a,b) reveals that $\mathbf{a}_{n+1/2}$ cannot be resolved explicitly since $\mathbf{x}_{n+1/2}$, which is a function of $\mathbf{x}_{n+1}$, is unknown. The terms $\mathbf{a}_{n+1/2}$ and $\mathbf{x}_{n+1/2}$ are linked by the time-stepping Equations (1.29a,b) and (1.31a,b), leading to a system of equations that have to be solved implicitly. There are a number of ways to approach this solution, but the most popular process splits the time step into a *predictor* phase followed by a *corrector* phase. The predictor phase is based on making a reasonable prediction of one of the two unknown variables, for instance $\mathbf{a}^{(0)}_{n+1/2} = \mathbf{a}_{n-1/2}$. This gives predictions for $\mathbf{x}$ and $\mathbf{v}$ as

$$\mathbf{v}^{(0)}_{n+1} = \mathbf{v}_n + \Delta t\, \mathbf{a}^{(0)}_{n+1/2}; \quad \mathbf{v}^{(0)}_{n+1/2} = \frac{1}{2}(\mathbf{v}_n + \mathbf{v}^{(0)}_{n+1}), \tag{1.32a,b}$$

$$\mathbf{x}^{(0)}_{n+1} = \mathbf{x}_n + \Delta t\, \mathbf{v}^{(0)}_{n+1/2}; \quad \mathbf{x}^{(0)}_{n+1/2} = \frac{1}{2}(\mathbf{x}_n + \mathbf{x}^{(0)}_{n+1}). \tag{1.32c,d}$$

A preliminary residual force $\mathbf{R}^{(0)}_{n+1/2}$ as a function of the predicted values $\mathbf{a}^{(0)}_{n+1/2}$ and $\mathbf{x}^{(0)}_{n+1/2}$ can be calculated from Equation (1.28) as

$$\mathbf{R}(\mathbf{x}^{(0)}_{n+1/2}, \mathbf{a}^{(0)}_{n+1/2}) = \mathbf{M}\,\mathbf{a}^{(0)}_{n+1/2} + \mathbf{T}(\mathbf{x}^{(0)}_{n+1/2}) - \mathbf{F}. \tag{1.33}$$

The corrector phase establishes a Newton–Raphson process to drive the above residual to zero by incrementing the position and acceleration at iteration step $(k-1)$ by $\Delta\mathbf{x}_{n+1/2}$ and $\Delta\mathbf{a}_{n+1/2}$ to give

$$\mathbf{x}^{(k)}_{n+1/2} = \mathbf{x}^{(k-1)}_{n+1/2} + \Delta\mathbf{x}_{n+1/2}; \quad \mathbf{a}^{(k)}_{n+1/2} = \mathbf{a}^{(k-1)}_{n+1/2} + \Delta\mathbf{a}_{n+1/2}. \tag{1.34a,b}$$

The corrector phase employs the iterative Newton–Raphson method which requires the linearization of Equation (1.28) at iteration step $(k-1)$ and time interval $n+1/2$ to give

$$\mathbf{R}(\mathbf{x}^{(k)}_{n+1/2}, \mathbf{a}^{(k)}_{n+1/2}) \approx \mathbf{R}\,(\mathbf{x}^{(k-1)}_{n+1/2}, \mathbf{a}^{(k-1)}_{n+1/2}) + D(\mathbf{M}\mathbf{a}^{(k-1)}_{n+1/2})[\Delta\mathbf{a}_{n+1/2}]$$
$$+\, D\mathbf{T}(\mathbf{x}^{(k-1)}_{n+1/2})[\Delta\mathbf{x}_{n+1/2}] = \mathbf{0}. \tag{1.35}$$

Since the mass matrix is constant the first directional derivative is simply $\mathbf{M}\,\Delta\mathbf{a}_{n+1/2}$. However, the second directional derivative involving the internal force $\mathbf{T}(\mathbf{x})$ is a function of the length l and the orientation given by the unit vector $\boldsymbol{n}$, both being a function of the position $\mathbf{x}$ at any iteration (k) time step n. This is the source of the geometrically nonlinear nature of the problem. Consideration of this second directional derivative requires the increments in $\mathbf{a}_{n+1/2}$ and $\mathbf{x}_{n+1/2}$ to be linked through the mid-point rule given by Equations (1.29a,b) and (1.31a,b) to give

$$\mathbf{x}_{n+1/2} = \mathbf{x}_n + \frac{\Delta t}{2}\mathbf{v}_n + \frac{\Delta t^2}{4}\mathbf{a}_{n+1/2}; \tag{1.36}$$

consequently, the increment $\Delta\mathbf{x}_{n+1/2}$ is given by the directional derivative

$$\Delta\mathbf{x}_{n+1/2} = D(\mathbf{x}_{n+1/2})[\Delta\mathbf{a}_{n+1/2}] = \frac{\Delta t^2}{4}\Delta\mathbf{a}_{n+1/2}. \tag{1.37}$$

Equation (1.35) is now rewritten in terms of the mass matrix $\mathbf{M}$ and a tangent stiffness matrix $\mathbf{K}$ as

$$\mathbf{R}(\mathbf{x}^{(k)}_{n+1/2}, \mathbf{a}^{(k)}_{n+1/2}) \approx \mathbf{R}(\mathbf{x}^{(k-1)}_{n+1/2}, \mathbf{a}^{(k-1)}_{n+1/2}) + \mathbf{M}\,\Delta\mathbf{a}_{n+1/2}$$
$$+\,\mathbf{K}(\mathbf{x}^{(k-1)}_{n+1/2})\,\Delta\mathbf{x}_{n+1/2} = \mathbf{0}, \tag{1.38a}$$

where the tangent stiffness matrix $\mathbf{K}$, which is considered in Section 1.3.5 below, is derived as

$$\mathbf{K}(\mathbf{x}^{(k-1)}_{n+1/2})\,\Delta\mathbf{x}_{n+1/2} = D\mathbf{T}(\mathbf{x}^{(k-1)}_{n+1/2})[\Delta\mathbf{x}_{n+1/2}]. \tag{1.38b}$$

Substituting for $\Delta\mathbf{x}_{n+1/2}$ from Equation (1.37) into Equation (1.38b) enables the corrector phase of the mid-point algorithm to be established solely in terms of $\Delta\mathbf{a}_{n+1/2}$ to give

$$\left[\mathbf{M} + \frac{\Delta t^2}{4}\mathbf{K}(\mathbf{x}^{(k-1)}_{n+1/2})\right]\Delta\mathbf{a}_{n+1/2} = -\mathbf{R}(\mathbf{x}^{(k-1)}_{n+1/2}, \mathbf{a}^{(k-1)}_{n+1/2}). \tag{1.39}$$

Solving the system of nonlinear Equations (1.39) enables the mid-point corrector phase to be completed to yield the position $\mathbf{x}_{n+1}$ and velocity $\mathbf{v}_{n+1}$ as

$$\mathbf{a}^{(k)}_{n+1/2} = \mathbf{a}^{(k-1)}_{n+1/2} + \Delta\mathbf{a}_{n+1/2}, \tag{1.40a}$$

$$\mathbf{x}^{(k)}_{n+1/2} = \mathbf{x}^{(k-1)}_{n+1/2} + \frac{\Delta t^2}{4}\Delta\mathbf{a}_{n+1/2}, \tag{1.40b}$$

$$\mathbf{x}^{(k)}_{n+1} = \mathbf{x}^{(k-1)}_{n+1} + \frac{\Delta t^2}{2}\Delta\mathbf{a}_{n+1/2}, \tag{1.40c}$$

$$\mathbf{v}^{(k)}_{n+1} = \mathbf{v}^{(k-1)}_{n+1} + \Delta t\Delta\mathbf{a}_{n+1/2}. \tag{1.40d}$$

The Newton–Raphson procedure involves repeated application of Equations (1.39) and (1.40) until the residual $\mathbf{R}(\mathbf{x}, \mathbf{a})$ is less than a given tolerance. Note that had the problem been geometrically linear then it would only be necessary to apply these equations once. Once convergence is achieved the predictor phase, given by Equation (1.32a,b), can be initiated for the next time step.

EXAMPLE 1.2: Discrete conservation of energy

Example 1.1 demonstrated the principle of conservation of energy for the spring-mass system. It is possible to maintain the energy conservation principle despite the use of time discretization when using certain types of time integration such as the mid-point rule. However, this requires subtle modifications to the algorithm which are described below. To begin, using the mid-point rule the equilibrium Equation (1.13a,b) at time $t_{n+1/2}$ can be expressed as

$$m\mathbf{a}_{n+1/2} + \frac{k(l_{n+1/2} - L)}{l_{n+1/2}}\mathbf{x}_{n+1/2} = m\boldsymbol{g},$$

where

$$\mathbf{a}_{n+1/2} = \frac{\mathbf{v}_{n+1} - \mathbf{v}_n}{\Delta t};\ \mathbf{x}_{n+1/2} = \frac{1}{2}(\mathbf{x}_{n+1} + \mathbf{x}_n);\ l_{n+1/2} = \|\mathbf{x}_{n+1/2}\|.$$

Multiplying the discrete "mid-point" equilibrium equation by $\mathbf{v}_{n+1/2} = \frac{1}{2}(\mathbf{v}_{n+1} + \mathbf{v}_n) = (\mathbf{x}_{n+1} - \mathbf{x}_n)/\Delta t$ gives

$$\frac{1}{2\Delta t}m(\mathbf{v}_{n+1} \cdot \mathbf{v}_{n+1} - \mathbf{v}_n \cdot \mathbf{v}_n) + k\frac{(l_{n+1/2} - L)}{l_{n+1/2}}\frac{1}{2\Delta t}(\mathbf{x}_{n+1} \cdot \mathbf{x}_{n+1}$$
$$-\mathbf{x}_n \cdot \mathbf{x}_n) + \frac{1}{\Delta t}mg(x_{2,n+1} - x_{2,n}) = 0.$$

Multiplying the above equation by Δt, and after some simple algebra, gives

$$\left(\frac{1}{2}mv_{n+1}^2 - \frac{1}{2}mv_n^2\right) + \frac{1}{2}k\frac{(l_{n+1/2} - L)}{l_{n+1/2}}(l_{n+1} + l_n)(l_{n+1} - l_n)$$
$$+ mg(x_{2,n+1} - x_{2,n}) = 0.$$

The first and last terms in the preceding equation give the change in the kinetic and potential energies. In order to transform the center term into a change of elastic energy, it is necessary to evaluate $l_{n+1/2}$ as $(l_{n+1} + l_n)/2$ rather than $\|\mathbf{x}_{n+1/2}\|$. In this case the elastic term becomes

$$\frac{1}{2}k\frac{(l_{n+1/2} - L)}{l_{n+1/2}}(l_{n+1} + l_n)(l_{n+1} - l_n)$$
$$= \frac{1}{2}k\frac{((l_{n+1} + l_n)/2 - L)}{(l_{n+1} + l_n)/2}(l_{n+1} + l_n)(l_{n+1} - l_n)$$

(continued)

Example 1.2: *(cont.)*

$$= \frac{1}{2}k\big[(l_{n+1} - L) + (l_n - L)\big]\big[(l_{n+1} - L) - (l_n - L)\big]$$
$$= \frac{1}{2}k(l_{n+1} - L)^2 - \frac{1}{2}k(l_n - L)^2.$$

Hence the discrete equilibrium equation implies the conservation of total energy as

$$\frac{1}{2}mv_{n+1}^2 + \frac{1}{2}k(l_{n+1} - L)^2 + mgx_{2,n+1}$$
$$= \frac{1}{2}mv_n^2 + \frac{1}{2}k(l_n - L)^2 + mgx_{2,n}.$$

Note that the evaluation of $l_{n+1/2}$ as the average between l_n and l_{n+1} will coincide with its evaluation as $\|\mathbf{x}_{n+1/2}\|$ when there is no rotation between $\mathbf{x}_n$ and $\mathbf{x}_{n+1}$. In more general cases, like the one shown in the figure below, this will not be exactly the case. For very small time steps, the differences, however, are minimal. Nevertheless, obtaining $l_{n+1/2}$ as the average ensures that energy is conserved and can have a beneficial effect in terms of the stability of the solution when large time steps are used. Since the average length is obtained under the assumption that the configurations at n and $n+1$ are aligned, this formulation is sometimes called "co-rotational" by some authors.

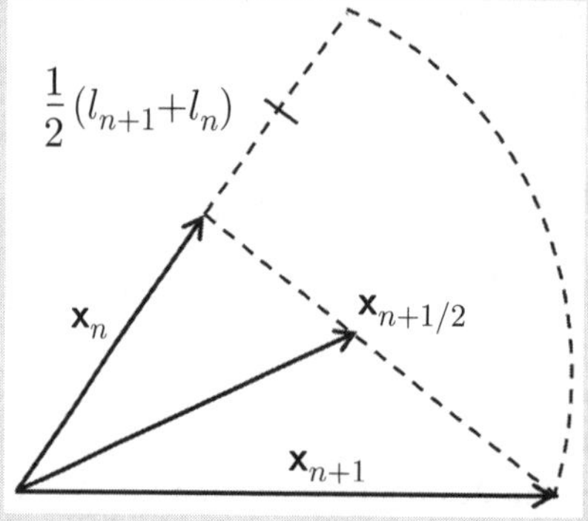

1.3.5 Tangent Stiffness Matrix

In order to determine the directional derivative in Equation (1.38b) it is necessary to find the linearization of a number of geometrical descriptors. For notational convenience and in recognition that the increment in mid-time-step acceleration given by Equation (1.37) represents a displacement, let $\Delta\mathbf{x}_{n+\frac{1}{2}} = \mathbf{u}$. Linearization is provided by the directional derivative for any position $\mathbf{x}$ as

$$D(\mathbf{x})[\mathbf{u}] = \frac{d}{d\epsilon}\bigg|_{\epsilon=0} (\mathbf{x} + \epsilon \mathbf{u}) = \mathbf{u}; \quad \mathbf{u} = \begin{bmatrix} u_1 \\ u_2 \end{bmatrix}. \tag{1.41}$$

A further example is the directional derivative of the length l^2 at time t, given as

$$\begin{aligned} Dl^2(\mathbf{x})[\mathbf{u}] &= \frac{d}{d\epsilon}\bigg|_{\epsilon=0} (\mathbf{x} + \epsilon \mathbf{u}) \cdot (\mathbf{x} + \epsilon \mathbf{u}) \\ &= \frac{d}{d\epsilon}\bigg|_{\epsilon=0} \mathbf{x} \cdot \mathbf{x} + 2\epsilon \mathbf{u} \cdot \mathbf{x} + \epsilon^2 \cdot \mathbf{u} \\ &= 2\mathbf{u} \cdot \mathbf{x}. \end{aligned} \tag{1.42}$$

The above equation can now be used to find the directional derivative of l as follows:

$$Dl^2(\mathbf{x})[\mathbf{u}] = 2l Dl(\mathbf{x})[\mathbf{u}]. \tag{1.43}$$

Substituting Equation (1.42) into equation yields

$$Dl(\mathbf{x})[\mathbf{u}] = \boldsymbol{n} \cdot \mathbf{u}. \tag{1.44}$$

The directional derivative $D(l^{-1})[\mathbf{u}]$ can now be found as

$$Dl^{-1}(\mathbf{x})[\mathbf{u}] = -l^{-2} Dl(\mathbf{x})[\mathbf{u}] = -l^{-2} \boldsymbol{n} \cdot \mathbf{u}. \tag{1.45}$$

Finally, the development of the tangent stiffness $\mathbf{K}$ requires the directional derivative of the unit normal $\boldsymbol{n}$ given in Equation (1.10c,d,e)$_d$. Observe that $\boldsymbol{n}$ is a function of l which itself is a function of $\mathbf{x}$. The directional derivative of $\boldsymbol{n}$ is found as

$$\begin{aligned} D\boldsymbol{n}(\mathbf{x})[\mathbf{u}] &= D\left(\frac{\mathbf{x}}{l}\right)[\mathbf{u}] \\ &= \mathbf{x} Dl^{-1}(\mathbf{x})[\mathbf{u}] + \frac{1}{l} D\mathbf{x}[\mathbf{u}] \\ &= -\frac{1}{l^2} (\boldsymbol{n} \cdot \mathbf{u}) \mathbf{x} + \frac{\mathbf{u}}{l} \\ &= -\frac{1}{l} (\boldsymbol{n} \cdot \mathbf{u}) \boldsymbol{n} + \frac{\mathbf{u}}{l}. \end{aligned} \tag{1.46}$$

The tangent stiffness $\mathbf{K}(\mathbf{x})$ can now be found using Equations (1.12a,b)$_a$, (1.35), (1.44), and (1.46) as

$$\begin{aligned} D\mathbf{T}(\mathbf{x})[\mathbf{u}] &= D\big(T\boldsymbol{n}(\mathbf{x})\big)[\mathbf{u}] \\ &= kD(l - L)[\mathbf{u}]\boldsymbol{n} + TD\big(\boldsymbol{n}(\mathbf{x})\big)[\mathbf{u}] \\ &= k(\boldsymbol{n} \cdot \boldsymbol{u})\boldsymbol{n} - T\frac{1}{l}(\boldsymbol{n} \cdot \mathbf{u})\boldsymbol{n} + T\frac{\mathbf{u}}{l} \end{aligned}$$

$$= \left[\left(k - \frac{T}{l}\right)(\boldsymbol{n} \otimes \boldsymbol{n})_{2\times 2} + \frac{T}{l}\boldsymbol{I}_{2\times 2} \right] \mathbf{u}$$

$$= \mathbf{K}(\mathbf{x})\mathbf{u}, \tag{1.47}$$

where the tangent stiffness is

$$\mathbf{K}(\mathbf{x}) = \left[\left(k - \frac{T}{l}\right)(\boldsymbol{n} \otimes \boldsymbol{n})_{2\times 2} + \frac{T}{l}\boldsymbol{I}_{2\times 2} \right]. \tag{1.48}$$

1.3.6 Mid-Point Rule Examples

It is expected that the implicit mid-point time integration should be able to produce reliable results using a much larger time step than the leap-frog explicit scheme for the same accuracy. For $m = 10$, $\mathbf{x}^{(0)} = [5, 5]^T$, $c = 0$, and $k = 50\,000$ Figures 1.9 and 1.10 show that the mid-point scheme, with a time step $\Delta t = 0.1$, produces the same results as the leap-frog scheme, with a time step $\Delta t = 0.02$. In both cases the high spring stiffness results in a predictable circular motion.

Furthermore, a correct formulation of a Newton–Raphson iterative solution should result in quadratic convergence of the residual norm. For the two-degrees-of-freedom spring-pendulum system this is clearly demonstrated in Figure 1.11.

In Example 1.2 it was shown that, in order to conserve energy for the spring pendulum simulation, it was necessary to calculate the length of the spring as $l_{n+1/2} = (l_{n+1} + l_n)/2$ as opposed to the more obvious $\|\mathbf{x}_{n+1/2}\|$. Figure 1.12

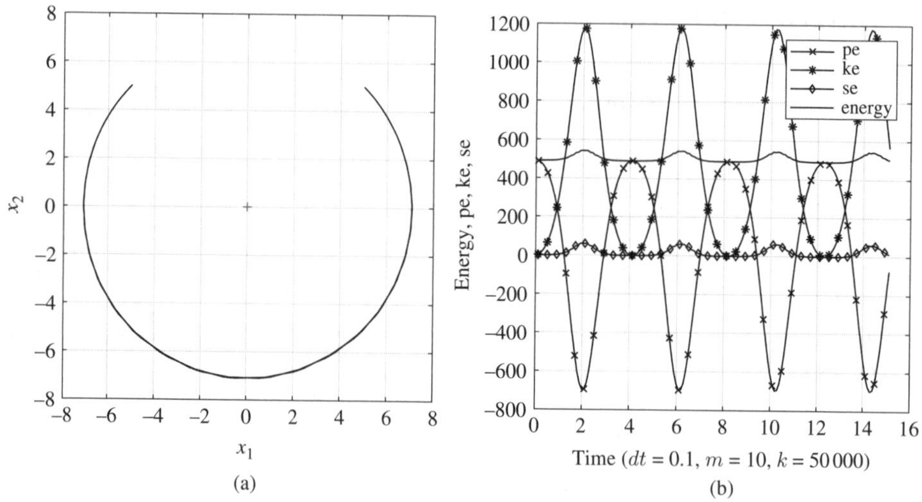

FIGURE 1.9 Spring-pendulum, mid-point, $\Delta t = 0.1$, $m = 10$, $k = 50\,000$: (a) Motion; (b) Energy–time.

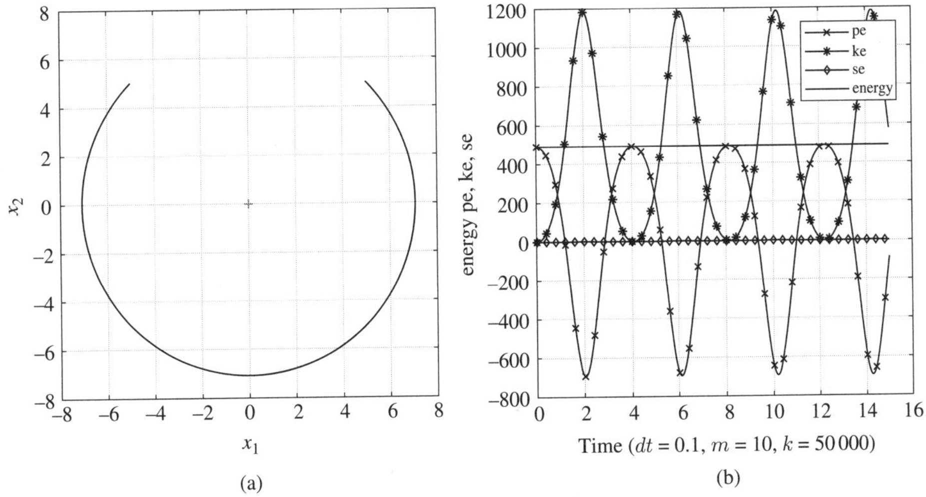

FIGURE 1.10 Spring-pendulum, leap-frog, $\Delta t = 0.02$, $m = 10$, $k = 50\,000$: (a) Motion; (b) Energy–time.

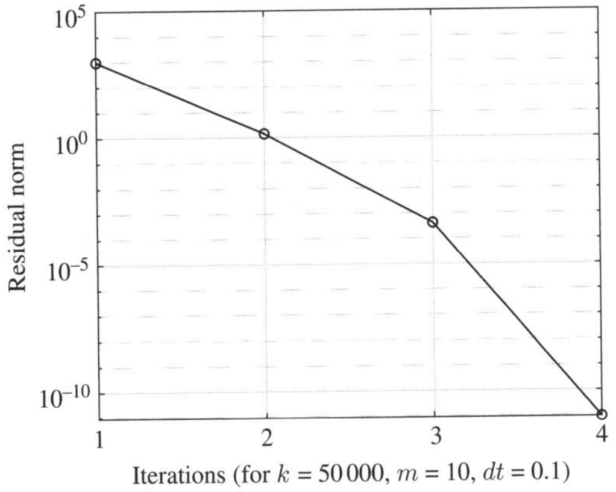

FIGURE 1.11 Mid-point integration: quadratic convergence for time $t = 15$.

shows the result of using the two alternatives, where it is clear that the average approach maintains the energy. It should be noted that, although the Newton–Raphson procedure converged, the algorithm used the tangent stiffness matrix given by Equation (1.48). However, this is not exactly correct since Equation (1.48) employed the directional derivative given by Equation (1.44), that is, $D(l)[\boldsymbol{u}]$ and not $D((l_{n+1} + l_n)/2))[\boldsymbol{u}]$.

Finally, Figure 1.13 shows a less predictable mid-point simulation with a time step of $\Delta t = 0.1$ and where the lower stiffness $k = 50$ results in a somewhat chaotic motion. The length calculation given in Example 1.2 is used, which makes the total energy remain constant. The mid-point algorithm is shown in Box 1.2.

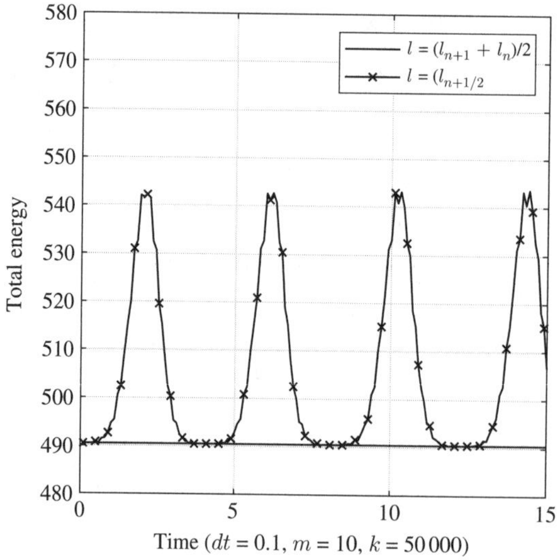

FIGURE 1.12 Spring-pendulum, mid-point, alternative length calculations: $\Delta t = 0.1$, $m = 10$, $k = 50\,000$.

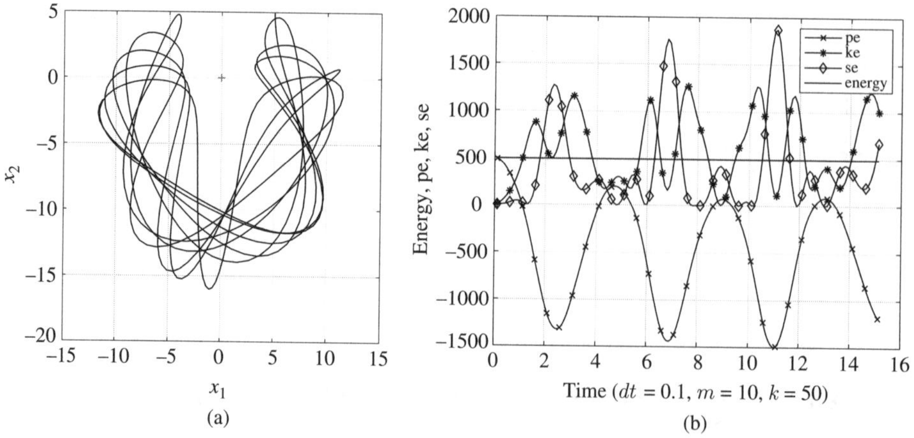

FIGURE 1.13 Spring-pendulum, mid-point (with energy conserving modified length formula $l_{n+1/2} = (l_n + l_{n+1})/2$), $\Delta t = 0.1$, $m = 10$, $k = 50$: (a) Motion ($t_{max} = 50$); (b) Energy–time ($t_{max} = 15$).

For the two-degrees-of-freedom spring-pendulum the leap-frog and mid-point rule MATLAB programs are presented at the end of the chapter in Boxes 1.3 and 1.4 respectively. Observe that in the leap-frog program energy calculation the velocity is adjusted as $\boldsymbol{v} - \boldsymbol{a}dt/2$ to ensure that the displacement and velocity are synchronized to the same time step.

BOX 1.2: Mid-point algorithm

- INPUT geometry $\mathbf{x}$, initial velocity $\mathbf{v}$, material properties m and k, and solution parameters
- INITIALIZE $\mathbf{v}_{n+1/2} = \mathbf{v}$, $\mathbf{a}_{n+1/2} = \mathbf{0}$
- WHILE $t < t_{max}$ (time steps)
  - PREDICT $\mathbf{v}$, $\mathbf{v}_{n+1/2}$, $\mathbf{x}$, and $\mathbf{x}_{n+1/2}$ (1.32a–d)
  - DO WHILE ($\|\mathbf{R}\|/\|\mathbf{F}\| >$ tolerance)
    - FIND $\mathbf{T}$ (1.12a,b)
    - FIND $\mathbf{R}$ (1.39)
    - SOLVE $(\mathbf{M}+\mathbf{K})\Delta\mathbf{a} = -\mathbf{R}$ (1.39)
    - CORRECT $\mathbf{a}^k_{n+1/2}$, $\mathbf{x}^k_{n+1/2}$, $\mathbf{x}^k_{n+1}$, $\mathbf{v}^k_{n+1}$ (1.40a–d)
    - SET $\mathbf{x}^k_{n+1} = \mathbf{x}^{k-1}_{n+1}$
    - (FIND kinetic, strain, and potential energies)
  - ENDDO
- ENDLOOP

Exercises

1. For the column problem given in Section 1.2.3, devise a numerical procedure to show that the static solution is $\theta \approx 167.42°$.
2. For the constrained spring-mass system shown in Figure 1.7,
   (a) show that the equations of motion are

$$ma_1 + k(x_1 - X_1) = -mg\cos\theta\sin\theta,$$

$$ma_2 + k(x_2 - X_2) = -mg\sin^2\theta;$$

   (b) for the initial conditions $\mathbf{x} = \mathbf{X}$ and velocity $\mathbf{v} = \mathbf{0}$ show that the solution to the above equations, in terms of the angular frequency ω, is

$$x_1(t) = X_1 + u_1^{\text{static}}(1 - \cos(\omega t)),$$

$$x_2(t) = X_2 + u_2^{\text{static}}(1 - \cos(\omega t)),$$

   where $\omega = (k/m)^{\frac{1}{2}}$ and the static displacements are

$$u_1^{\text{static}} = \frac{-mg\cos\theta\sin\theta}{k},$$

$$u_2^{\text{static}} = \frac{-mg\sin^2\theta}{k}.$$

3. (a) Referring to Example 1.2, show that the tangent stiffness matrix for the case when the length is calculated at the average $l_{n+1/2} = (l_{n+1} + l_n)/2$ is

$$\bar{\mathbf{K}}(\mathbf{x}_{n+1/2}) = \left[\left(k - \frac{T}{l_{n+1/2}}\right)(\boldsymbol{n} \otimes \boldsymbol{m})_{2\times 2} + \frac{T}{l_{n+1/2}}\boldsymbol{I}_{2\times 2}\right],$$

where $T = k(l_{n+1/2} - L)$, $\boldsymbol{n} = \boldsymbol{x}_{n+1/2}/l_{n+1/2}$, and $\boldsymbol{m} = \boldsymbol{x}_{n+1}/l_{n+1}$.

(b) Change the mid-point program to incorporate $l_{n+1/2} = (l_{n+1}+l_n)/2$ and $\bar{\mathbf{K}}(\mathbf{x})$, and confirm that the total energy is constant and that convergence is quadratic.

BOX 1.3: Leap-frog time integration

```
function SpringpendulumLeapFrog
%Leap Frog time integration
%------------------------
clear;clf;clc;cla;whitebg('white');
% mass coords
x=[5;5];L=norm(x);xprev=x;
% initial velocities and accelerations
v=[0;0];vprev=v;ahalf=[0;0];
% rod stiffness, mass,force
springK=50000;mass=10;g=[0;-9.81];
M=[mass,0;0,mass];F=M*g;
% control data
tmax=30;dt=0.02;t=0;count=0;
% start velocity
a=accel(x,L,F,M,springK);
v=v+a*dt/2;
%time loop
while t<tmax
    count=count+1;
% leap frog time integration
    x=x+v*dt;
    a=accel(x,L,F,M,springK);
    v=v+a*dt;
% update time, data for output
   t=t+dt;
   tt(count)=t;xx(count)=x(1);yy(count)=x(2);
% energy calculation
   l=norm(x);vel=norm(v-a*dt/2);
   ke(count)=0.5*mass*vel^2;
   se(count)=0.5*springK*(L-l)^2;
   pe(count)=mass*-x'*g;
   energy(count)=(ke(count)+se(count)+pe(count));
end
% Graphics
   plot(0,0,'+r'); hold on
```

(continued)

Box 1.3: Leap-frog time integration *(cont.)*

```
   plot(xx,yy,'LineWidth',1,'Color','k');
   axis square;xlabel('x(1)');ylabel('x(2)');
   axis([-8, 8, -8, 8]);grid on
end
function a=accel(x,L,F,M,springK)
%geometry
l=norm(x);n=x/l;
Fint=springK*(l-L);
T=Fint*n;
%accelerations
a=M\(F-T);
end
```

BOX 1.4: Mid-point time integration

```
% Mid-point time integration
%------------------------
clear;clf;clc;cla;whitebg('white');
% mass coords
x=[5;5];L=norm(x);xprev=x;
% initial velocities and accelerations
v=[0;0];vprev=v;ahalf=[0;0];
% rod stiffness, mass,force
springK=50;mass=10;g=[0;-9.81];
M=[mass,0;0,mass];F=M*g;
% control data
tmax=30;dt=0.1;t=0;count=0;
cnorm=1e-6;miter=50;
% time loop
while t<tmax
  t=t+dt;rnorm=10^6;
  count=count+1;niter=0;
% predictor
  v=vprev+dt*ahalf;vhalf=(vprev+v)/2;
  x=xprev+dt*vhalf;xhalf=(xprev+x)/2;
% corrector
  while ((rnorm>cnorm)&&(niter<miter))
   niter=niter+1;
% residual
% find length
   l=norm(xhalf);n=xhalf/l;
   Fint=springK*(l-L);T=Fint*n;
   resid=M*ahalf+T-F;rnorm=norm(resid);
%'stiffness'
   Ka=M+0.25*dt*dt*((springK-Fint/l)*(n*n')...
       + (Fint/l)*eye(2));
   da=-Ka\resid;
% update
```

(continued)

Box 1.4: Mid-point time integration *(cont.)*

```
    ahalf=ahalf+da;
    xhalf=xhalf+0.25*dt*dt*da;
    x=x+0.5*dt*dt*da;
    v=v+dt*da;
  end
  xprev=x;vprev=v;
  tt(count)=t;xx(count)=x(1);yy(count)=x(2);
% energy calculation
  l=norm(x);vel=norm(v);
  ke(count)=0.5*mass*vel^2;
  se(count)=0.5*springK*(L-l)^2;
  pe(count)=mass*-x'*g;
  energy(count)=(ke(count)+se(count)+pe(count));
end
% graphics
  plot(0,0,'+r'); hold on
  plot(xx,yy,'LineWidth',1,'Color','k');
  axis square;xlabel('x(1)');ylabel('x(2)');
  grid on
```

CHAPTER TWO

DYNAMIC ANALYSIS OF THREE-DIMENSIONAL TRUSSES

2.1 INTRODUCTION

This chapter describes the dynamic behavior of nonlinear three-dimensional pin-jointed trusses. In doing so, the chapter aims to introduce in the simpler context of trusses a number of concepts that will be used in the more complex context of solids. The chapter follows closely the presentation of pin-jointed trusses given in Chapter 3 of NL-Statics. However, a number of formulas and derivations will be repeated for completeness, although the reader will occasionally be referred to the derivations in the NL-Statics volume.

The chapter starts with a kinematic and dynamic description of an individual pin-jointed axial truss member. In particular, the assumption of concentrated mass at the ends of the rod will be introduced at this stage. This is followed by the assembly of individual rod equations into the global dynamic equilibrium equations of a complete truss.

The global dynamic equilibrium equations will be recast in Section 2.3 in the form of a variational principle, first in the form of the principle of virtual work and then as the more advanced Hamilton principle, in which concepts such as the Lagrangian, the action integral, phase space, and the simplectic product are introduced. These concepts will be generalized in Chapter 3 for three-dimensional solids.

The set of dynamic equilibrium equations require the use of a discrete time-stepping scheme in order to advance the solution in time, a process also known as the *numerical time integration* of the dynamic equilibrium equations. The same leap-frog and mid-point time integration presented in Chapter 1 can be used in the context of pin-jointed trusses. In the case of the leap-frog scheme, this leads to an explicit solution procedure whereby the next positions at each time step can be evaluated from the previous positions and current external forces without the solution of a nonlinear set of equations involving the evaluation of a tangent operator (called the tangent matrix). The explicit nature of this time-stepping scheme makes it very attractive from a computational view point but it requires very small time steps in order to remain stable, as already discussed in Chapter 1.

The mid-point time integration scheme is free from stability constraints but requires the solution of a nonlinear system of equations at each time step. A Newton–Raphson solution of this system will be presented which will require the evaluation of a dynamic tangent operator at each iteration. Finally, in order to provide a preliminary introduction to a commonly used type of time integration scheme, namely the Newmark family, the trapezoidal time-stepping scheme will be described. Similarly to the mid-point rule, this scheme leads to a set of nonlinear equations at each time step but has the advantage of being unconditionally stable for any time-step size.

The simplicity of the dynamic equilibrium equations for the pin-jointed truss can also be used to introduce some important concepts, such as conservation of linear and angular momentum and energy. This will be discussed both before and after the introduction of discrete time integration. Moreover, the concept of simplectic integrators as those that preserve the simplectic product in the phase space will be explained in some detail.

2.2 DYNAMIC EQUILIBRIUM

2.2.1 Truss Member – Kinematics

A general pin-jointed three-dimensional truss comprises an assembly of individual truss members, or, in finite element terms, individual truss elements. The kinematics of a single truss member, shown in Figure 2.1, is determined by the initial (at time $t = 0$) and current (at time t) positions of the nodal end points a and b. Capital letters $\boldsymbol{X}_a$ and $\boldsymbol{X}_b$ describe the initial positions and lower case $\boldsymbol{x}_a$ and $\boldsymbol{x}_b$ denote the current positions. The velocity and acceleration of the nodes are given by the first and second time derivatives of $\boldsymbol{x}_a$ and $\boldsymbol{x}_b$ as

$$\boldsymbol{v}_a = \frac{d\boldsymbol{x}_a}{dt}\;;\; \boldsymbol{a}_a = \frac{d\boldsymbol{v}_a}{dt} = \frac{d^2\boldsymbol{x}_a}{dt^2}, \tag{2.1a,b}$$

$$\boldsymbol{v}_b = \frac{d\boldsymbol{x}_b}{dt}\;;\; \boldsymbol{a}_b = \frac{d\boldsymbol{v}_b}{dt} = \frac{d^2\boldsymbol{x}_b}{dt^2}. \tag{2.1c,d}$$

The length of the truss member at the initial and current configurations can be evaluated as the norm of the vector joining nodes a to b as

$$L = \sqrt{(\boldsymbol{X}_b - \boldsymbol{X}_a)\cdot(\boldsymbol{X}_b - \boldsymbol{X}_a)}\;;\;\; l(t) = \sqrt{(\boldsymbol{x}_b - \boldsymbol{x}_a)\cdot(\boldsymbol{x}_b - \boldsymbol{x}_a)}. \tag{2.2a,b}$$

The change in length will typically be accompanied by a change in cross-sectional area from A to $a(t)$, the volume from V to $v(t)$, and the density from ρ_R to $\rho(t)$.

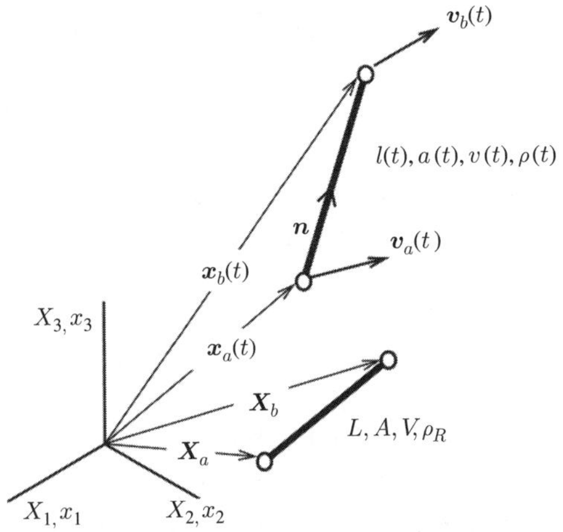

FIGURE 2.1 Truss element: Kinematics.

These changes are related by simple geometric considerations and the fact that the total mass of the truss member must remain constant, hence

$$v(t) = a(t)l(t) \; ; \quad V = AL \; ; \quad m = \rho v = \rho_R V \; ; \quad \rho = \frac{\rho_R}{J}, \qquad \text{(2.3a,b,c,d)}$$

where $J = v/V$ denotes the volume ratio between current and initial configurations.

2.2.2 Truss Member – Forces

Any change in truss member length will produce axial stress determined by the constitutive relation between the resulting strain and stress. In large strain nonlinear situations, a number of alternative constitutive relations are available. These are studied in some detail in NL-Statics; see Sections 1.3 or 3.3. A simple relationship between the logarithmic strain and the Kirchhoff stress will be adopted here whereby the strain is measured as the natural logarithm* of the current stretch $\lambda = l/L$ as

$$\varepsilon = \ln \lambda = \ln \frac{l}{L}. \tag{2.4}$$

A linear constitutive relationship between the nonlinear strain measure and the Kirchhoff stress τ will be assumed as

$$\tau = E\varepsilon, \tag{2.5}$$

* The natural logarithm strain makes engineering sense as it is the integration of the engineering strains dl/l between L and l.

where E is the well-known Young's modulus and the Kirchhoff stress is related to the Cauchy stress σ by the volume ratio as

$$\tau = J\sigma. \tag{2.6}$$

The magnitude of the tensile force in the truss member can now be obtained by multiplying the Cauchy stress by the cross-sectional area. Using Equations (2.3a,b,c,d)$_{a,b}$ to (2.6) it is easy to show that

$$T = \sigma a = E\frac{V}{l}\varepsilon. \tag{2.7}$$

Observe that in the small strain case where $l \approx L$ the logarithmic strain ε above would simply be the engineering strain $\Delta l/l$ and the ratio V/l would be the cross-sectional area A, thereby leading to the common expression for the tension $T = EA\Delta l/l$.

The tensile force vector at nodes a and b can be found using the unit vector $\boldsymbol{n}$ (see Figure 2.1), given as

$$\boldsymbol{n} = \frac{\boldsymbol{x}_b - \boldsymbol{x}_a}{l}. \tag{2.8}$$

Assuming that T is positive in tension, the corresponding forces vectors at a and b are (see Figure 2.2)

$$\boldsymbol{T}_a = -T\boldsymbol{n}\,; \quad \boldsymbol{T}_b = T\boldsymbol{n} = -\boldsymbol{T}_a. \tag{2.9a,b}$$

In addition to the tensile forces caused by deformation, the dynamic equilibrium equations will require consideration of the inertial forces due to the mass and acceleration. A commonly used simplified but effective technique is to consider the entire mass $m = \rho_R V = \rho v$ of the truss member to be divided into two equal amounts, each assigned to the nodes a and b as shown in Figure 2.2.

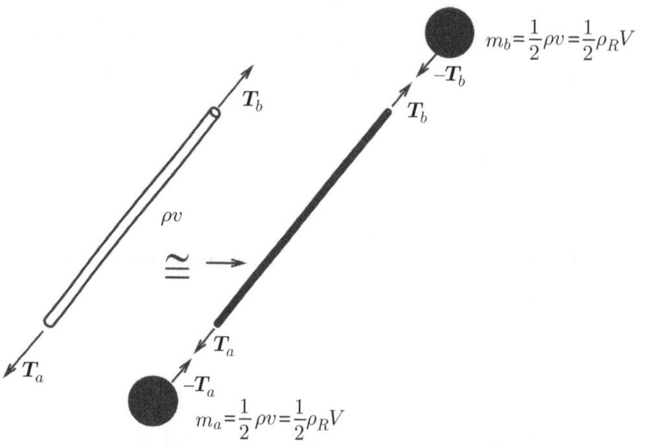

FIGURE 2.2 Truss element: Lumped-mass model.

This technique is known as a *lumped-mass* model and will be justified at greater length in Chapter 4. Essentially the lumped-mass model separates the mass of the system, which is concentrated at the nodes, from the stiffness, which is provided by the joining members. In this way, the lumped-mass model replaces the forces acting on the distributed mass of the system by forces acting on masses concentrated at the nodes. These forces are provided by the stiffness of the members joining the nodes and by any external forces applied to the nodes. This leads to a particularly simple form of the dynamic equilibrium equations that are formally identical to the equilibrium of a system of particles. Note that in Figure 2.2 the member itself, which is now massless, is already in equilibrium by virtue of the fact that $\boldsymbol{T}_a + \boldsymbol{T}_b = \boldsymbol{0}$. From Newton's third law the forces that the truss member exerts on the nodes are equal and opposite to $\boldsymbol{T}_a$ and $\boldsymbol{T}_b$, as shown in Figure 2.2.

2.2.3 Dynamic Equilibrium of a Pin-Jointed Truss

Consider a general truss, as shown in Figure 2.3(b) comprising members $e = 1, 2, \ldots, M$ and nodes $a = 1, 2, \ldots, N$. The total mass of node a, namely m_a, is obtained by adding the contributions, $m_e/2$, from all members (e) (1 to M_a) containing node a ($e \ni a$) as

$$m_a = \sum_{\substack{e=1 \\ e \ni a}}^{M_a} \frac{1}{2} m_e. \tag{2.10}$$

The forces applied to node a will comprise contributions from the tensions $\boldsymbol{T}_a^e$ from all members e connected to node a in addition to a possible external force $\boldsymbol{F}_a$. The dynamic equilibrium equation for node a is simply given by Newton's second law

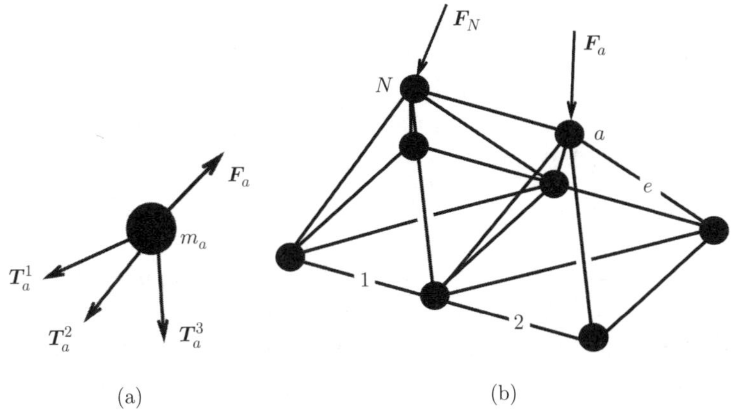

FIGURE 2.3 Pin-jointed truss: (a) Nodal forces; (b) Global forces.

$$m_a \boldsymbol{a}_a = \boldsymbol{F}_a - \boldsymbol{T}_a \,; \quad \boldsymbol{T}_a = \sum_{\substack{e=1 \\ e \ni a}}^{M_a} \boldsymbol{T}_a^e. \tag{2.11a,b}$$

In this expression, $\boldsymbol{T}_a$ now represents the resultant of all the member tensile forces connected to node a, see Figure 2.3(a). Equation (2.11a,b) is a system of ordinary differential equations in time for the position vector $\boldsymbol{x}_a(t)$ at node a, which can be rewritten as

$$m_a \frac{d^2 \boldsymbol{x}_a}{dt^2} = \boldsymbol{F}_a - \boldsymbol{T}_a \,; \quad a = 1, 2, \ldots, N. \tag{2.12}$$

Observe that in the above equations $\boldsymbol{T}_a$ is a function of $\boldsymbol{x}_b$ for all the nodes $\boldsymbol{x}_b$ to which node a is connected. Alternatively the above equations can be re-expressed using a global vector notation as

$$\mathbf{Ma} = \mathbf{F} - \mathbf{T}(\mathbf{x}), \tag{2.13}$$

where the global vectors $\mathbf{a}$, $\mathbf{T}$, and $\mathbf{F}$ contain nodal entries,

$$\mathbf{x} = \begin{bmatrix} \boldsymbol{x}_1 \\ \boldsymbol{x}_2 \\ \vdots \\ \boldsymbol{x}_N \end{bmatrix} ; \ \mathbf{a} = \begin{bmatrix} \boldsymbol{a}_1 \\ \boldsymbol{a}_2 \\ \vdots \\ \boldsymbol{a}_N \end{bmatrix} ; \ \mathbf{T}(\mathbf{x}) = \begin{bmatrix} \boldsymbol{T}_1 \\ \boldsymbol{T}_2 \\ \vdots \\ \boldsymbol{T}_N \end{bmatrix} ; \ \mathbf{F} = \begin{bmatrix} \boldsymbol{F}_1 \\ \boldsymbol{F}_2 \\ \vdots \\ \boldsymbol{F}_N \end{bmatrix}, \tag{2.14a,b,c,d}$$

where $\mathbf{a} = d^2\mathbf{x}/dt^2$. The so-called mass matrix $\mathbf{M}$ is a diagonal matrix defined as

$$\mathbf{M} = \begin{bmatrix} m_1 \boldsymbol{I} & \mathbf{0} & \cdots & \mathbf{0} \\ \mathbf{0} & m_2 \boldsymbol{I} & \cdots & \mathbf{0} \\ \vdots & \vdots & & \vdots \\ \mathbf{0} & \mathbf{0} & \cdots & m_N \boldsymbol{I} \end{bmatrix} ; \ \boldsymbol{I} = \begin{bmatrix} 1\,0\,0 \\ 0\,1\,0 \\ 0\,0\,1 \end{bmatrix} ; \ \mathbf{0} = \begin{bmatrix} 0\,0\,0 \\ 0\,0\,0 \\ 0\,0\,0 \end{bmatrix}. \tag{2.15a,b,c}$$

The time integration of Equation (2.13) using various time-stepping schemes will be discussed in Section 2.4. However, before doing so it is helpful to recast the dynamic equilibrium equation in terms of a variational principle, known as the least action principle or Hamilton's principle, as described in the next section.

2.3 DYNAMIC VARIATIONAL PRINCIPLES

It is well known that the static equilibrium of an elastic solid or indeed a pin-jointed truss can be formulated as a stationary point of the total potential energy. This interpretation of equilibrium can be beneficial when deploying discretization processes for the case of a solid. More generally, the examination of the stationary conditions of the total potential energy provide an understanding of the phenomenon of structural stability. A similar variational formulation is possible in the case of dynamic equilibrium by employing the concepts of kinetic energy, potential energy, and the Lagrangian function.

The introduction of the Lagrangian function, defined as the kinetic energy minus the sum of the internal and external potential energies, leads to the formulation of the Lagrange equations, which provide an alternative description of the dynamic equilibrium equations. However, of greater consequence is that the stationary condition of the integral of the Lagrangian over the time span of the dynamic system, known as the action integral, provides not only the Lagrange equations and thus an expression of dynamic equilibrium, but also a consistent methodology for the development of time-stepping numerical integration schemes.

These topics are explored analytically in this section and in a time-discrete manner in Section 2.4 for the simple pin-jointed truss case; will be further developed for solids in Chapter 3.

2.3.1 Kinetic Energy, Potential Energy, and the Lagrangian

A detailed discussion of the total potential energy due to internal strain energy and external potential energy for a pin-jointed truss has been presented in NL-Statics, Section 3.5; however, the essential derivations are repeated below for completeness. Insofar as the total energy of a dynamic system comprises the addition of kinetic energy and total potential energy, it is necessary to begin by considering the kinetic energy.

As explained above, the distributed mass of the pin-jointed truss is modeled by assuming the mass to be concentrated at the nodes of the truss. This enables the total kinetic energy K to be found as the sum of the kinetic energies of the nodes as

$$K(\mathbf{v}) = \sum_{a=1}^{N} \frac{1}{2} m_a \boldsymbol{v}_a \cdot \boldsymbol{v}_a \,; \quad \boldsymbol{v}_a = \frac{d\boldsymbol{x}_a}{dt}, \tag{2.16a,b}$$

or in matrix/vector form as

$$K(\mathbf{v}) = \frac{1}{2}\mathbf{v}^T \mathbf{M}\mathbf{v}\,; \quad \mathbf{v} = \begin{bmatrix} \boldsymbol{v}_1 \\ \boldsymbol{v}_2 \\ \vdots \\ \boldsymbol{v}_N \end{bmatrix} = \frac{d\mathbf{x}}{dt}. \tag{2.17a,b}$$

To derive an expression for the total elastic strain energy in the truss, recall from NL-Statics, Section 3.3, that the Kirchhoff stress τ, for a typical member e, see Equation (2.5), can be derived from an elastic potential energy $\Psi(\varepsilon_e)$ per unit initial volume as

$$\tau_e = E\varepsilon_e = \frac{d\Psi(\varepsilon)}{d\varepsilon_e}\,; \quad \Psi(\varepsilon_e) = \frac{1}{2}E\varepsilon_e^2\,; \quad \varepsilon_e = \ln\frac{l_e}{L_e}. \tag{2.18a,b,c}$$

If the initial volume of the member e is V_e, the total elastic strain energy (internal energy) for the complete truss can be defined as

$$\Pi_{\text{int}}(\mathbf{x}) = \sum_{e=1}^{M} \Psi(\varepsilon_e)V_e = \sum_{e=1}^{M} \frac{1}{2}E\,\varepsilon_e^2\,V_e. \tag{2.19}$$

For the case of constant external forces, such as a self weight (gravity force), it is also possible to define an external potential energy as

$$\Pi_{\text{ext}}(\mathbf{x}) = -\sum_{a=1}^{N} \boldsymbol{F}_a \cdot \boldsymbol{x}_a = -\mathbf{F}^T\mathbf{x}. \tag{2.20}$$

For instance, in the simple case where $\boldsymbol{F}_a$ is the gravity force acting downward on the mass m_a at node a with a gravity constant g, the force is

$$\boldsymbol{F}_a = m_a\,g \begin{bmatrix} 0 \\ 0 \\ -1 \end{bmatrix}, \tag{2.21}$$

then $\Pi_{\text{ext}}(\mathbf{x})$ becomes the standard gravity potential energy,

$$\Pi_{\text{ext}}(\mathbf{x}) = -\sum_{a=1}^{N} m_a\,g\boldsymbol{x}_a \cdot \begin{bmatrix} 0 \\ 0 \\ -1 \end{bmatrix} = \sum_{a=1}^{N} m_a\,g\,h_a, \tag{2.22}$$

where h_a is the height of node a with reference to the appropriate coordinate. The total energy of the dynamic system can now be defined as the sum of the kinetic energy, the elastic internal energy, and the external potential energy as

$$\Pi(\mathbf{x}, \mathbf{v}) = K(\mathbf{v}) + \Pi_{\text{int}}(\mathbf{x}) + \Pi_{\text{ext}}(\mathbf{x}), \tag{2.23}$$

where it is explicitly noted that K is a function of the nodal velocities $\mathbf{v} = \dot{\mathbf{x}}$, whereas Π_{int} and Π_{ext} are functions of the nodal positions $\mathbf{x}$.

Following on from Equation (2.23), the concept of the *Lagrangian* function $\mathcal{L}$ can now be considered. This function is usually applied to a system of particles or a general mechanical system comprising a discrete number of degrees of freedom, usually represented by a vector of variables $\boldsymbol{q}(t)$, which for the pin-jointed truss are the positions of the nodes $\mathbf{x}(t)$. The Lagrangian $\mathcal{L}$ is generally defined as the kinetic energy minus the sum of the internal and external potential energies as

$$\mathcal{L}(\mathbf{x},\mathbf{v}) = K(\mathbf{v}) - \Pi_{\text{int}}(\mathbf{x}) - \Pi_{\text{ext}}(\mathbf{x}). \tag{2.24}$$

The following section will employ the above definition to rewrite the dynamic equilibrium equations in the form of the so-called Lagrange equations and to derive an equivalent variational principle known as Hamilton's principle, or the least action principle.

2.3.2 Lagrange Equations

Using the Lagrangian function given by Equation (2.24) it is possible to show that the dynamic equilibrium Equation (2.13) can be expressed in the form of the equations of Lagrange as

$$\frac{d}{dt}\frac{\partial \mathcal{L}(\mathbf{x},\mathbf{v})}{\partial \mathbf{v}} - \frac{\partial \mathcal{L}(\mathbf{x},\mathbf{v})}{\partial \mathbf{x}} = \mathbf{0}. \tag{2.25}$$

To show this, recall the definition of the Lagrangian given in Equation (2.24) and Equation (2.16a,b), which enables the first term in Equation (2.25) to yield the inertial forces as

$$\begin{aligned}\frac{d}{dt}\frac{\partial \mathcal{L}(\mathbf{x},\mathbf{v})}{\partial \mathbf{v}} &= \frac{d}{dt}\frac{\partial K(\mathbf{v})}{\partial \mathbf{v}}\\ &= \frac{d}{dt}\mathbf{M}\mathbf{v}\\ &= \mathbf{M}\mathbf{a}.\end{aligned} \tag{2.26}$$

The second term in Equation (2.25) gives rise to the internal and external forces; this can be shown by noting that

$$-\frac{\partial \mathcal{L}(\mathbf{x},\mathbf{v})}{\partial \mathbf{x}} = \frac{\partial \Pi_{\text{int}}(\mathbf{x})}{\partial \mathbf{x}} + \frac{\partial \Pi_{\text{ext}}(\mathbf{x})}{\partial \mathbf{x}}. \tag{2.27}$$

The derivative of the external energy term can easily be obtained from Equation (2.20) to give

$$\frac{\partial \Pi_{\text{ext}}(\mathbf{x})}{\partial \mathbf{x}} = \frac{\partial}{\partial \mathbf{x}}\left(-\mathbf{x}^T\mathbf{F}\right) = -\mathbf{F}. \tag{2.28}$$

The derivative of the internal energy term gives the internal force vector $\mathbf{T}$ as

$$\frac{\partial \Pi_{\text{int}}(\mathbf{x})}{\partial \mathbf{x}} = \mathbf{T}. \tag{2.29}$$

For now, the proof of this equation will be delayed but it is given in Example 2.3 below. Substituting Equations (2.26)–(2.29) into Equation (2.25) gives the dynamic equilibrium equations as

$$\mathbf{Ma} = \mathbf{F} - \mathbf{T}, \tag{2.30}$$

which shows that the equations of Lagrange (2.25) are equivalent to the set of dynamic equilibrium equations derived from Newton's second law in Section 2.2 and in particular Equation (2.13). The significance of this equivalence will become clear in Section 2.3.3, where the Lagrange equations are derived from a variational statement known as Hamilton's principle.

EXAMPLE 2.3: Prove Equation (2.29)

Recall Equations (2.18a,b,c) and (2.19) as $\Pi_{\text{int}}(\mathbf{x}) = \sum_{e=1}^{M} \Psi(\varepsilon_e(\mathbf{x}))V_e$, where $\varepsilon_e = \ln l_e/L_e$ and the Kirchhoff stress for member e is $\tau_e = d\Psi_e(\varepsilon)/d\varepsilon_e$; consequently,

$$\begin{aligned}\frac{\partial \Pi_{\text{int}}}{\partial \mathbf{x}} &= \sum_{e=1}^{M} V_e \frac{d\Psi_e}{d\varepsilon_e}\frac{d\varepsilon_e}{dl_e}\frac{\partial l_e}{\partial \mathbf{x}} \\ &= \sum_{e=1}^{M} V_e \tau_e \frac{1}{l_e}\frac{\partial l_e}{\partial \mathbf{x}} \\ &= \sum_{e=1}^{M} T_e \frac{\partial l_e}{\partial \mathbf{x}},\end{aligned} \tag{a}$$

where Equation (2.7) for the member tension T_e has been used. From Equation (2.2a,b)$_b$ for member e joining nodes a and b, $l_e^2 = (\boldsymbol{x}_b - \boldsymbol{x}_a)\cdot(\boldsymbol{x}_b - \boldsymbol{x}_a)$, and therefore

$$\frac{\partial l_e^2}{\partial \boldsymbol{x}_a} = 2l_e \frac{\partial l_e}{\partial \boldsymbol{x}_a}\,; \quad \frac{\partial l_e^2}{\partial \boldsymbol{x}_b} = 2l_e \frac{\partial l_e}{\partial \boldsymbol{x}_b},$$

$$\frac{\partial l_e^2}{\partial \boldsymbol{x}_a} = \frac{\partial}{\partial \boldsymbol{x}_a}\left[(\boldsymbol{x}_b - \boldsymbol{x}_a)\cdot(\boldsymbol{x}_b - \boldsymbol{x}_a)\right] = -2(\boldsymbol{x}_b - \boldsymbol{x}_a),$$

$$\frac{\partial l_e^2}{\partial \boldsymbol{x}_b} = \frac{\partial}{\partial \boldsymbol{x}_b}\left[(\boldsymbol{x}_b - \boldsymbol{x}_a)\cdot(\boldsymbol{x}_b - \boldsymbol{x}_a)\right] = 2(\boldsymbol{x}_b - \boldsymbol{x}_a).$$

Combining the above equations gives

$$\frac{\partial l_e}{\partial \boldsymbol{x}_a} = \frac{-(\boldsymbol{x}_b - \boldsymbol{x}_a)}{l_e} = -\boldsymbol{n}_e\,; \quad \frac{\partial l_e}{\partial \boldsymbol{x}_b} = \frac{(\boldsymbol{x}_b - \boldsymbol{x}_a)}{l_e} = \boldsymbol{n}_e,$$

(continued)

Example 2.3: *(cont.)*

where $\boldsymbol{n}_e$ is the unit vector defining the orientation of the member (see Equation (2.8)) and the member forces are given as $\boldsymbol{T}_a^e = -T_e\boldsymbol{n}_e$ and $\boldsymbol{T}_b^e = T_e\boldsymbol{n}_e$; see Equation (2.9a,b).

It is now evident that the $\boldsymbol{x}_a$ component of the derivative with respect to $\mathbf{x}$ in Equation (a) above becomes

$$\frac{\partial \Pi_{\text{int}}}{\partial \boldsymbol{x}_a} = \sum_{e=1}^{M} \boldsymbol{T}_a^e = \boldsymbol{T}_a,$$

or in block vector notation

$$\frac{\partial \Pi_{\text{int}}}{\partial \mathbf{x}} = \mathbf{T}.$$

2.3.3 Action Integral and Hamilton's Principle

The Lagrange equations derived in the previous section are in fact the stationary condition of a variational principle due to Hamilton. In order to demonstrate this it is first necessary to introduce the *action integral function* as the integral of the Lagrangian taken between two fixed time steps t_1 and t_2 as

$$\mathcal{A}\big(\mathbf{x}(t)\big) = \int_{t_1}^{t_2} \mathcal{L}\big(\mathbf{x}(t), \dot{\mathbf{x}}(t)\big)\, dt. \tag{2.31}$$

Note that $\mathcal{A}$ is a functional of the time-dependent function $\mathbf{x}(t)$. This dependence is further emphasized by observing that in the above equation the Lagrangian $\mathcal{L}$ has been written as a function of $\mathbf{x}(t)$ and $\dot{\mathbf{x}}(t)$ rather than $\mathbf{x}$ and $\mathbf{v}$ as in Section 2.3.2, although, of course, $\mathbf{v} = \dot{\mathbf{x}}(t)$.

Hamilton's principle, also known as the *least action principle*, states that the path $\mathbf{x}(t)$ followed by a system moving according to the dynamic equilibrium equations between two positions at times t_1 and t_2 represents a stationary point of the action integral as defined by Equation (2.31). In other words, the stationary conditions of the integral in Equation (2.31) are the equations of Lagrange. Figure 2.4 provides an illustration of this principle, showing alternative paths from $\mathbf{x}(t_1)$ to $\mathbf{x}(t_2)$. The correct path that satisfies dynamic equilibrium during this journey is obtained by determining the stationary point, that is, the minimum or maximum, of the action integral. Consequently, in order to show that Hamilton's principle leads to the Lagrange equations it is necessary to find the stationary conditions of the functional $\mathcal{A}\big(\mathbf{x}(t)\big)$ with respect to the paths in time of all nodal positions $\mathbf{x}$. This implies finding the path $\mathbf{x}(t)$ from t_1 to t_2 such that the directional derivative of $\mathcal{A}$ vanishes. Using the definition of the directional derivative given in Section 2.3

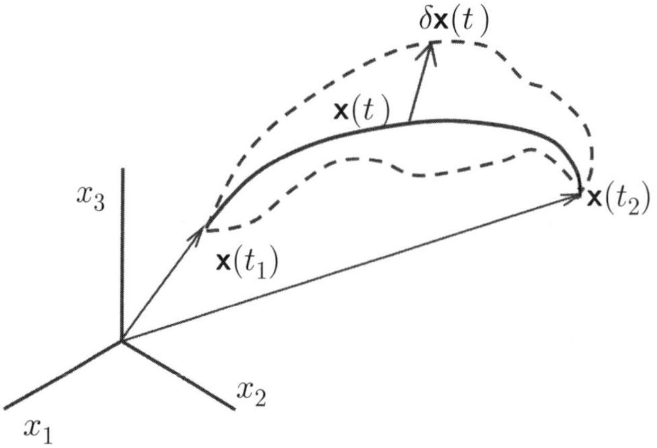

FIGURE 2.4 Alternative paths from $\mathbf{x}(t_1)$ to $\mathbf{x}(t_2)$ for an individual node.

of NL-Statics, this is expressed as

$$D\mathcal{A}\big(\mathbf{x}(t)\big)[\delta\mathbf{x}(t)] = \frac{d}{d\epsilon}\bigg|_{\epsilon=0} \mathcal{A}\big(\mathbf{x}(t) + \epsilon\delta\mathbf{x}(t)\big) = 0, \tag{2.32}$$

where $\delta\mathbf{x}(t)$ is an arbitrary variation of $\mathbf{x}(t)$ as shown in Figure 2.4 but subject to the start and end conditions

$$\delta\mathbf{x}(t_1) = \delta\mathbf{x}(t_2) = 0. \tag{2.33}$$

Substituting Equation (2.31) for $\mathcal{A}\big(\mathbf{x}(t)\big)$ into Equation (2.32) yields the stationary conditions of $\mathcal{A}$ as

$$\begin{aligned} D\mathcal{A}\big(\mathbf{x}(t)\big)[\delta\mathbf{x}(t)] &= \frac{d}{d\epsilon}\bigg|_{\epsilon=0} \int_{t_1}^{t_2} \mathcal{L}\big(\mathbf{x}(t) + \epsilon\delta\mathbf{x}(t), \dot{\mathbf{x}}(t)\big) + \epsilon\delta\dot{\mathbf{x}}(t)\big)\, dt \\ &= \int_{t_1}^{t_2} \left(\delta\mathbf{x}^T \frac{\partial \mathcal{L}}{\partial \mathbf{x}} + \delta\dot{\mathbf{x}}^T \frac{\partial \mathcal{L}}{\partial \dot{\mathbf{x}}}\right) dt, \end{aligned} \tag{2.34}$$

where $\delta\dot{\mathbf{x}} = d(\delta\mathbf{x})/dt$. Integrating the second term in this expression by parts gives

$$\begin{aligned} D\mathcal{A}\big(\mathbf{x}(t)\big)[\delta\mathbf{x}(t)] &= \int_{t_1}^{t_2} \delta\mathbf{x}^T \frac{\partial \mathcal{L}}{\partial \mathbf{x}}\, dt + \delta\mathbf{x}^T \frac{\partial \mathcal{L}}{\partial \dot{\mathbf{x}}}\bigg|_{t_1}^{t_2} - \int_{t_1}^{t_2} \delta\mathbf{x}^T \frac{d}{dt}\frac{\partial \mathcal{L}}{\partial \dot{\mathbf{x}}}\, dt \\ &= -\int_{t_1}^{t_2} \delta\mathbf{x}^T \left(\frac{d}{dt}\frac{\partial \mathcal{L}}{\partial \dot{\mathbf{x}}} - \frac{\partial \mathcal{L}}{\partial \mathbf{x}}\right) = 0. \end{aligned} \tag{2.35}$$

Note that the middle term in the first line of Equation (2.35) vanishes due to the start and end conditions for $\delta\mathbf{x}(t)$ given by Equation (2.33). Since the integral in the last line of Equation (2.35) must vanish for any arbitrary $\delta\mathbf{x}(t)$, it follows that the stationary conditions of $\mathcal{A}(\mathbf{x}(t))$ lead to the Lagrange equations as

$$\frac{d}{dt}\frac{\partial \mathcal{L}}{\partial \dot{\mathbf{x}}} - \frac{\partial \mathcal{L}}{\partial \mathbf{x}} = 0, \tag{2.36}$$

which have already been shown in Section 2.3.2 to be equivalent to the dynamic equilibrium equations expressed as

$$\mathbf{Ma} = \mathbf{F} - \mathbf{T}. \tag{2.37}$$

The fact that equilibrium paths render the action integral stationary will be used in Section 2.4 to derive time-stepping schemes that satisfy the same principle but in a discrete manner.

Finally, before concluding this section, it is worth observing that the product of the equilibrium equations, expressed either in Lagrangian form or equivalently in Newtonian form, by an arbitrary set of variations of $\mathbf{x}(t)$, as shown in the final line of Equation (2.35), represents the principle of virtual work as follows:

$$\delta\mathbf{x}^T\left(\frac{d}{dt}\frac{\partial \mathcal{L}}{\partial \dot{\mathbf{x}}} - \frac{\partial \mathcal{L}}{\partial \mathbf{x}}\right) = 0, \tag{2.38a}$$

$$\delta\mathbf{x}^T(\mathbf{Ma} - \mathbf{F} + \mathbf{T}) \quad = 0. \tag{2.38b}$$

Mathematically this expression does not provide any further information than the equilibrium equation itself. Physically, it represents the balance between the virtual work done by external forces and the work done by internal and inertial forces during an arbitrary motion $\delta\mathbf{x}$ occurring at any position $\mathbf{x}(t)$:

$$\delta\mathbf{x}^T\mathbf{Ma} + \delta\mathbf{x}^T\mathbf{T} = \delta\mathbf{x}^T\mathbf{F}. \tag{2.39}$$

The term $\delta\mathbf{x}^T$ is often replaced by a virtual velocity $\delta\mathbf{v}^T$ or virtual displacement $\delta\mathbf{u}^T$ without any significant change to the physical implication of Equation (2.39), except for the fact that if velocity is used, one should more strictly refer to virtual power rather than virtual work. In addition, the use of virtual velocity obviates the need to qualify the magnitude of either $\delta\mathbf{x}^T$ or $\delta\mathbf{u}^T$. The principle of virtual work (or power) will become more relevant when dealing with solids as it provides a useful tool with which to introduce the discretization of the body into finite elements. In the case of the truss, the body is already a collection of discrete members; consequently, the use of the principle of virtual work is unnecessary.

2.4 TIME INTEGRATION SCHEMES

The set of dynamic equilibrium equations derived in this and the previous chapter need to be integrated in time in order to advance the solution from an initial known state to a configuration at time t. This cannot be done analytically for nonlinear problems and consequently a numerical process is required. Two methods, namely the leap-frog scheme and the mid-point rule, have already been presented in Chapter 1 in the context of one- and two-degrees-of-freedom examples. This section will revisit these schemes as applied to the more general pin-jointed truss containing many degrees of freedom. Moreover, these two schemes will be re-derived from a discrete approximation of the least action variational principle. Such schemes are known as *variational time integration schemes*. In addition, a third,very commonly used, time integration scheme, known as the trapezoidal rule, will be presented for completeness. This scheme belongs to a family of time integrators known as the Newmark method, of which the leap-frog method is also a particular case.

2.4.1 Discrete Action Integral

The motion of the pin-jointed truss from its initial state to its final position of interest can be described by a sequence of positions $\mathbf{x}_n$ at times $t_0, t_1, \ldots, t_n, t_{n+1}, \ldots, t_{N_t}$ separated by a fixed time step Δt so that $t_{n+1} = t_n + \Delta t$ for $n = 0, 1, \ldots, N_t$. The action integral can similarly be broken down into the sum of discrete time segments as

$$\mathcal{A}(\mathbf{x}(t)) = \sum_{n=0}^{N_t-1} \int_{t_n}^{t_{n+1}} \mathcal{L}(\mathbf{x}(t), \dot{\mathbf{x}}(t))\, dt. \tag{2.40}$$

Discrete time integration schemes can be derived by providing suitable approximations to the integral of the Lagrangian Equation (2.40) as

$$L_n(\mathbf{x}_n, \mathbf{x}_{n+1}) \approx \int_{t_n}^{t_{n+1}} \mathcal{L}(\mathbf{x}(t), \dot{\mathbf{x}}(t))\, dt. \tag{2.41}$$

Note that it is assumed that L_n will only depend upon the starting and end positions $\mathbf{x}_n$ and $\mathbf{x}_{n+1}$. The action integral may now be approximated by the sum of discrete Lagrangians as

$$\mathcal{A}(\mathbf{x}(t)) \approx \mathcal{S}(\mathbf{x}_0, \mathbf{x}_1, \ldots, \mathbf{x}_{N_t}) = \sum_{n=0}^{N_t-1} L_n(\mathbf{x}_n, \mathbf{x}_{n+1}). \tag{2.42}$$

The stationary conditions of the action integral now become the stationary conditions of $\mathcal{S}$ with respect to each of the nodal variables $\mathbf{x}_n$, as

$$D\mathcal{A}[\delta\mathbf{x}] = \sum_{n=0}^{N_t-1} \delta\mathbf{x}_n^T \frac{\partial \mathcal{S}(\mathbf{x}_0, \mathbf{x}_1, \ldots, \mathbf{x}_{N_t})}{\partial \mathbf{x}_n} = 0. \tag{2.43}$$

Now observe that only $L_{n-1}(\mathbf{x}_{n-1}, \mathbf{x}_n)$ and $L_n(\mathbf{x}_n, \mathbf{x}_{n+1})$ are functions of $\mathbf{x}_n$ in the summation that defines $\mathcal{S}$ in Equation (2.42). This, together with the fact that each of the virtual variations $\delta \mathbf{x}_n$ is arbitrary and independent of each other, yields a discrete version of the least action integral principle at step n as

$$\frac{\partial \mathcal{S}}{\partial \mathbf{x}_n} = \frac{\partial L_{n-1}(\mathbf{x}_{n-1}, \mathbf{x}_n)}{\partial \mathbf{x}_n} + \frac{\partial L_n(\mathbf{x}_n, \mathbf{x}_{n+1})}{\partial \mathbf{x}_n} = \mathbf{0}. \tag{2.44}$$

Given a particular definition of the discrete Lagrangian L_n, this expression will provide an equation from which it is possible to obtain $\mathbf{x}_{n+1}$ knowing $\mathbf{x}_{n-1}$ and $\mathbf{x}_n$. To initiate the sequential evaluation of positions $\mathbf{x}_n$ as the number of time steps progresses, it will be necessary in Equation (2.44) to start with $\mathbf{x}_0$ and $\mathbf{x}_1$. Whereas $\mathbf{x}_0$ will be known from initial conditions, a procedure to obtain $\mathbf{x}_1$ from $\mathbf{x}_0$ and the initial velocity $\mathbf{v}_0$, for instance $\mathbf{x}_1 = \mathbf{x}_0 + \Delta t \mathbf{v}_0$, will be required. However, more accurate techniques will be presented below in the context of specific time integration schemes derived from particular choices of the discrete Lagrangian function $L_n(\mathbf{x}_n, \mathbf{x}_{n+1})$.

2.4.2 Leap-Frog Time Integration

Consider the following approximation to the integrated Lagrangian $L_n(\mathbf{x}_n, \mathbf{x}_{n+1})$:

$$\begin{aligned} L_n^{LF}(\mathbf{x}_n, \mathbf{x}_{n+1}) &\approx \int_{t_n}^{t_{n+1}} \left(\frac{1}{2}\mathbf{v}^T \mathbf{M} \mathbf{v} - \Pi_{\text{int}}(\mathbf{x}) - \Pi_{\text{ext}}(\mathbf{x})\right) dt \\ &\approx \frac{\Delta t}{2} \left(\frac{\mathbf{x}_{n+1} - \mathbf{x}_n}{\Delta t}\right)^T \mathbf{M} \left(\frac{\mathbf{x}_{n+1} - \mathbf{x}_n}{\Delta t}\right) \\ &\quad - \frac{\Delta t}{2}\left(\Pi_{\text{int}}(\mathbf{x}_n) + \Pi_{\text{int}}(\mathbf{x}_{n+1})\right) \\ &\quad - \frac{\Delta t}{2}\left(\Pi_{\text{ext}}(\mathbf{x}_n) + \Pi_{\text{ext}}(\mathbf{x}_{n+1})\right). \end{aligned} \tag{2.45}$$

This expression simply implies an approximation of the velocity vector by the increment in $\mathbf{x}$ over the time step Δt and a trapezoidal integration of the internal and external energy terms. Recall that the derivative of the internal energy with respect to $\mathbf{x}$ gives the internal forces $\mathbf{T}$, and similarly the derivative of the external energy gives the negative of the applied force, that is, $-\mathbf{F}$; see Equations (2.29) and (2.28). Consequently, Equation (2.44) can be obtained via the following intermediate steps:

$$\frac{\partial L_{n-1}^{LF}(\mathbf{x}_{n-1}, \mathbf{x}_n)}{\partial \mathbf{x}_n} = \mathbf{M}\left(\frac{\mathbf{x}_n - \mathbf{x}_{n-1}}{\Delta t}\right) - \frac{\Delta t}{2}\mathbf{T}(\mathbf{x}_n) + \frac{\Delta t}{2}\mathbf{F}(t_n), \tag{2.46a}$$

$$\frac{\partial L_n^{LF}(\mathbf{x}_n, \mathbf{x}_{n+1})}{\partial \mathbf{x}_n} = -\mathbf{M}\left(\frac{\mathbf{x}_{n+1} - \mathbf{x}_n}{\Delta t}\right) - \frac{\Delta t}{2}\mathbf{T}(\mathbf{x}_n) + \frac{\Delta t}{2}\mathbf{F}(t_n). \tag{2.46b}$$

Adding these two expressions, substituting into the general variational time-stepping Equation (2.44), and rearranging gives

$$\mathbf{M}\left(\frac{\mathbf{x}_{n+1} - 2\mathbf{x}_n + \mathbf{x}_{n-1}}{\Delta t^2}\right) + \mathbf{T}(\mathbf{x}_n) = \mathbf{F}(t_n). \tag{2.47}$$

This equation can be rewritten in more familiar terms for time step t_n as

$$\mathbf{M}\mathbf{a}_n + \mathbf{T}(\mathbf{x}_n) = \mathbf{F}(t_n); \quad \mathbf{a}_n = \frac{\mathbf{x}_{n+1} - 2\mathbf{x}_n + \mathbf{x}_{n-1}}{\Delta t^2}, \tag{2.48a,b}$$

where $\mathbf{a}_n$ is known as the central difference approximation of the acceleration at time step t_n.† Since $\mathbf{M}$ is a diagonal matrix for the pin-jointed truss under consideration, the acceleration can be obtained explicitly from Equation (2.48a,b) for time t_n as

$$\mathbf{a}_n = \mathbf{M}^{-1}\big(\mathbf{F}(t_n) - \mathbf{T}(\mathbf{x}_n)\big). \tag{2.49}$$

Assuming that both $\mathbf{x}_{n-1}$ and $\mathbf{x}_n$ are known, the geometry of the truss can be advanced to time t_{n+1} by using Equation (2.48a,b)$_b$ to give

$$\mathbf{x}_{n+1} = 2\mathbf{x}_n - \mathbf{x}_{n-1} + \Delta t^2 \mathbf{a}_n. \tag{2.50}$$

To show that this expression is simply the leap-frog scheme derived in Chapter 1, note first that Equation (2.50) can be rewritten as

$$\frac{\mathbf{x}_{n+1} - \mathbf{x}_n}{\Delta t} = \frac{\mathbf{x}_n - \mathbf{x}_{n-1}}{\Delta t} + \Delta t \mathbf{a}_n. \tag{2.51}$$

Defining the intermediate velocity variables $\mathbf{v}_{n+1/2}$ and $\mathbf{v}_{n-1/2}$ for $n = 0, 1, \ldots, N_t$ as

$$\mathbf{v}_{n+1/2} = \frac{\mathbf{x}_{n+1} - \mathbf{x}_n}{\Delta t}; \quad \mathbf{v}_{n-1/2} = \frac{\mathbf{x}_n - \mathbf{x}_{n-1}}{\Delta t} \tag{2.52a,b}$$

gives the two stages of the leap-frog scheme as

$$\mathbf{v}_{n+1/2} = \mathbf{v}_{n-1/2} + \Delta t \mathbf{a}_n, \tag{2.53a}$$

$$\mathbf{x}_{n+1} = \mathbf{x}_n + \Delta t \mathbf{v}_{n+1/2}, \tag{2.53b}$$

where the acceleration $\mathbf{a}_n$ is calculated using the explicit Equation (2.49). For a direct application of the leap-frog algorithm from $n = 1$ onward it is necessary to know the value of $\mathbf{x}_0$ and $\mathbf{v}_{1/2}$. The initial position of the truss will determine $\mathbf{x}_0$, whereas $\mathbf{v}_{1/2}$ can be obtained using the initial velocity $\mathbf{v}_0$ and acceleration $\mathbf{a}_0$ as

$$\mathbf{v}_{1/2} = \mathbf{v}_0 + \frac{\Delta t}{2}\mathbf{a}_0; \quad \mathbf{a}_0 = \mathbf{M}^{-1}\big(\mathbf{T}(\mathbf{x}_0) - \mathbf{F}(t_0)\big). \tag{2.54a,b}$$

† The acceleration $\mathbf{a}_n$ can be found directly as $(\mathbf{v}_{n+1/2} - \mathbf{v}_{n-1/2})/\Delta t$, hence the term "central difference."

The leap-frog algorithm described above is a simple and efficient time integration scheme. However, it is subject to the same time-step stability considerations described in Chapter 1 for the spring-mass system which led to a time-step restriction of $\Delta t \leq 2/\omega$. The evaluation of the natural frequency ω for a nonlinear elastic truss member and the resulting time-step restriction is shown in Example 2.3.

2.4.3 Mid-Point Time Integration

The mid-point time-stepping rule was introduced in Chapter 1 as an alternative time integration scheme that does not suffer from the same time-step limitations of the leap-frog method. However, it is an implicit scheme because to advance a time step requires the solution of a nonlinear set of equations using the Newton–Raphson method and hence the evaluation of the tangent stiffness matrix. This has already been evaluated for the pin-jointed truss in Chapter 3 of NL-Statics, but will be briefly revisited in Example 2.2 below.

The mid-point rule is derived by using the following approximate integrated Lagrangian:

$$L_n^{MP}(\mathbf{x}_n, \mathbf{x}_{n+1}) = \frac{\Delta t}{2}\left(\frac{\mathbf{x}_{n+1} - \mathbf{x}_n}{\Delta t}\right)^T \mathbf{M}\left(\frac{\mathbf{x}_{n+1} - \mathbf{x}_n}{\Delta t}\right) - \Delta t\left[\Pi_{\text{int}}(\mathbf{x}_{n+1/2}) + \Pi_{\text{ext}}(\mathbf{x}_{n+1/2})\right], \tag{2.55}$$

where by definition

$$\mathbf{x}_{n+1/2} = \frac{1}{2}(\mathbf{x}_n + \mathbf{x}_{n+1}). \tag{2.56}$$

Observe that the only difference between this integrated Lagrangian and Equation (2.45) for the leap-frog scheme is that the internal and external energy terms are now integrated via the mid-point method rather than by the trapezoidal rule. Both integrations are second-order accurate‡ and equivalent for quadratic energy functions, but will differ for more complex energy expressions, such as the squared logarithmic expression used in the constitutive equation for the truss member; see Equation (2.19).

The application of Equation (2.44) for the above discrete Lagrangian is obtained from the derivatives of L_n and L_{n-1} as

$$\frac{\partial L_{n-1}^{MP}(\mathbf{x}_{n-1}, \mathbf{x}_n)}{\partial \mathbf{x}_n} = \mathbf{M}\left(\frac{\mathbf{x}_n - \mathbf{x}_{n-1}}{\Delta t}\right) - \frac{\Delta t}{2}\mathbf{T}(\mathbf{x}_{n-1/2}) + \frac{\Delta t}{2}\mathbf{F}(t_{n-1/2}), \tag{2.57a}$$

‡ Second-order accurate means that the time integration scheme is exact if the variables are at most quadratic in time, otherwise there will be errors that are proportional to Δt^2. For example, the central difference acceleration term in Equations (2.48a,b) and (2.58) can alternatively be obtained by assuming a parabolic distribution of $\mathbf{x}$ through the steps $n-1$, n, and $n+1$; hence, second-order accurate.

$$\frac{\partial L_n^{MP}(\mathbf{x}_n, \mathbf{x}_{n+1})}{\partial \mathbf{x}_n} = -\mathbf{M}\left(\frac{\mathbf{x}_{n+1} - \mathbf{x}_n}{\Delta t}\right) - \frac{\Delta t}{2}\mathbf{T}(\mathbf{x}_{n+1/2}) + \frac{\Delta t}{2}\mathbf{F}(t_{n+1/2}). \tag{2.57b}$$

Adding the above two expressions and performing simple algebraic manipulations (dividing by Δt and changing signs), the following equation is obtained:

$$\mathbf{M}\left(\frac{\mathbf{x}_{n+1} - 2\mathbf{x}_n + \mathbf{x}_{n-1}}{\Delta t^2}\right) + \frac{1}{2}\left[\mathbf{T}(\mathbf{x}_{n+1/2}) - \mathbf{F}(t_{n+1/2})\right] + \frac{1}{2}\left[\mathbf{T}(\mathbf{x}_{n-1/2}) - \mathbf{F}(t_{n-1/2})\right] = \mathbf{0}. \tag{2.58}$$

To rewrite this equilibrium equation as a recognizable mid-point rule as defined in Chapter 1, a number of algebraic transformations and definitions of intermediate variables are needed. To begin, nodal velocities at each step are defined in such a way that

$$\frac{\mathbf{v}_{n+1} + \mathbf{v}_n}{2} = \frac{\mathbf{x}_{n+1} - \mathbf{x}_n}{\Delta t}, \tag{2.59a}$$

$$\mathbf{x}_{n+1} = \mathbf{x}_n + \frac{\Delta t}{2}(\mathbf{v}_{n+1} + \mathbf{v}_n); \quad n = 0, 1, \ldots, N_t. \tag{2.59b}$$

For instance, for $n = 0$, Equation (2.59a) would give a definition of $\mathbf{v}_1$ in terms of $\mathbf{x}_1$ as

$$\mathbf{v}_1 = \frac{2}{\Delta t}(\mathbf{x}_1 - \mathbf{x}_0) - \mathbf{v}_0, \tag{2.60}$$

where $\mathbf{v}_0$ and $\mathbf{x}_0$ would emerge from the initial conditions. The repeated application of Equation (2.59a) enables $\mathbf{v}_{n+1}$ to be found sequentially, knowing $\mathbf{v}_n$ and the nodal positions. With these definitions, the central difference approximation to the acceleration can be re-expressed as

$$\begin{aligned}
\frac{\mathbf{x}_{n+1} - 2\mathbf{x}_n + \mathbf{x}_{n-1}}{\Delta t^2} &= \frac{1}{\Delta t}\left[\frac{\mathbf{x}_{n+1} - \mathbf{x}_n}{\Delta t} - \frac{\mathbf{x}_n - \mathbf{x}_{n-1}}{\Delta t}\right] \\
&= \frac{1}{\Delta t}\left[\frac{\mathbf{v}_{n+1} + \mathbf{v}_n}{2} - \frac{\mathbf{v}_n + \mathbf{v}_{n-1}}{2}\right] \\
&= \frac{1}{2}\left[\frac{\mathbf{v}_{n+1} - \mathbf{v}_n}{\Delta t} + \frac{\mathbf{v}_n - \mathbf{v}_{n-1}}{\Delta t}\right] \\
&= \frac{1}{2}\left[\mathbf{a}_{n+1/2} + \mathbf{a}_{n-1/2}\right],
\end{aligned} \tag{2.61}$$

where the intermediate accelerations are *defined* as

$$\mathbf{a}_{n+1/2} = \frac{\mathbf{v}_{n+1} - \mathbf{v}_n}{\Delta t}; \quad \mathbf{a}_{n-1/2} = \frac{\mathbf{v}_n - \mathbf{v}_{n-1}}{\Delta t}. \tag{2.62a,b}$$

Substituting Equation (2.61) into Equation (2.58) gives

$$\frac{1}{2}\left[\mathbf{M}\mathbf{a}_{n+1/2} + \mathbf{T}(\mathbf{x}_{n+1/2}) - \mathbf{F}(t_{n+1/2})\right] + \frac{1}{2}\left[\mathbf{M}\mathbf{a}_{n-1/2} + \mathbf{T}(\mathbf{x}_{n-1/2}) - \mathbf{F}(t_{n-1/2})\right] = \mathbf{0}. \tag{2.63}$$

Recall that the above equation not only enshrines the mid-point time integration scheme, but also ensures satisfaction of the least action principle, which, in a discrete manner, is equivalent to the dynamic equilibrium equation which is enforced above as the average of the equilibrium equations at times $t_{n-1/2}$ and $t_{n+1/2}$. It provides a means of evaluating $\mathbf{a}_{n+1/2}$ and hence $\mathbf{v}_{n+1}$ and $\mathbf{x}_{n+1}$, knowing $\mathbf{x}_{n-1/2}$ and $\mathbf{a}_{n-1/2}$. However, to evaluate the acceleration at the first half time step, that is, $\mathbf{a}_{1/2}$, in the absence of $\mathbf{a}_{-1/2}$, it is logical to simply enforce equilibrium at $t_{1/2}$ as

$$\mathbf{M}\mathbf{a}_{1/2} + \mathbf{T}(\mathbf{x}_{1/2}) - \mathbf{F}(t_{1/2}) = \mathbf{0}. \tag{2.64}$$

In so doing, this ensures that when Equation (2.63) is enforced at subsequent time steps equilibrium is satisfied at each half time step $n + 1/2$. To elaborate a little further, by obtaining the acceleration $\mathbf{a}_{1/2}$ from Equation (2.64) at time step $n = 1$, the second term in Equation (2.63) has already been satisfied. At time step $n = 2$ the second term in square brackets in Equation (2.63) again has already been satisfied from $n = 1$, and so on in a sequential manner. Consequently, it is only necessary to consider the solution to the first term in Equation (2.63) as

$$\mathbf{M}\mathbf{a}_{n+1/2} + \mathbf{T}(\mathbf{x}_{n+1/2}) - \mathbf{F}(t_{n+1/2}) = \mathbf{0}. \tag{2.65}$$

This equation is the standard mid-point rule already introduced in Chapter 1. It is complemented by the update equations for positions and velocities given by Equations (2.62a,b)$_a$, (2.59a), and (2.56), which are grouped below for convenience:

$$\mathbf{v}_{n+1} = \mathbf{v}_n + \Delta t \mathbf{a}_{n+1/2}, \tag{2.66a}$$

$$\mathbf{x}_{n+1} = \mathbf{x}_n + \frac{\Delta t}{2}(\mathbf{v}_{n+1} + \mathbf{v}_n), \tag{2.66b}$$

$$\mathbf{x}_{n+1/2} = \frac{1}{2}(\mathbf{x}_n + \mathbf{x}_{n+1}). \tag{2.66c}$$

In these relationships, $\mathbf{x}_{n+1/2}$ is directly related to $\mathbf{a}_{n+1/2}$, and therefore Equation (2.65) is a system of nonlinear equations to be solved for either $\mathbf{x}_{n+1/2}$ or $\mathbf{a}_{n+1/2}$. This is usually achieved using a Newton–Raphson process, starting from an initial guess, known as a predictor, followed by a linear iteration, called the corrector. This procedure has already been presented in Chapter 1, Section 1.3.4.

Assuming as an initial prediction of the acceleration $\mathbf{a}^{(0)}_{n+1/2} = \mathbf{a}_{n-1/2}$, from a previous time step $t_{n-1/2}$, a predictor phase gives velocities and positions as,[§]

$$\mathbf{v}^{(0)}_{n+1} = \mathbf{v}_n + \Delta t \mathbf{a}^{(0)}_{n+1/2}, \tag{2.67a}$$

$$\mathbf{x}^{(0)}_{n+1} = \mathbf{x}_n + \frac{\Delta t}{2}(\mathbf{v}^{(0)}_{n+1} + \mathbf{v}_n), \tag{2.67b}$$

$$\mathbf{x}^{(0)}_{n+1/2} = \frac{1}{2}(\mathbf{x}^{(0)}_{n+1} + \mathbf{x}_n). \tag{2.67c}$$

To initiate the corrector iterations a residual (or out-of-balance force) is evaluated using the above predictions or later iterative values as

$$\mathbf{R}(\mathbf{x}^{(k)}_{n+1/2}, \mathbf{a}^{(k)}_{n+1/2}) = \mathbf{M}\mathbf{a}^{(k)}_{n+1/2} + \mathbf{T}(\mathbf{x}^{(k)}_{n+1/2}) - \mathbf{F}(t_{n+1/2}), \tag{2.68}$$

where k represents the iteration number, starting from the predicted values at $k = 0$ given by Equations (2.67).

In order to drive the residual force to zero, thereby satisfying the dynamic equilibrium equations, a correction to the half-time-step acceleration $\Delta\mathbf{a}_{n+1/2}$ and position $\Delta\mathbf{x}_{n+1/2}$ is introduced into a linearization of the residual force $\mathbf{R}$ as

$$\mathbf{R}(\mathbf{x}^{(k+1)}_{n+1/2}, \mathbf{a}^{(k+1)}_{n+1/2}) \approx \mathbf{R}(\mathbf{x}^{(k)}_{n+1/2}, \mathbf{a}^{(k)}_{n+1/2}) + \mathbf{M}\Delta\mathbf{a}_{n+1/2} + D\mathbf{T}(\mathbf{x}^{(k)}_{n+1/2})[\Delta\mathbf{x}_{n+1/2}] = \mathbf{0}. \tag{2.69}$$

The linearization of the internal force term $\mathbf{T}$ with respect to a general increment in position $\Delta\mathbf{x}$ is provided by the tangent stiffness matrix $\mathbf{K}$ developed in Example 2.2 as

$$D\mathbf{T}(\mathbf{x})[\Delta\mathbf{x}] = \mathbf{K}(\mathbf{x})\Delta\mathbf{x}. \tag{2.70}$$

In particular, increments in position $\Delta\mathbf{x}_{n+1/2}$ are related to increments in acceleration $\Delta\mathbf{a}_{n+1/2}$ via the mid-point rule, Equations (2.66), to give

$$\Delta\mathbf{v}_{n+1} = \Delta t \Delta\mathbf{a}_{n+1/2}, \tag{2.71a}$$

$$\Delta\mathbf{x}_{n+1} = \frac{\Delta t}{2}\Delta\mathbf{v}_{n+1} = \frac{\Delta t^2}{2}\Delta\mathbf{a}_{n+1/2}, \tag{2.71b}$$

$$\Delta\mathbf{x}_{n+1/2} = \frac{1}{2}\Delta\mathbf{x}_{n+1} = \frac{\Delta t^2}{4}\Delta\mathbf{a}_{n+1/2}. \tag{2.71c}$$

[§] For time-stepping schemes involving Newton–Raphson iteration, the superscripts $^{(*)}$ refer to iteration number.

Introducing Equations (2.71c) and (2.70) into the linearized residual Equation (2.69) gives the corrector equation for $\Delta\mathbf{a}_{n+1/2}$ as

$$\left[\mathbf{M} + \frac{\Delta t^2}{4}\mathbf{K}\right]\Delta\mathbf{a}_{n+1/2} = -\mathbf{R}(\mathbf{x}^{(k)}_{n+1/2}, \mathbf{a}^{(k)}_{n+1/2}). \tag{2.72}$$

Solving this linear system of equations for $\Delta\mathbf{a}_{n+1/2}$ enables the new iterative positions, velocities, and accelerations to be updated as

$$\mathbf{a}^{(k+1)}_{n+1/2} = \mathbf{a}^{(k)}_{n+1/2} + \Delta\mathbf{a}_{n+1/2}, \tag{2.73a}$$

$$\mathbf{v}^{(k+1)}_{n+1} = \mathbf{v}^{(k)}_{n+1} + \Delta t\Delta\mathbf{a}_{n+1/2}, \tag{2.73b}$$

$$\mathbf{x}^{(k+1)}_{n+1} = \mathbf{x}^{(k)}_{n+1} + \frac{\Delta t^2}{2}\Delta\mathbf{a}_{n+1/2}, \tag{2.73c}$$

$$\mathbf{x}^{(k+1)}_{n+1/2} = \mathbf{x}^{(k)}_{n+1/2} + \frac{\Delta t^2}{4}\Delta\mathbf{a}_{n+1/2}. \tag{2.73d}$$

This process needs to be repeated until the magnitude of the residual force is sufficiently small. Usually the larger the time step, the more iterations are required for convergence.

EXAMPLE 2.2: Tangent stiffness matrix

The development of the tangent stiffness operator or matrix is necessary for the implementation of implicit time integration schemes where a Newton–Raphson process needs to be applied to solve the nonlinear dynamic equilibrium equations. The tangent stiffness operator $\mathbf{K}^{(e)}(\mathbf{x})$ for truss element (e) derives from the directional derivative of the internal force vector $\mathbf{T}^{(e)}(\mathbf{x})$ in the direction of an arbitrary change in nodal positions $\Delta\mathbf{x}$ as

$$D\mathbf{T}^{(e)}(\mathbf{x})[\mathbf{u}] = \mathbf{K}\mathbf{u}; \quad \mathbf{u} = \begin{bmatrix} u_a \\ u_b \end{bmatrix}. \tag{a}$$

The substitution $\mathbf{u} = \Delta\mathbf{x}$ is used for convenience and simply indicates a displacement; also the element designation (e) is omitted for clarity. Note that in both the mid-point and trapezoidal schemes the change in iterative position is a function of the change in acceleration, see Equations (2.73) and (2.80), and consequently a final substitution as $\Delta\mathbf{x} = (\Delta t^2/4)\Delta\mathbf{a}_{n+1/2}$ or $\Delta\mathbf{x} = (\Delta t^2/4)\Delta\mathbf{a}_{n+1}$ respectively will need to be introduced in the Newton–Raphson equations; see for example Equation (2.79). To evaluate

(continued)

Example 2.2: *(cont.)*

K consider a single truss element joining typical nodes a and b for which the internal forces vector is given by Equations (2.4) and (2.7)–(2.9a,b) as

$$\mathbf{T} = \begin{bmatrix} \boldsymbol{T}_a \\ \boldsymbol{T}_b \end{bmatrix} = \begin{bmatrix} -T\boldsymbol{n} \\ T\boldsymbol{n} \end{bmatrix}, \tag{b}$$

$$T = E\frac{V}{l}\ln\frac{l}{L};\ \boldsymbol{n} = \frac{\boldsymbol{x}_b - \boldsymbol{x}_a}{l};\ l^2 = (\boldsymbol{x}_b - \boldsymbol{x}_a)\cdot(\boldsymbol{x}_b - \boldsymbol{x}_a). \tag{c,d,e}$$

Recall from Section 2.2.1 that $\boldsymbol{n}$ is the unit vector (see Figure 2.1), l is the current length, L is the initial length, V is the initial volume, and E is Young's modulus. To obtain the directional derivative in Equation (a) it is first necessary to determine the directional derivative of the current length l. In other words, l being a function of the nodal positions $\boldsymbol{x}_a$ and $\boldsymbol{x}_b$, how does l change if these positions change by $\boldsymbol{u}_a$ and $\boldsymbol{u}_b$? The directional derivative of the current length, l, is most easily derived by noting that

$$Dl^2(\boldsymbol{x})[\boldsymbol{u}] = 2l\,Dl(\boldsymbol{x})[\boldsymbol{u}], \quad \text{hence } Dl(\boldsymbol{x})[\boldsymbol{u}] = \frac{1}{2l}Dl^2(\boldsymbol{x})[\boldsymbol{u}]. \tag{f}$$

Substituting for l^2 from Equation (e) gives

$$\begin{aligned} Dl^2(\boldsymbol{x})[\boldsymbol{u}] &= \frac{d}{d\epsilon}\bigg|_{\epsilon=0}\left[(\boldsymbol{x}_b + \epsilon\boldsymbol{u}_b - \boldsymbol{x}_a - \epsilon\boldsymbol{u}_a)\cdot(\boldsymbol{x}_b + \epsilon\boldsymbol{u}_b - \boldsymbol{x}_a - \epsilon\boldsymbol{u}_a)\right] \\ &= 2(\boldsymbol{x}_b - \boldsymbol{x}_a)\cdot(\boldsymbol{u}_b - \boldsymbol{u}_a), \end{aligned} \tag{g}$$

$$\text{from which} \quad Dl(\boldsymbol{x})[\boldsymbol{u}] = \boldsymbol{n}\cdot(\boldsymbol{u}_b - \boldsymbol{u}_a). \tag{h}$$

The directional derivative of the $T\boldsymbol{n}$ term in Equations (b)–(d) can be obtained as

$$\begin{aligned} D(T\boldsymbol{n})[\boldsymbol{u}] &= D\left(E\frac{V}{l^2}\left(\ln\frac{l}{L}\right)(\boldsymbol{x}_b - \boldsymbol{x}_a)\right)[\boldsymbol{u}] \\ &= E\frac{V}{l^2}\left(\ln\frac{l}{L}\right)(\boldsymbol{u}_b - \boldsymbol{u}_a) + \frac{d}{dl}\left(E\frac{V}{l^2}\ln\frac{l}{L}\right)Dl(\boldsymbol{x})[\boldsymbol{u}](\boldsymbol{x}_b - \boldsymbol{x}_a) \\ &= \frac{T}{l}(\boldsymbol{u}_b - \boldsymbol{u}_a) + \frac{d}{dl}\left(E\frac{V}{l^2}\ln\frac{l}{L}\right)\boldsymbol{n}\cdot(\boldsymbol{u}_b - \boldsymbol{u}_a)(l\boldsymbol{n}) \\ &= \left[\frac{T}{l}\boldsymbol{I}_{3\times 3} + k_e\boldsymbol{n}\otimes\boldsymbol{n}\right](\boldsymbol{u}_b - \boldsymbol{u}_a), \end{aligned} \tag{i}$$

(continued)

Example 2.2: *(cont.)*

where the effective truss member stiffness k_e is defined and evaluated as

$$k_e = l\frac{d}{dl}\left(E\frac{V}{l^2}\ln\frac{l}{L}\right) = \frac{V}{l^2}\frac{d}{d\ln l}\left(E\ln\frac{l}{L}\right) + \left(VE\ln\frac{l}{L}\right)l\frac{d}{dl}\left(\frac{1}{l^2}\right)$$

$$= \frac{EV}{l^2} - 2\frac{V\tau}{l^2} = \frac{V}{l^2}(E - 2\tau).$$

Substitution of Equation (i) into Equation (a) and re-arranging terms yields the tangent stiffness matrix operator $\mathbf{K}^{(e)}$ for a single truss element (e) as[†]

$$\mathbf{K}^{(e)} = \begin{bmatrix} \mathbf{K}^{(e)}_{aa} & \mathbf{K}^{(e)}_{ab} \\ \mathbf{K}^{(e)}_{ba} & \mathbf{K}^{(e)}_{bb} \end{bmatrix}; \quad \mathbf{K}^{(e)}_{aa} = \mathbf{K}^{(e)}_{bb} = k_e\boldsymbol{n}\otimes\boldsymbol{n} + \frac{T}{l}\boldsymbol{I},$$

$$\mathbf{K}^{(e)}_{ab} = \mathbf{K}^{(e)}_{ba} = -\mathbf{K}^{(e)}_{aa},$$

hence

$$D\mathbf{T}^{(e)}(\mathbf{x})[\mathbf{u}] = \begin{bmatrix} \mathbf{K}^{(e)}_{aa} & \mathbf{K}^{(e)}_{ab} \\ \mathbf{K}^{(e)}_{ba} & \mathbf{K}^{(e)}_{bb} \end{bmatrix}\begin{bmatrix} u_a \\ u_b \end{bmatrix}.$$

† An alternative derivation of this tangent stiffness matrix, involving extensive use of the directional derivative, is given in Section 3.4.3 of NL-Statics.

EXAMPLE 2.3: Critical time step for a truss element

In Chapter 1 it was shown that the leap-frog scheme becomes unstable for time steps larger than a critical value given as $2/\omega$, where ω is the natural frequency of the spring-mass system. The more general case of the truss can be considered as an assembly of nonlinear spring-mass systems comprising the individual truss members. While being nonlinear, a linearized natural frequency of vibration of each truss member can be obtained by considering the mass stiffness eigenvalue problem

$$\mathbf{K}^{(e)}\mathbf{v} = \omega^2\mathbf{M}^{(e)}\mathbf{v}, \tag{a}$$

where the mass and stiffness matrix (see Example 2.2) for a single truss member are

(continued)

Example 2.3: *(cont.)*

$$\mathbf{M}^{(e)} = \begin{bmatrix} \frac{m_e}{2}\boldsymbol{I} & 0 \\ 0 & \frac{m_e}{2}\boldsymbol{I} \end{bmatrix}; \ \mathbf{K}^{(e)} = \begin{bmatrix} \mathbf{K} & -\mathbf{K} \\ -\mathbf{K} & \mathbf{K} \end{bmatrix},$$

$$m_e = \rho_R V; \qquad \mathbf{K} = \frac{V}{l^2}(E - 2\tau)\boldsymbol{n}_e \otimes \boldsymbol{n}_e + \frac{T}{l}\boldsymbol{I}.$$

The first and fundamental eigenvalue for the truss member is a pure extension given by the eigenvector $\mathbf{v}$ as

$$\mathbf{v} = \begin{bmatrix} -\boldsymbol{n}_e \\ \boldsymbol{n}_e \end{bmatrix}.$$

Substituting this eigenvector into Equation (a) and multiplying by $\mathbf{v}^T$ gives, after employing Equation (2.7) and some simple algebra, the frequency of vibration as

$$\omega^2 = \frac{\mathbf{v}^T\mathbf{K}^{(e)}\mathbf{v}}{\mathbf{v}^T\mathbf{M}^{(e)}\mathbf{v}} = \frac{4(\frac{V}{l^2})(E-2\tau) + 4\frac{T}{l}}{m_e}$$

$$= \frac{4V(E-\tau)}{l^2\rho_R V} = \frac{1}{l^2}\frac{(E-\tau)}{\rho_R}.$$

Consequently, the frequency of vibration ω is

$$\omega = \frac{2c}{l}; \quad c = \sqrt{\frac{E-\tau}{\rho_R}}, \tag{b}$$

where c is the speed of a linear sound wave traveling along the truss member. The resulting critical time step is finally written in what is known as the Courant–Friedrichs–Lewy form as

$$\Delta t \leq \Delta t_{\text{crit}} = \frac{2}{\omega} = \frac{l}{c}.$$

The time a sound wave takes to traverse the length of the truss member is given by the expression l/c. This implies that the critical time step is determined by the time taken by a sound wave to traverse the shortest member. This conclusion extends to the general finite element case where the critical time step is determined by the highest natural frequency occurring amongst all elements. The resulting time-step values are typically very small compared to the natural period of vibration of the overall truss or solid; consequently, a very large number of time steps are invariably needed to complete an analysis using the leap-frog method.

2.4.4 Trapezoidal Time Integration

The final time-stepping scheme to be presented in this chapter is the so-called trapezoidal scheme. This is not a variational time integrator and hence does not follow from Equation (2.44) via an approximate integrated Lagrangian. Instead the trapezoidal rule can be derived from a simple trapezoidal integration of the nodal velocities as

$$\mathbf{v}_{n+1} = \mathbf{v}_n + \Delta t \frac{(\mathbf{a}_{n+1} + \mathbf{a}_n)}{2}, \tag{2.74}$$

where $(\mathbf{a}_{n+1} + \mathbf{a}_n)/2$ is a trapezoidal approximation of the average acceleration over the time step n to $n+1$. The nodal position at $n+1$ can now be approximated by a Taylor series expansion employing the average acceleration to give

$$\mathbf{x}_{n+1} = \mathbf{x}_n + \Delta t \mathbf{v}_n + \frac{1}{2}\Delta t^2 \left(\frac{\mathbf{a}_{n+1} + \mathbf{a}_n}{2} \right) + \ldots \tag{2.75}$$

Equations (2.74) and (2.75) are combined with the dynamic equilbrium equations at each time step to provide $\mathbf{a}_{n+1}$ and consequently advance the motion of the system (the truss, for example). At time step $n+1$ the accelerations and positions must satisfy

$$\mathbf{M}\mathbf{a}_{n+1} + \mathbf{T}(\mathbf{x}_{n+1}) = \mathbf{F}(t_{n+1}). \tag{2.76}$$

Observe from Equation (2.75) that the implicit nature of the above equation is obvious since $\mathbf{x}_{n+1}$ depends upon $\mathbf{a}_{n+1}$ and consequently a predictor–corrector process similar to that employed in the mid-point scheme will be required. The predictor can be established as

$$\mathbf{a}_{n+1}^{(0)} = \mathbf{a}_n, \tag{2.77a}$$

$$\mathbf{v}_{n+1}^{(0)} = \mathbf{v}_n + \frac{\Delta t}{2}(\mathbf{a}_n + \mathbf{a}_{n+1}^{(0)}), \tag{2.77b}$$

$$\mathbf{x}_{n+1}^{(0)} = \mathbf{x}_n + \Delta t \mathbf{v}_n + \frac{\Delta t^2}{4}(\mathbf{a}_n + \mathbf{a}_{n+1}^{(0)}). \tag{2.77c}$$

The residual (or out-of-balance) force required for the Newton–Raphson iteration process is now

$$\mathbf{R}_{n+1}(\mathbf{x}_{n+1}^{(k)}, \mathbf{a}_{n+1}^{(k)}) = \mathbf{M}\mathbf{a}_{n+1}^{(k)} + \mathbf{T}(\mathbf{x}_{n+1}^{(k)}) - \mathbf{F}(t_{n+1}). \tag{2.78}$$

To solve the above dynamic equilibrium equation, that is, to find $\mathbf{x}_{n+1}^{(k)}$ and $\mathbf{a}_{n+1}^{(k)}$ such that $\mathbf{R}_{n+1}(\mathbf{x}_{n+1}^{(k)}, \mathbf{a}_{n+1}^{(k)}) \rightarrow 0$, a corrector phase similar to that used in the mid-point scheme must establish the linearized residual equation for $\Delta\mathbf{a}_{n+1}$ as

$$\left[\mathbf{M} + \frac{\Delta t^2}{4}\mathbf{K} \right] \Delta\mathbf{a}_{n+1} = -\mathbf{R}_{n+1}(\mathbf{x}_{n+1}^{(k)}, \mathbf{a}_{n+1}^{(k)}), \tag{2.79}$$

where the derivation of the tangent stiffness matrix $\mathbf{K}$ is given in Example 2.2 above. The system of linear equations (2.79) is solved for $\Delta\mathbf{a}_{n+1}$, allowing the acceleration, velocity, and position to be updated as

$$\mathbf{a}_{n+1}^{(k+1)} = \mathbf{a}_{n+1}^{(k)} + \Delta\mathbf{a}_{n+1}, \tag{2.80a}$$

$$\mathbf{v}_{n+1}^{(k+1)} = \mathbf{v}_{n+1}^{(k)} + \frac{\Delta t}{2}\Delta \mathbf{a}_{n+1}, \tag{2.80b}$$

$$\mathbf{x}_{n+1}^{(k+1)} = \mathbf{x}_{n+1}^{(k)} + \frac{\Delta t^2}{4}\Delta \mathbf{a}_{n+1}. \tag{2.80c}$$

The factor $\Delta t^2/4$ above is the same as that used in the update for $\mathbf{x}_{n+1/2}^{(k+1)}$ in Equation (2.73) in the mid-point time integration. In the Newton–Raphson procedure, Equations (2.79) and (2.80) are repeatedly employed until the residual $\mathbf{R}_{n+1}(\mathbf{x}_{n+1}^{(k)}, \mathbf{a}_{n+1}^{(k)})$ is less than a given tolerance.

The trapezoidal rule is a widely used integration scheme which belongs to a group of schemes known as the Newmark family. These are derived from a Taylor series and are generally written in terms of two arbitrary parameters γ and β as follows:

$$\mathbf{x}_{n+1} = \mathbf{x}_n + \Delta t \mathbf{v}_n + \frac{\Delta t}{2}\big[(1-2\beta)\mathbf{a}_n + 2\beta \mathbf{a}_{n+1}\big], \tag{2.81a}$$

$$\mathbf{v}_{n+1} = \mathbf{v}_n + \Delta t\big[(1-\gamma)\mathbf{a}_n + \gamma \mathbf{a}_{n+1}\big]. \tag{2.81b}$$

The choice of $\gamma = 1/2$ and $\beta = 1/4$ corresponds to the trapezoidal rule. In fact, it is easy to show that the choice of $\gamma = 1/2$ and $\beta = 0$ corresponds to the leap-frog method.

EXAMPLE 2.4: Viscous effects

So far this chapter has assumed that the tensile force in the truss member is purely a function of the relative positions of the nodes. Indeed, this is the case for the hyperelastic material models where stresses are derived from an elastic potential energy; see for example Equation (2.19). However, real materials are subject to dissipative effects resulting from plasticity or viscosity. Moreover, the introduction of viscous effects is useful from a "numerical point of view" as a simple way to converge to a static solution through the addition of artificial damping. A very simple damping effect can be added to the truss member model described in this chapter by augmenting the tensile force T given in Equation (2.7) with a velocity-dependent viscous term as

$$T = \frac{EV}{l}\varepsilon + c_v(\boldsymbol{v}_b - \boldsymbol{v}_a) \cdot \boldsymbol{n},$$

where c_v is the viscous damping coefficient. This equation represents a simple linear damper as shown overleaf.

(continued)

Example 2.4: *(cont.)*

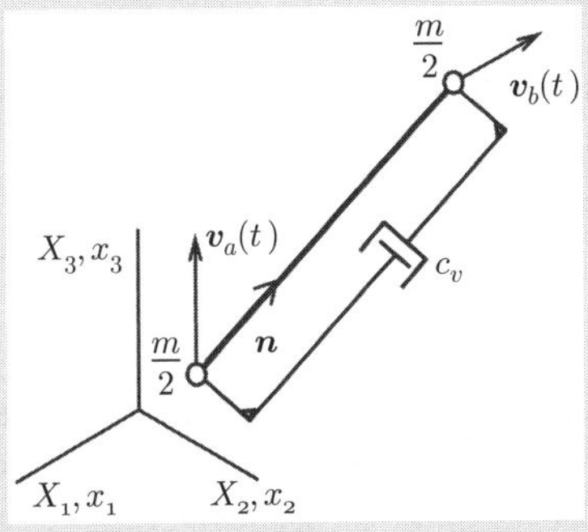

Truss member with viscous damper.

The value of the damping coefficient c can be related to the nondimensional critical damping coefficient ξ, the natural frequency of vibration ω, and the mass m as (refer to any standard dynamics textbook)

$$\xi = \frac{c_v}{2\omega m}$$

Using Equation (b) from Example 2.3 provides a useful expression for c_v in terms of the nondimensional damping coefficient ξ as

$$c_v = 4\xi\rho c a,$$

where a is the current truss member cross-sectional area, ρ is the current mass density, and c is the speed of sound in the member. For simplicity, c_v can be set initially at time $t = 0$ in the time-stepping process and maintained as constant throughout the motion; however, this will imply a varying amount of damping relative to the critical value. The addition of viscous effects will require a reformulation of the directional derivative of the residual force $\mathbf{R}$ for implicit time-stepping schemes.

2.5 GLOBAL CONSERVATION LAWS

In solid dynamics there are a number of conservation principles that have important practical consequences and applications. In particular, the conservation of linear momentum, angular momentum, and energy are conditions satisfied by an elastic solid in motion that have simple counterparts in the case of a pin-jointed truss.

Conservation principles will be presented both in the context of a time continuum system, that is, in rate form, and in an incremental or discrete setting following the application of a time integration scheme.

2.5.1 Conservation of Linear Momentum

In the pin-jointed truss that has been considered in this chapter, the mass is concentrated at each node as the sum of half the masses of each truss member connected to the node. Consequently, linear momentum of each node is $m_a \boldsymbol{v}_a$ and the total linear momentum of the truss, for all N nodes, is

$$\boldsymbol{L} = \sum_{a=1}^{N} m_a \boldsymbol{v}_a. \tag{2.82}$$

The rate of change of this total linear momentum vector can be obtained from the dynamic equilibrium Equation (2.12) as

$$\begin{aligned}\frac{d\boldsymbol{L}}{dt} &= \sum_{a=1}^{N} m_a \frac{d\boldsymbol{v}_a}{dt} \\ &= \sum_{a=1}^{N} (\boldsymbol{F}_a - \boldsymbol{T}_a) \\ &= \sum_{a=1}^{N} \boldsymbol{F}_a - \sum_{a=1}^{N} \boldsymbol{T}_a.\end{aligned} \tag{2.83}$$

The first term, the sum of $\boldsymbol{F}_a$, is the resultant of the external forces applied to the truss. The second term, that is, the sum of the internal forces $\boldsymbol{T}_a$, can easily be shown to equal zero by noting that the sum of the two internal force vectors associated with each truss member is zero since they are equal and opposite. Mathematically this implies that

$$\begin{aligned}\sum_{a=1}^{N} \boldsymbol{T}_a &= \sum_{a=1}^{N} \sum_{e \ni a}^{M_a} \boldsymbol{T}_a^e = \sum_{e=1}^{M} \sum_{a=1}^{2} \boldsymbol{T}_a^e \\ &= \sum_{e=1}^{M} (\boldsymbol{T}_a^e + \boldsymbol{T}_b^e) = \sum_{e=1}^{M} \boldsymbol{0} = \boldsymbol{0},\end{aligned} \tag{2.84}$$

where N is the number of nodes, M is the number of members, and M_a is the number of members containing node a. Consequently, the rate of change of linear momentum is given by the resultant external force as

$$\frac{d\boldsymbol{L}}{dt} = \sum_{a=1}^{N} \boldsymbol{F}_a. \tag{2.85}$$

In the absence of external forces the total linear momentum is conserved. It is also relatively easy to show that linear momentum is conserved after the introduction of leap-frog or mid-point rule time integration schemes. For example, in the case of the leap-frog scheme given by a combination of Equations (2.49) and (2.53), the change in total linear momentum at a typical step can be calculated as

$$\begin{aligned}
\boldsymbol{L}_{n+1/2} - \boldsymbol{L}_{n-1/2} &= \sum_{a=1}^{N} m_a \boldsymbol{v}_{a,n+1/2} - \sum_{a=1}^{N} m_a \boldsymbol{v}_{a,n-1/2} \\
&= \sum_{a=1}^{N} \Delta t \, m_a \boldsymbol{a}_{a,n} \\
&= \sum_{a=1}^{N} \Delta t \boldsymbol{F}_a(t_n) - \sum_{a=1}^{N} \Delta t \boldsymbol{T}_a(\mathbf{x}_n) \\
&= \Delta t \sum_{a=1}^{N} \boldsymbol{F}_a(t_n).
\end{aligned} \tag{2.86}$$

Note that $\boldsymbol{T}_a(\mathbf{x}_n)$ is a function not only of $\boldsymbol{x}_{a,n}$, but also of all $\boldsymbol{x}$ coordinates having nodes connected to node a. The increment in the total linear momentum between the half-time-step velocities is again given by the resultant force, but now multiplied by the time step Δt. Again, in the absence of extenal forces, linear momentum is conserved. A similar derivation is possible for the mid-point rule using Equations (2.65) and (2.66a) to give the change in linear momentum from steps n to $n+1$ as

$$\begin{aligned}
\boldsymbol{L}_{n+1} - \boldsymbol{L}_{n} &= \sum_{a=1}^{N} m_a \boldsymbol{v}_{a,n+1} - \sum_{a=1}^{N} m_a \boldsymbol{v}_{a,n} \\
&= \sum_{a=1}^{N} \Delta t \, m_a \boldsymbol{a}_{a,n+1/2} \\
&= \sum_{a=1}^{N} \Delta t \boldsymbol{F}_a(t_{n+1/2}) - \sum_{a=1}^{N} \Delta t \boldsymbol{T}_a(\mathbf{x}_{n+1/2}) \\
&= \Delta t \sum_{a=1}^{N} \boldsymbol{F}_a(t_{n+1/2}).
\end{aligned} \tag{2.87}$$

In both the above cases (leap-frog and mid-point) conservation is a consequence of the total sum of the nodal internal forces at t_n or $t_{n+1/2}$ being zero since the nodal truss forces for an individual truss member are equal and opposite.

Generally it is the case that any time integration scheme obtained by means of a discrete approximation of the Lagrangian as described in Section 2.4.1 will maintain the momentum conservation properties of the time-dependent equilibrium equations.

2.5.2 Conservation of Angular Momentum

Conservation of angular momentum is particularly important in problems involving large rotations, especially so for long term simulations. The total angular momentum of the truss is given by the sum of the angular momenta of each node where the mass of each truss member is concentrated. Taking the origin of the coordinate system as the reference point for the calculation of angular momentum gives

$$\boldsymbol{A} = \sum_{a=1}^{N} \boldsymbol{x}_a(t) \times m_a \boldsymbol{v}_a(t); \quad \boldsymbol{v}_a = \frac{d\boldsymbol{x}_a}{dt}. \tag{2.88a,b}$$

The rate of change of angular momentum is obtained by differentiating Equation (2.88a,b) to give

$$\begin{aligned}\frac{d\boldsymbol{A}}{dt} &= \sum_{a=1}^{N} \frac{d\boldsymbol{x}_a}{dt} \times m_a \boldsymbol{v}_a + \sum_{a=1}^{N} \boldsymbol{x}_a \times m_a \frac{d\boldsymbol{v}_a}{dt} \\ &= \sum_{a=1}^{N} \boldsymbol{x}_a \times m_a \boldsymbol{a}_a,\end{aligned} \tag{2.89}$$

where the first summation involving the cross product of parallel vectors is zero. Substituting the dynamic equilibrium Equation (2.11a,b) into Equation (2.89) yields the rate of change of angular momentum as

$$\begin{aligned}\frac{d\boldsymbol{A}}{dt} &= \sum_{a=1}^{N} \boldsymbol{x}_a \times (\boldsymbol{F}_a - \boldsymbol{T}_a) \\ &= \sum_{a=1}^{N} \boldsymbol{x}_a \times \boldsymbol{F}_a - \sum_{a=1}^{N} \boldsymbol{x}_a \times \boldsymbol{T}_a.\end{aligned} \tag{2.90}$$

The first term is the total moment due to the external forces acting on the truss, which should be the only source of changes in angular momentum. In the absence of external forces, or when the total moment of the nonzero external forces equals zero, then the angular momentum should remain constant, i.e. conserved. This implies that the total moment of the internal forces must vanish. To show this, note that for each truss member the internal forces $\boldsymbol{T}_a$ and $\boldsymbol{T}_b$ are equal and opposite in direction, that is, $\boldsymbol{T}_a = -\boldsymbol{T}_b$. This allows the internal force term in Equation (2.90) to be written as

$$\sum_{a=1}^{N} \boldsymbol{x}_a \times \boldsymbol{T}_a = \sum_{a=1}^{N} \boldsymbol{x}_a \times \sum_{e \ni a}^{M_a} \boldsymbol{T}_a^e = \sum_{e=1}^{M} \sum_{a=1}^{2} \boldsymbol{x}_a \times \boldsymbol{T}_a^e$$
$$= \sum_{e=1}^{M} (\boldsymbol{x}_a \times \boldsymbol{T}_a^e + \boldsymbol{x}_b \times \boldsymbol{T}_b^e)$$
$$= \sum_{e=1}^{M} (\boldsymbol{x}_b - \boldsymbol{x}_a) \times \boldsymbol{T}_b^e. \tag{2.91}$$

Equations (2.8) and (2.9a,b) give the force $\boldsymbol{T}_b^e$ in the truss member (e) in terms of the unit vector $\boldsymbol{n}_e = (\boldsymbol{x}_b - \boldsymbol{x}_a)/\|\boldsymbol{x}_b - \boldsymbol{x}_a\|$ and the magnitude of the force $T^e \equiv T$. Substituting into the above equation and again noting that the cross product of parallel vectors is zero gives

$$\sum_{a=1}^{N} \boldsymbol{x}_a \times \boldsymbol{T}_a = \sum_{e=1}^{M} (\boldsymbol{x}_b - \boldsymbol{x}_a) \times (T^e \boldsymbol{n}_e) = \boldsymbol{0}. \tag{2.92}$$

The term within the summation in the above equation is simply the moment due to the equal and opposite and co-linear end forces acting on the truss member (see Figure 2.5), which is equal to zero, that is,

$$\boldsymbol{x}_b \times T^e \boldsymbol{n}_e + \boldsymbol{x}_a \times (-T^e \boldsymbol{n}_e) = T^e (\boldsymbol{x}_b - \boldsymbol{x}_a) \times \boldsymbol{n}_e = \boldsymbol{0}. \tag{2.93}$$

The same conclusion can be reached for any mechanical system that can be represented by a set of particles with discrete masses and interacting forces acting

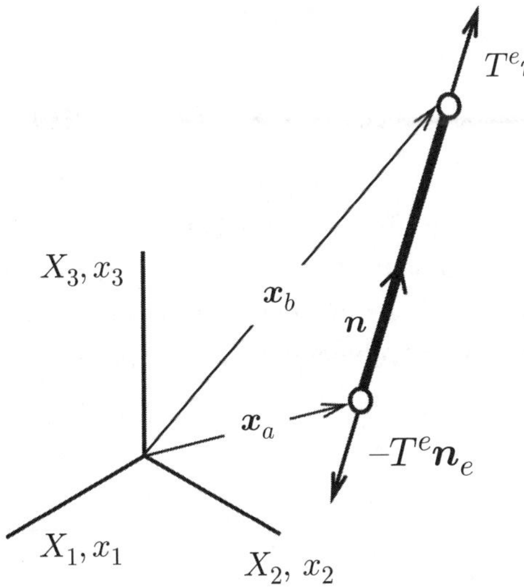

FIGURE 2.5 Total moment due to internal forces on a single truss member.

in the direction of the vector joining each pair of particles. Such is the case, for example, in molecular dynamics.

Combining Equations (2.90) and (2.92) gives the total rate of change of angular momentum as

$$\frac{d\boldsymbol{A}}{dt} = \sum_{a=1}^{N} \boldsymbol{x}_a \times \boldsymbol{F}_a. \tag{2.94}$$

As previously stated, if external forces are absent or the total moment due to external forces is zero then the total angular momentum of the truss is conserved.

The same angular conservation result can be achieved after time discretization using either the leap-frog or the mid-point rule. For brevity, only the mid-point rule will be developed, but the reader can follow similar steps for the leap-frog scheme. In fact, it is possible to prove that any time-stepping scheme derived from the variational process described in Section 2.4 will maintain the linear and angular momentum conservation properties expressed in rate form.

To prove the discrete conservation of angular momentum in the context of the mid-point rule, consider the change of angular momentum from time step n to $n+1$ as

$$\begin{aligned}\boldsymbol{A}_{n+1} - \boldsymbol{A}_n &= \sum_{a=1}^{N} \boldsymbol{x}_{a,n+1} \times m_a \boldsymbol{v}_{a,n+1} - \sum_{a=1}^{N} \boldsymbol{x}_{a,n} \times m_a \boldsymbol{v}_{a,n} \\ &= \sum_{a=1}^{N} (\boldsymbol{x}_{a,n+1} \times m_a \boldsymbol{v}_{a,n+1} - \boldsymbol{x}_{a,n} \times m_a \boldsymbol{v}_{a,n}) \\ &= \sum_{a=1}^{N} \Big[(\boldsymbol{x}_{a,n+1} - \boldsymbol{x}_{a,n}) \times \frac{1}{2} (\boldsymbol{v}_{a,n+1} + \boldsymbol{v}_{a,n}) m_a \\ &\qquad + \frac{1}{2} (\boldsymbol{x}_{a,n+1} + \boldsymbol{x}_{a,n}) \times m_a (\boldsymbol{v}_{a,n+1} - \boldsymbol{v}_{a,n}) \Big]. \end{aligned} \tag{2.95}$$

The expanded expression between the second and last equalities in Equation (2.95) is justified by invoking the distributive property of the cross product to show, after a simple algebraic manipulation, that the last equality can be re-expressed as the second equality. Introducing the mid-point rule, see Equations (2.66), into Equation (2.95), enables the discrete change in angular momentum to be expressed in terms of the mid-point acceleration as

$$\begin{aligned}\boldsymbol{A}_{n+1} - \boldsymbol{A}_n &= \sum_{a=1}^{N} \boldsymbol{v}_{a,n+1/2} \times m_a \boldsymbol{v}_{a,n+1/2} + \sum_{a=1}^{N} \boldsymbol{x}_{a,n+1/2} \times m_a \boldsymbol{a}_{a,n+1/2} \\ &= \sum_{a=1}^{N} \boldsymbol{x}_{a,n+1/2} \times m_a \boldsymbol{a}_{a,n+1/2}. \end{aligned} \tag{2.96}$$

Finally, substituting the nodal dynamic equilibrium equation at time $t_{n+1/2}$ and noting that the resultant moment of the internal forces at the mid-step is zero, see Equation (2.93) and Figure 2.5, gives the discrete angular momentum conservation property as

$$\begin{aligned}\boldsymbol{A}_{n+1} - \boldsymbol{A}_n &= \sum_{a=1}^{N} \boldsymbol{x}_{a,n+1/2} \times m_a \boldsymbol{a}_{a,n+1/2}\\ &= \sum_{a=1}^{N} \boldsymbol{x}_{a,n+1/2} \times \left[\boldsymbol{F}_a(t_{n+1/2}) - \boldsymbol{T}_a(\mathbf{x}_{n+1/2})\right]\\ &= \sum_{a=1}^{N} \boldsymbol{x}_{a,n+1/2} \times \boldsymbol{F}_a(t_{n+1/2}).\end{aligned} \tag{2.97}$$

Note that $\boldsymbol{T}_a(\mathbf{x}_{n+1/2})$ is a function not only of $\boldsymbol{x}_{a,n+1/2}$, but also of all $\boldsymbol{x}$ coordinates of nodes connected to node a. As mentioned above, a similar result can be arrived at for the leap-frog scheme, but not for the trapezoidal time integration scheme given in Section 2.4.4 since it does not derive from the least action principle. Consequently, the trapezoidal scheme is not advised for long term simulations involving large rotations.

2.5.3 Conservation of Energy

To conclude this section on conservation properties, it is possible to show that the equilibrium Equation (2.30) implies that the rate of change of total energy of an elastic truss (or solid) with constant external forces is zero. This is achieved by taking the time derivative of the total energy given by Equation (2.23) together with Equations (2.28) and (2.29) to yield

$$\begin{aligned}\frac{d\Pi}{dt} &= \frac{d}{dt}\left(K(\mathbf{v}) + \Pi_{\text{int}}(\mathbf{x}) + \Pi_{\text{ext}}(\mathbf{x})\right)\\ &= \frac{d}{dt}\left(\frac{1}{2}\mathbf{v}^T\mathbf{M}\mathbf{v}\right) + \left(\frac{\partial \Pi_{\text{int}}}{\partial \mathbf{x}}\right)^T \frac{d\mathbf{x}}{dt} + \left(\frac{\partial \Pi_{\text{ext}}}{\partial \mathbf{x}}\right)^T \frac{d\mathbf{x}}{dt}\\ &= \mathbf{v}^T\mathbf{M}\mathbf{a} + \mathbf{v}^T\mathbf{T}(\mathbf{x}) - \mathbf{v}^T\mathbf{F}\\ &= \mathbf{v}^T\left(\mathbf{M}\mathbf{a} + \mathbf{T}(\mathbf{x}) - \mathbf{F}\right) = 0.\end{aligned} \tag{2.98}$$

In more complex cases, where the external force $\mathbf{F}$ is not constant and cannot be derived from a potential $\Pi_{\text{ext}}(\mathbf{x})$, it is easy to show that the rate of change of total energy coincides with the rate of work done by the external forces.

An important question that emerges in the simulation of large strain solid dynamics is the preservation of energy conservation after the application of a numerical time-stepping scheme. Unlike linear and angular momentum, the use of a time integration scheme derived from a discrete approximation of the Lagrangian via a discrete variational principle does not, in general, guarantee discrete conservation of energy. This was already manifest in Chapter 1 in the case of a simple spring-mass system, where the evaluation of the spring force had to be modified in an ad hoc manner to preserve strict conservation of energy. The same outcome applies to the more complex case of the nonlinear pin-jointed truss described in this chapter, or, indeed, in the general continuum situation discussed in future chapters.

In a manner similar to that employed in Chapter 1, it is possible to modify the evaluation of the internal truss member force so that energy is exactly conserved from one time step to the next when using a time integration scheme such as the mid-point rule. Such techniques are known as *energy-momentum time integrators* and are sometimes employed when the exact conservation of energy is either essential or has an important effect on ensuring the stability of the time-stepping scheme. Typical examples of this occur when simulating large long term rotations of very stiff systems.

To study the issues surrounding discrete energy conservation in the presence of large rotations and/or nonlinear material behavior, recall the mid-point expression of the dynamic equilibrium Equation (2.65) as

$$\mathbf{M}\left(\frac{\mathbf{v}_{n+1}-\mathbf{v}_n}{\Delta t}\right)+\mathbf{T}(\mathbf{x}_{n+1/2})-\mathbf{F}=\mathbf{0}, \tag{2.99}$$

where the external force term has been assumed constant and the mid-point acceleration has been replaced by the velocity increment; see Equations (2.66). Multiplying Equation (2.99) by the position increment $\mathbf{x}_{n+1}-\mathbf{x}_n=\Delta t(\mathbf{v}_{n+1}+\mathbf{v}_n)/2$ gives an energy expression as follows:

$$\begin{aligned}\frac{1}{2}(\mathbf{v}_{n+1}+\mathbf{v}_n)^T\mathbf{M}(\mathbf{v}_{n+1}-\mathbf{v}_n)\ &+(\mathbf{x}_{n+1}-\mathbf{x}_n)^T\mathbf{T}(\mathbf{x}_{n+1/2})\\ &+(-\mathbf{x}_{n+1}^T\mathbf{F}+\mathbf{x}_n^T\mathbf{F})=\mathbf{0}.\end{aligned} \tag{2.100}$$

The last term in this expression is simply the increment in external potential energy, this energy being defined in Equation (2.20) as $\Pi_{\text{ext}}(\mathbf{x})=-\mathbf{x}^T\mathbf{F}$. The first term can be shown to give the change in kinetic energy as

$$\begin{aligned}\frac{1}{2}(\mathbf{v}_{n+1}+\mathbf{v}_n)^T\mathbf{M}(\mathbf{v}_{n+1}-\mathbf{v}_n)&=\frac{1}{2}\big[\mathbf{v}_{n+1}^T\mathbf{M}\mathbf{v}_{n+1}-\mathbf{v}_{n+1}^T\mathbf{M}\mathbf{v}_n\\ &\qquad+\mathbf{v}_n^T\mathbf{M}\mathbf{v}_{n+1}-\mathbf{v}_n^T\mathbf{M}\mathbf{v}_n\big]\\ &=\frac{1}{2}\mathbf{v}_{n+1}^T\mathbf{M}\mathbf{v}_{n+1}-\frac{1}{2}\mathbf{v}_n^T\mathbf{M}\mathbf{v}_n\\ &=K(\mathbf{v}_{n+1})-K(\mathbf{v}_n),\end{aligned} \tag{2.101}$$

where the fact that $\mathbf{M}$ is symmetric has been used to cancel $\mathbf{v}_{n+1}^T \mathbf{M} \mathbf{v}_n$ with $\mathbf{v}_n^T \mathbf{M} \mathbf{v}_{n+1}$. Insofar as the first and last terms in Equation (2.100) have been identified as changes in kinetic and external potential energy, respectively, it therefore transpires that discrete energy conservation will be satisfied *if* the internal force term in Equation (2.100) satisfies the remaining change, that is, the change in internal energy as

$$(\mathbf{x}_{n+1} - \mathbf{x}_n)^T \mathbf{T}(\mathbf{x}_{n+1/2}) = \Pi_{\text{int}}(\mathbf{x}_{n+1}) - \Pi_{\text{int}}(\mathbf{x}_n), \tag{2.102}$$

whereby Equation (2.100) would then emerge as the discrete conservation expression

$$K(\mathbf{v}_{n+1}) + \Pi_{\text{int}}(\mathbf{x}_{n+1}) + \Pi_{\text{ext}}(\mathbf{x}_{n+1}) = K(\mathbf{v}_n) + \Pi_{\text{int}}(\mathbf{x}_n) + \Pi_{\text{ext}}(\mathbf{x}_n); \tag{2.103}$$

that is, the total energy remains unchanged from one time step to the next. Unfortunately, Equation (2.102) is not generally true. Such is the case in situations involving large rotations or nonlinear material behavior when, for example, the logarithmic strain–Kirchhoff stress relation is used for the truss member. It is useful to show that this is the case given that, in doing so, a correction to the evaluation of the truss member force can be derived so that Equation (2.102) can be satisfied. The resulting algebra is somewhat lengthy and tedious, but is basically simple. To begin, express the left side of Equation (2.102) as

$$(\mathbf{x}_{n+1} - \mathbf{x}_n)^T \mathbf{T}(\mathbf{x}_{n+1/2}) = \sum_{a=1}^{N} (\boldsymbol{x}_{a,n+1} - \boldsymbol{x}_{a,n}) \cdot \sum_{\substack{e=1 \\ e \ni a}}^{M_a} \boldsymbol{T}_a^e(\mathbf{x}_{n+1/2}). \tag{2.104}$$

Recalling Equations (2.2a,b)$_b$, (2.4), (2.18a,b,c), (2.8), (2.11a,b), and (2.7) for the internal force term $\boldsymbol{T}_a^e$ gives¶

$$\begin{aligned}
(\mathbf{x}_{n+1} - \mathbf{x}_n)^T \mathbf{T}(\mathbf{x}_{n+1/2}) &= \sum_{a=1}^{N} (\boldsymbol{x}_{a,n+1} - \boldsymbol{x}_{a,n}) \cdot \sum_{\substack{e=1 \\ e \ni a}}^{M_a} \boldsymbol{T}_a^e(\mathbf{x}_{n+1/2}) \\
&= \sum_{e=1}^{M} \sum_{a=1}^{2} (\boldsymbol{x}_{a,n+1} - \boldsymbol{x}_{a,n}) \cdot \boldsymbol{T}_a^e(\mathbf{x}_{n+1/2}) \\
&= \sum_{e=1}^{M} \left[(\boldsymbol{x}_{b,n+1} - \boldsymbol{x}_{b,n}) \cdot \boldsymbol{T}_e(\mathbf{x}_{n+1/2}) - (\boldsymbol{x}_{a,n+1} - \boldsymbol{x}_{a,n}) \cdot \boldsymbol{T}_e(\mathbf{x}_{n+1/2}) \right] \\
&= \sum_{e=1}^{M} \left[(\boldsymbol{x}_{b,n+1} - \boldsymbol{x}_{a,n+1}) - (\boldsymbol{x}_{b,n} - \boldsymbol{x}_{a,n}) \right] \\
&\quad \cdot \frac{V_e}{l_{e,n+1/2}} \tau_e(l_{e,n+1/2}) \frac{\boldsymbol{x}_{b,n+1/2} - \boldsymbol{x}_{a,n+1/2}}{l_{e,n+1/2}}
\end{aligned}$$

¶ (e) refers to an element (truss member) whether subscript or superscript as convenient.

$$
\begin{aligned}
&= \sum_{e=1}^{M} \frac{V_e}{l_{e,n+1/2}^2} \tau_e(l_{e,n+1/2}) \Big(\big[(\boldsymbol{x}_{b,n+1} - \boldsymbol{x}_{a,n+1}) - (\boldsymbol{x}_{b,n} - \boldsymbol{x}_{a,n})\big] \\
&\qquad \cdot \frac{1}{2}\big[(\boldsymbol{x}_{b,n+1} - \boldsymbol{x}_{a,n+1}) + (\boldsymbol{x}_{b,n} - \boldsymbol{x}_{a,n})\big] \Big) \\
&= \sum_{e=1}^{M} \frac{1}{2} \frac{V_e}{l_{e,n+1/2}^2} \tau_e(l_{e,n+1/2}) \big[(\boldsymbol{x}_{b,n+1} - \boldsymbol{x}_{a,n+1}) \cdot (\boldsymbol{x}_{b,n+1} - \boldsymbol{x}_{a,n+1}) \\
&\qquad - (\boldsymbol{x}_{b,n} - \boldsymbol{x}_{a,n}) \cdot (\boldsymbol{x}_{b,n} - \boldsymbol{x}_{a,n})\big] \\
&= \sum_{e=1}^{M} \frac{1}{2} \frac{V_e}{l_{e,n+1/2}^2} \tau_e(l_{e,n+1/2}) (l_{e,n+1}^2 - l_{e,n}^2) \\
&= \sum_{e=1}^{M} \frac{1}{2} \frac{V_e}{l_{e,n+1/2}^2} \left. \frac{d\Psi}{d\ln(\frac{l_e}{L_e})} \right|_{l_{e,n+1/2}} (l_{e,n+1} - l_{e,n})(l_{e,n+1} + l_{e,n}) \\
&= \sum_{e=1}^{M} V_e \left. \frac{d\Psi(l_e)}{dl_e} \right|_{l_{e,n+1/2}} (l_{e,n+1} - l_{e,n}) \left(\frac{\frac{1}{2}(l_{e,n+1} + l_{e,n})}{l_{e,n+1/2}} \right).
\end{aligned} \tag{2.105}
$$

To achieve the desired result that the change in internal energy is given by Equation (2.102), which is now re-expressed as the final equation in Equation (2.105), two conditions must be satisfied. First, it would be necessary for the length $l_{e,n+1/2}$ at the half time step to coincide with the average of the lengths at steps n and $n+1$; and, additionally, for the derivative of the strain energy to satisfy

$$
\left. \frac{d\Psi(l_e)}{dl_e} \right|_{l_{e,n+1/2}} (l_{e,n+1} - l_{e,n}) = \Psi(l_{e,n+1}) - \Psi(l_{e,n}). \tag{2.106}
$$

In this case, Equations (2.102) and (2.105) would give the total change in internal energy as

$$
\begin{aligned}
(\mathbf{x}_{n+1} - \mathbf{x}_n)^T \mathbf{T}(\mathbf{x}_{n+1/2}) &= \sum_{e=1}^{M} V_e \big(\Psi(l_{e,n+1}) - \Psi(l_{e,n})\big) \\
&= \Pi_{\text{int}}(\mathbf{x}_{n+1}) - \Pi_{\text{int}}(\mathbf{x}_n),
\end{aligned} \tag{2.107}
$$

thereby completing the conservation Equation (2.103).

Generally, however, Equation (2.107) is not satisfied since, in the presence of rotations, $l_{e,n+1/2} \neq \frac{1}{2}(l_{e,n+1} + l_{e,n})$, and Equation (2.106) is not satisfied for complex strain energy functions Ψ. Therefore energy is not exactly conserved. Nevertheless it is possible to amend the evaluation of the member force $\boldsymbol{T}_e$ to ensure that energy is conserved. This is achieved by replacing the calculation of $\boldsymbol{T}_e(\mathbf{x}_{n+1/2})$ by the following secant expression:

$$\boldsymbol{T}^*_{e,n+1/2} = V_e \left(\frac{\Psi(l_{e,n+1}) - \Psi(l_{e,n})}{l_{e,n+1} - l_{e,n}} \right) \left(\frac{\boldsymbol{x}_{b,n+1/2} - \boldsymbol{x}_{a,n+1/2}}{\frac{1}{2}(l_{e,n+1} + l_{e,n})} \right). \qquad (2.108)$$

It is a simple exercise to show that replacing $\boldsymbol{T}_e = (V_e \tau_e / l_e)\boldsymbol{n}_e$ in Equation (2.102) by the above expression gives by construction the exact energy increment as

$$(\mathbf{x}_{n+1} - \mathbf{x}_n)^T \mathbf{T}^*(\mathbf{x}_{n+1/2}) = \Pi_{\text{int}}(\mathbf{x}_{n+1}) - \Pi_{\text{int}}(\mathbf{x}_n). \qquad (2.109)$$

In essence, the evaluation of the Kirchhoff stress τ as a derivative with respect to the length of the strain energy function Ψ is replaced by a secant expression as the increment in strain energy per increment in member length. In addition, in a similar manner to Example 1.2, the evaluation of the intermediate length $l_{n+1/2}$ has been replaced by the average length between time steps n and $n+1$ in Equation (2.108). The use of Equation (2.108) leads to what is known as an "energy-momentum" integration scheme, which, by design, satisfies conservation of linear and angular momenta as well as energy. Replacing $\boldsymbol{T}_e$ by $\boldsymbol{T}^*_e$ does have one computational disadvantage. Employing Equation (2.108) leads to a tangent stiffness matrix that is unsymmetric, unlike the symmetric matrix derived in Example 2.2. This is because the modified force $\boldsymbol{T}^*_e$ is no longer formulated from an exact derivative of the strain energy.

2.6 HAMILTONIAN FORMULATIONS

The preceding sections presented a comprehensive description of the dynamics of a pin-jointed truss. All the concepts introduced can be extended to more complex settings, such as three-dimensional solid mechanics. From a practical computational standpoint, it is not necessary to develop the formulations any further. However, there is a particular additional theoretical development that brings interesting mathematical insights into the physics of the motion of an elastic body. These insights can be used to ensure that time integration schemes reproduce certain mathematical properties of the exact motion; these are discussed below. It must be emphasized, however, that the reader may choose to skip this section as the ideas are somewhat advanced and require a good understanding of the mapping concepts developed in Chapter 3 of NL-Statics.

The theoretical concepts discussed in this section come under the umbrella term "Hamiltonian" formulations insofar as they originate from the definition of the Hamiltonian as a scalar function measuring the energy of the system in terms of momentum variables and positions. The combination of positions and momentum variables defines a *phase space* where the dynamic equilibrium equations of motion imply the conservation of certain mathematical properties, such as the so-called simplectic product. The integration schemes that preserve such properties are known as simplectic integrators and have been demonstrated to be particularly adept at approximating complex situations. While the leap-frog method and the mid-point rule, and indeed any variational integrators, are simplectic, proof will only be provided for the mid-point rule.

2.6.1 Momentum Variables and the Hamiltonian

The dynamic equations of motion for a truss have been formulated in terms of the nodal positions, $\mathbf{x}$, and the associated velocities $\mathbf{v} = \dot{\mathbf{x}}$ and accelerations $\mathbf{a} = \dot{\mathbf{v}} = \ddot{\mathbf{x}}$. To progress with the Hamiltonian formulation requires the introduction of momentum variables labeled as $\mathbf{p}$. In the case of the truss, $\mathbf{p}$ is a global vector containing the linear momentum $\boldsymbol{p}_a$ of each node a, which is related to the nodal velocity as

$$\boldsymbol{p}_a = m_a \boldsymbol{v}_a; \quad \mathbf{p} = \begin{bmatrix} \boldsymbol{p}_1 \\ \boldsymbol{p}_2 \\ \vdots \\ \boldsymbol{p}_N \end{bmatrix} = \mathbf{M}\mathbf{v}. \tag{2.110}$$

In a more general setting, the momentum variable $\mathbf{p}$ can be defined as the derivative of the Lagrangian $\mathcal{L}(\mathbf{x}, \mathbf{v})$ with respect to the velocity variables, known as a *generalized momentum*. Using the definition of the Lagrangian given by Equation (2.24), it is easy to show that this general definition coincides with Equations (2.110) as

$$\begin{aligned} \mathbf{p} &= \frac{\partial \mathcal{L}(\mathbf{x}, \mathbf{v})}{\partial \mathbf{v}} \\ &= \frac{\partial K(\mathbf{v})}{\partial \mathbf{v}} \\ &= \frac{\partial}{\partial \mathbf{v}} \left(\frac{1}{2} \mathbf{v}^T \mathbf{M} \mathbf{v} \right) = \mathbf{M}\mathbf{v}. \end{aligned} \tag{2.111}$$

The kinetic energy is now expressed in terms of the momentum vector as

$$\begin{aligned} K(\mathbf{v}) &= \frac{1}{2} \mathbf{v}^T \mathbf{M} \mathbf{v} = \frac{1}{2} (\mathbf{M}^{-1}\mathbf{p})^T \mathbf{M} (\mathbf{M}^{-1}\mathbf{p}) \\ &= \frac{1}{2} \mathbf{p}^T \mathbf{M}^{-T} \mathbf{p} = \frac{1}{2} \mathbf{p}^T \mathbf{M}^{-1} \mathbf{p}, \end{aligned} \tag{2.112}$$

where the symmetry of the mass matrix $\mathbf{M}$ has been noted. Corresponding to the general momentum definition $\partial\mathcal{L}/\partial\mathbf{v}$, the velocity $\mathbf{v}$ can be derived from the "conjugate" differential relationship

$$\mathbf{v} = \mathbf{M}^{-1}\mathbf{p} = \frac{\partial K(\mathbf{p})}{\partial \mathbf{p}}. \tag{2.113}$$

For the simple truss system under consideration, the Hamiltonian $\mathcal{H}$ can now be defined as the energy of the system expressed in terms of the momentum $\mathbf{p}$ and position variables $\mathbf{x}$ as

$$\mathcal{H}(\mathbf{x}, \mathbf{p}) = K(\mathbf{p}) + \Pi_{\text{int}}(\mathbf{x}) + \Pi_{\text{ext}}(\mathbf{x}). \tag{2.114}$$

The actual numerical value of $\mathcal{H}$ is identical to the total energy $\Pi(\mathbf{x}, \mathbf{v})$ given by Equation (2.23), the only difference being that, whereas Π is expressed in terms of $\mathbf{x}$ and its time derivative $\mathbf{v} = \dot{\mathbf{x}}$, the Hamiltonian $\mathcal{H}$ is a function of $\mathbf{x}$ and $\mathbf{p}$ *considered as independent* of each other. This is discussed in more detail below.

The dynamic equilibrium equations given in Section 2.3.2 and expressed by Equation (2.25) can now be re-written using the Hamiltonian. As a start note that Equation (2.113) can be written as a rate equation for $\mathbf{x}$ as

$$\dot{\mathbf{x}} = \frac{\partial \mathcal{H}(\mathbf{x}, \mathbf{p})}{\partial \mathbf{p}}, \tag{2.115}$$

by virtue of the fact that $\mathcal{H}$ depends on $\mathbf{p}$ only through the kinetic energy $K(\mathbf{p})$. A very similar equation for the rate of the momentum $\mathbf{p}$ can be derived through simple algebra by combining the definition of $\mathbf{p}$ from Equation (2.111) with the Lagrangian equilibrium Equation (2.25) and the definition of the Hamiltonian from Equation (2.114) to give

$$\begin{aligned}
\dot{\mathbf{p}} &= \frac{d}{dt}\frac{\partial \mathcal{L}(\mathbf{x}, \mathbf{v})}{\partial \mathbf{v}} \\
&= \frac{\partial \mathcal{L}(\mathbf{x}, \mathbf{v})}{\partial \mathbf{x}} \\
&= -\frac{\partial}{\partial \mathbf{x}}\left(\Pi_{\text{int}}(\mathbf{x}) + \Pi_{\text{ext}}(\mathbf{x})\right) \\
&= -\frac{\partial \mathcal{H}(\mathbf{x}, \mathbf{p})}{\partial \mathbf{x}}.
\end{aligned} \tag{2.116}$$

At this juncture it is worth pausing to reflect on the difference between the equations of Lagrange (2.25) and the Hamiltonian Equations (2.115) and (2.116). The Lagrangian $\mathcal{L}(\mathbf{x}, \mathbf{v})$ is a function of $\mathbf{x}$ and $\mathbf{v}$, where $\mathbf{v} = \dot{\mathbf{x}}$, so in effect $\mathcal{L}$ is a functional of the function $\mathbf{x}(t)$. In contrast, the Hamiltonian $\mathcal{H}(\mathbf{x}, \mathbf{p})$ is a function of $\mathbf{x}$ and $\mathbf{p}$ considered as independent variables. If, for example, in Equation (2.116) it is assumed a priori that $\mathbf{p} = \mathbf{M}\dot{\mathbf{x}}$, then the equation will deliver the dynamic equilibrium Equation (2.13). In addition, this latter circumstance transforms the first-order Hamiltonian Equation (2.116) into the second-order Newtonian Equation (2.13). Given that in $\mathcal{H}(\mathbf{x},\mathbf{p})$ the quantity $\dot{\mathbf{x}}$ cannot be deduced from $\mathbf{p}$, a means must be found to find $\dot{\mathbf{x}}$ from $\mathcal{H}(\mathbf{x},\mathbf{p})$, and this is provided by Equation (2.115). The necessary coupling between $\dot{\mathbf{x}}$ and $\mathbf{p}$ is provided through $\mathcal{H}(\mathbf{x},\mathbf{p})$, in effect ensuring $\mathbf{p} = \mathbf{M}\dot{\mathbf{x}}$. Finally, observe that the solution to the dynamic equilibrium problem is obtained by integrating either the second-order Equation (2.13) or the two first-order Equations (2.115) and (2.116).

The above digression is consolidated by Equations (2.115) and (2.116) being conveniently grouped together into a single evolution expression as

$$\frac{d}{dt}\begin{bmatrix}\mathbf{x}\\ \mathbf{p}\end{bmatrix} = \boldsymbol{J}\begin{bmatrix}\partial\mathcal{H}/\partial\mathbf{x}\\ \partial\mathcal{H}/\partial\mathbf{p}\end{bmatrix}; \quad \boldsymbol{J} = \begin{bmatrix}\mathbf{0}_{3N\times 3N} & \mathbf{I}_{3N\times 3N}\\ -\mathbf{I}_{3N\times 3N} & \mathbf{0}_{3N\times 3N}\end{bmatrix}, \tag{2.117a,b}$$

where N is the number of free nodes in the truss and the $6N \times 6N$ matrix $\boldsymbol{J}$ is known as the *simplectic operator*. Equation (2.117a,b) constitutes the *Hamiltonian dynamic equilibrium equation*. A useful physical interpretation of this equation and the operator $\boldsymbol{J}$ is discussed in the next section.

2.6.2 Phase Space and Hamiltonian Maps

Equation (2.117a,b) can be simplified (both notationally and conceptually) by defining a phase vector z as

$$z = \begin{bmatrix}\mathbf{x}\\ \mathbf{p}\end{bmatrix}, \tag{2.118}$$

which for a three-dimensional truss belongs to a $6 \times N$-dimensional phase space. With this new notation the Hamiltonian dynamic equilibrium equation becomes

$$\dot{z} = \boldsymbol{J}\boldsymbol{\nabla}_{z}\mathcal{H}(z), \tag{2.119}$$

where $\boldsymbol{\nabla}_{z}$ is the gradient with respect to z as shown in Equation (2.117a,b)$_a$. The equation above can be integrated in time from a set of initial conditions $\boldsymbol{Z} = z(t=0) = (\mathbf{x}_0^T, \mathbf{p}_0^T)^T$. The relationship between $\boldsymbol{Z}$ and $z(t)$ can be interpreted as a mapping in the phase space from an initial state to a current position and expressed as

$$z = \phi_{\mathcal{H}}(\boldsymbol{Z}, t). \tag{2.120}$$

This mapping is known, not unsurprisingly, as the Hamiltonian map‖ from an initial configuration $\boldsymbol{Z}$ in the phase space to its current state $z(t)$ at time t. This is

‖ The Hamiltonian map is often called the Hamiltonian flow due to its dependence upon time.

mathematically similar to the mapping $\boldsymbol{x} = \phi(\boldsymbol{X}, t)$ used to define the motion of a particle $\boldsymbol{X}$ to the current position $\boldsymbol{x}(t)$ at time t introduced in Section 4.2 of NL-Statics. The main difference is that $\boldsymbol{z}$ resides in a space with twice as many dimensions as degrees of freedom in the system and $\phi_{\mathcal{H}}$ is determined by the time integration of Equation (2.119), that is, its associated "velocity" satisfies

$$\dot{\boldsymbol{z}} = \frac{\partial \phi_{\mathcal{H}}(\boldsymbol{Z}, t)}{\partial t} = \boldsymbol{J}\boldsymbol{\nabla}_{\boldsymbol{Z}}\mathcal{H}(\boldsymbol{z}). \tag{2.121}$$

The mapping $\phi_{\mathcal{H}}$ can be represented graphically for the simple one-degree-of-freedom case, in which case the phase space is two dimensional. This is illustrated in Figure 2.6, where the trajectory followed from a given set of initial conditions $\boldsymbol{Z}_P$ is shown together with alternative trajectories from neighboring initial conditions $\boldsymbol{Z}_{Q_1}$ and $\boldsymbol{Z}_{Q_2}$. As a result of the motion, the initial point $\boldsymbol{Z}_P$ moves to $\boldsymbol{z}_p$ at time t, whereas $\boldsymbol{Z}_{Q_1}$ moves to $\boldsymbol{z}_{q_1}$ and $\boldsymbol{Z}_{Q_2}$ to $\boldsymbol{z}_{q_2}$. The elemental vectors joining $\boldsymbol{Z}_P$ to $\boldsymbol{Z}_{Q_1}$ and $\boldsymbol{Z}_{Q_2}$, namely $d\boldsymbol{Z}_1$ and $d\boldsymbol{Z}_2$, respectively, become vectors $d\boldsymbol{z}_1$ and $d\boldsymbol{z}_2$ at time t, now joining $\boldsymbol{z}_p$ to $\boldsymbol{z}_{q_1}$ and $\boldsymbol{z}_{q_2}$; see Figure 2.6. The relationship between $d\boldsymbol{Z}_1$ and $d\boldsymbol{z}_1$ and $d\boldsymbol{Z}_2$ and $d\boldsymbol{z}_2$, see Equation (2.120), is given by the deformation gradient $\boldsymbol{F}_{\boldsymbol{Z}}$ of the mapping $\phi_{\mathcal{H}}$ as

$$\begin{aligned} d\boldsymbol{z}_i &= \phi_{\mathcal{H}}(\boldsymbol{Z}_{Qi}, t) - \phi_{\mathcal{H}}(\boldsymbol{Z}_P, t) \\ &\approx \left.\frac{\partial \phi_{\mathcal{H}}(\boldsymbol{Z}, t)}{\partial \boldsymbol{Z}}\right|_{\boldsymbol{Z}_P, t} d\boldsymbol{Z}_i \\ &= \boldsymbol{F}_{\boldsymbol{Z}}(\boldsymbol{Z}_P, t)\, d\boldsymbol{Z}_i; \quad i = 1, 2, \end{aligned} \tag{2.122a}$$

where

$$\boldsymbol{F}_{\boldsymbol{Z}} = \frac{\partial \phi_{\mathcal{H}}(\boldsymbol{Z}, t)}{\partial \boldsymbol{Z}}. \tag{2.122b}$$

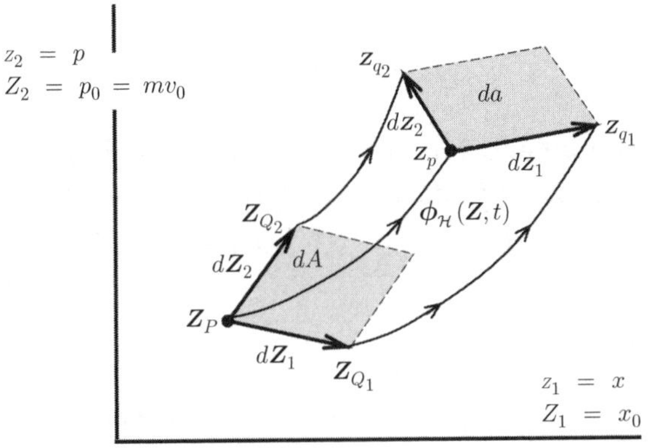

FIGURE 2.6 Hamiltonian map, or flow, in the phase space showing trajectories from different starting positions $\boldsymbol{Z}_P$, $\boldsymbol{Z}_{Q_1}$, and $\boldsymbol{Z}_{Q_2}$.

The interpretation of the equations of motion as a map makes it possible to deduce some important results. For instance, it is easy to show that the Hamiltonian flow, in the phase space, is volume conserving, which is known as *Liouville's theorem*. To show this, recall from Section 4.14 of NL-Statics that the rate of change of volume is given by the divergence of the velocity, which, using Equation (2.121), yields

$$\begin{aligned} \operatorname{div} \dot{z} &= \operatorname{tr}(\boldsymbol{\nabla}_z \dot{z}) \\ &= \operatorname{tr}\left(\frac{\partial}{\partial z}\left(\boldsymbol{J}\frac{\partial \mathcal{H}}{\partial z}\right)\right) \\ &= \operatorname{tr}\left(\boldsymbol{J}\frac{\partial^2 \mathcal{H}}{\partial z \partial z}\right) \\ &= \boldsymbol{J} : \frac{\partial^2 \mathcal{H}}{\partial z \partial z} = 0. \end{aligned} \tag{2.123}$$

Volume is conserved as a consequence of $\boldsymbol{J}$ in Equation (2.117a,b)$_b$ being a skew symmetric tensor and the second derivative of $\mathcal{H}$ being symmetric, thus rendering $\operatorname{div} \dot{z} = 0$. In the context of the motion depicted in Figure 2.6, this implies that $da = dA$.

The conservation of phase space volume of Hamiltonian maps is an interesting result. However, a far more profound property of this map relates to the simplectic operator defined above. This is the subject of the next section.

EXAMPLE 2.5: Phase space for the simple column

As an illustration of phase space, consider the simple column shown in Figure 1.1, Section 1.2.1. The phase vector given by Equation (2.118) simplifies to

$$z = \begin{bmatrix} \mathsf{x} \\ \mathsf{p} \end{bmatrix} = \begin{bmatrix} \theta \\ mL^2\dot{\theta} \end{bmatrix}.$$

For this example, the diagram in Figure 2.6 can be constructed using the leap-frog time integration with the following data: $\theta_0 = 1$, $L = 2$, $k = 100$, $c = 0$, $g = 9.81$, $m = 1$, $\Delta t = 0.0001$, and $t_{max} = 1.1$, together with three starting values for θ and $\dot{\theta}$ as

$\theta_P = 1$	$\dot{\theta}_P = 1$
$\theta_{Q_1} = 1.02$	$\dot{\theta}_{Q_1} = 1$
$\theta_{Q_2} = 1$	$\dot{\theta}_{Q_2} = 1.1$

(continued)

Example 2.5: *(cont.)*

With reference to Figure 2.6, the simplectic nature of the phase space is demonstrated by the area da remaining constant at all time steps.

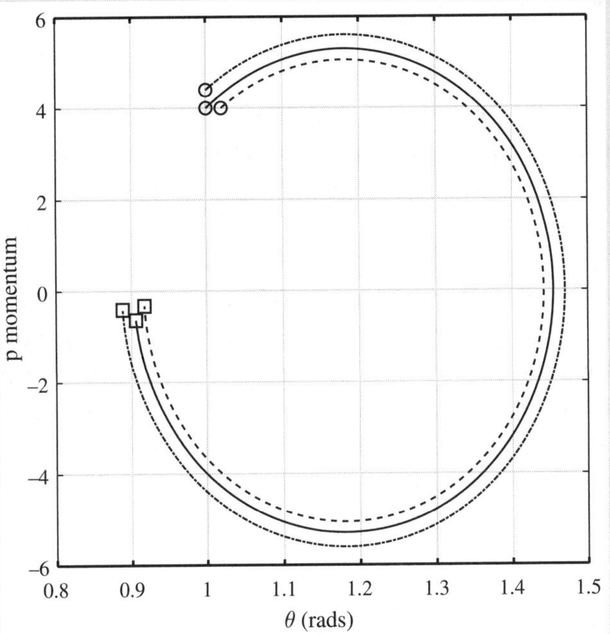

Simple column – phase space map from initial to current configuration.

2.6.3 The Simplectic Product

The simplectic operator $\boldsymbol{J}$ defined in Equation (2.117a,b)$_b$ satisfies a number of important properties. It is easy to check by direct multiplication the following relationships:

$$\boldsymbol{J}^T\boldsymbol{J} = \boldsymbol{I}; \quad \boldsymbol{J}\boldsymbol{J}^T = \boldsymbol{I}; \quad \boldsymbol{J}^T = -\boldsymbol{J}; \quad \boldsymbol{J}\boldsymbol{J} = -\boldsymbol{I}. \tag{2.124a,b,c,d}$$

These properties imply that $\boldsymbol{J}$ is in fact a rotation operator; Equation (2.124a,b,c,d)$_d$ signifies that $\boldsymbol{J}$ rotates vectors by $90°$ so that when applied twice, as in $\boldsymbol{J}\boldsymbol{J}$, it returns the same vector but in the opposite direction. This can be demonstrated by applying $\boldsymbol{J}$ to any arbitrary vector in the phase space, for instance $d\boldsymbol{z}_1$ shown in Figure 2.6, to give

$$\boldsymbol{J}d\boldsymbol{z}_1 = \begin{bmatrix} \mathbf{0} & \mathbf{I} \\ -\mathbf{I} & \mathbf{0} \end{bmatrix} \begin{bmatrix} d\mathbf{x}_1 \\ d\mathbf{p}_1 \end{bmatrix} = \begin{bmatrix} d\mathbf{p}_1 \\ -d\mathbf{x}_1 \end{bmatrix}. \tag{2.125}$$

That $\boldsymbol{J}d\boldsymbol{z}_1$ is perpendicular to $d\boldsymbol{z}_1$ can be checked by taking the scalar product:

$$\begin{aligned} d\boldsymbol{z}_1 \cdot \boldsymbol{J}d\boldsymbol{z}_1 &= \begin{bmatrix} d\mathbf{x}_1^T & d\mathbf{p}_1^T \end{bmatrix} \begin{bmatrix} d\mathbf{p}_1 \\ -d\mathbf{x}_1 \end{bmatrix} \\ &= d\mathbf{x}_1^T\, d\mathbf{p}_1 - d\mathbf{p}_1^T d\mathbf{x}_1 \\ &= 0. \end{aligned} \quad (2.126)$$

In general, given two arbitrary vectors $d\boldsymbol{z}_1$ and $d\boldsymbol{z}_2$ in the phase space, their *simplectic product* is defined as $d\boldsymbol{z}_1 \cdot \boldsymbol{J}d\boldsymbol{z}_2$, and, since $\boldsymbol{J}$ is skew symmetric, the simplectic product changes sign if the vectors are commuted, that is,

$$d\boldsymbol{z}_1 \cdot \boldsymbol{J}d\boldsymbol{z}_2 = -d\boldsymbol{z}_2 \cdot \boldsymbol{J}d\boldsymbol{z}_1. \quad (2.127)$$

Equation (2.126) has shown that the simplectic product of a vector by itself is zero, a property which has the following important consequence. The value of the Hamiltonian, that is, the total energy of the system, remains constant whenever $\mathcal{H}$ is only a function of $\boldsymbol{z}$ and does not explicitly depend on the time t. In practice, this implies that the external forces are not a function of t. In order to prove this important property, use Equation (2.119) to give the rate of change of the Hamiltonian $\dot{\mathcal{H}}$ as

$$\begin{aligned} \dot{\mathcal{H}}(\boldsymbol{z}) &= \boldsymbol{\nabla}_{\boldsymbol{z}}\mathcal{H} \cdot \dot{\boldsymbol{z}} \\ &= \boldsymbol{\nabla}_{\boldsymbol{z}}\mathcal{H} \cdot \boldsymbol{J}\boldsymbol{\nabla}_{\boldsymbol{z}}\mathcal{H} \\ &= 0. \end{aligned} \quad (2.128)$$

For the one-degree-of-freedom situation shown in Figure 2.6, the simplectic product $d\boldsymbol{z}_1 \cdot \boldsymbol{J}d\boldsymbol{z}_2$ can be interpreted as the magnitude of the area da as

$$\begin{aligned} da &= d\boldsymbol{z}_1 \cdot \boldsymbol{J}d\boldsymbol{z}_2 \\ &= \|d\boldsymbol{z}_1 \times d\boldsymbol{z}_2\| \\ &= dx_1 dp_2 - dx_2 dp_1. \end{aligned} \quad (2.129)$$

For problems with more degrees of freedom the simplectic product gives the sum total of the areas corresponding to each pair of $(\mathbf{x}, \mathbf{p})$ variables. The key feature of the Hamiltonian map is that it preserves the simplectic product of arbitrary elemental vectors, hence

$$d\boldsymbol{z}_1 \cdot \boldsymbol{J}d\boldsymbol{z}_2 = d\boldsymbol{Z}_1 \cdot \boldsymbol{J}d\boldsymbol{Z}_2, \quad (2.130)$$

or in rate form (to be proved below)

$$\frac{d}{dt}(d\boldsymbol{z}_1 \cdot \boldsymbol{J}d\boldsymbol{z}_2) = 0. \quad (2.131)$$

For a one-degree-of-freedom situation this statement is equivalent to Liouville's theorem, that is, the conservation of the area da. This is demonstrated for the one-degree-of-freedom situation in Example 2.5. As mentioned above, for multiple degrees of freedom Equation (2.130) implies that individual elements of area da are conserved as well as the combined hypervolume in $6N$ space.

Proof of Equation (2.131), which implies Equation (2.130), is reasonably straightforward and is based on the relationship between $d\boldsymbol{z}_i$ and $d\boldsymbol{Z}_i$ $(i = 1, 2)$ given by the deformation gradient $\boldsymbol{F}_Z$ as

$$\begin{aligned}\frac{d}{dt}(d\boldsymbol{z}_1 \cdot \boldsymbol{J} d\boldsymbol{z}_2) &= \frac{d}{dt}\big[(\boldsymbol{F}_Z d\boldsymbol{Z}_1) \cdot \boldsymbol{J}(\boldsymbol{F}_Z d\boldsymbol{Z}_2)\big] \\ &= d\boldsymbol{Z}_1 \cdot \frac{d}{dt}\big[\boldsymbol{F}_Z^T \boldsymbol{J} \boldsymbol{F}_Z\big] d\boldsymbol{Z}_2 \\ &= d\boldsymbol{Z}_1 \cdot \big[\dot{\boldsymbol{F}}_Z^T \boldsymbol{J} \boldsymbol{F}_Z + \boldsymbol{F}_Z^T \boldsymbol{J} \dot{\boldsymbol{F}}_Z\big] d\boldsymbol{Z}_2.\end{aligned} \tag{2.132}$$

The time derivative of $\boldsymbol{F}_Z$ can be obtained from Equations (2.121) and (2.122) and repeated use of the chain rule as

$$\begin{aligned}\dot{\boldsymbol{F}}_Z &= \frac{\partial^2 \boldsymbol{\phi}_{\mathcal{H}}}{\partial t \partial \boldsymbol{Z}} = \frac{\partial \dot{\boldsymbol{z}}}{\partial \boldsymbol{Z}} \\ &= \frac{\partial \dot{\boldsymbol{z}}}{\partial \boldsymbol{z}} \left(\frac{\partial \boldsymbol{z}}{\partial \boldsymbol{Z}} \right) = \boldsymbol{J} \frac{\partial^2 \mathcal{H}}{\partial \boldsymbol{z} \partial \boldsymbol{z}} \boldsymbol{F}_Z.\end{aligned} \tag{2.133}$$

Substituting Equation (2.133) into (2.132) and the use of Equations (2.124a,b,c,d)$_{b,d}$ for the simplectic operator gives (after some simple algebra) the rate of change of the simplectic product as

$$\begin{aligned}\frac{d}{dt}(d\boldsymbol{z}_1 \cdot \boldsymbol{J} d\boldsymbol{z}_2) &= d\boldsymbol{Z}_1 \cdot \left[\boldsymbol{F}_Z^T \frac{\partial^2 \mathcal{H}}{\partial \boldsymbol{z} \partial \boldsymbol{z}} \boldsymbol{J}^T \boldsymbol{J} \boldsymbol{F}_Z + \boldsymbol{F}_Z^T \boldsymbol{J} \boldsymbol{J} \frac{\partial^2 \mathcal{H}}{\partial \boldsymbol{z} \partial \boldsymbol{z}} \boldsymbol{F}_Z \right] d\boldsymbol{Z}_2 \\ &= (\boldsymbol{F}_Z d\boldsymbol{Z}_1) \cdot \left[\frac{\partial^2 \mathcal{H}}{\partial \boldsymbol{z} \partial \boldsymbol{z}} - \frac{\partial^2 \mathcal{H}}{\partial \boldsymbol{z} \partial \boldsymbol{z}} \right] (\boldsymbol{F}_Z d\boldsymbol{Z}_2) \\ &= d\boldsymbol{z}_1 \cdot \boldsymbol{0}\, d\boldsymbol{z}_2 \\ &= 0.\end{aligned} \tag{2.134}$$

This shows that the exact Hamiltonian map preserves the simpletic product, or, in short, "is simplectic." Unfortunately, the analytical integration of the Hamiltonian map from the rate Equation (2.121) is rarely possible. Inevitably this equation needs to be integrated in time via a time-stepping procedure. Some discrete time-stepping schemes are able to maintain the preservation of the simplectic product and are thus known as simplectic. Such simplectic schemes are able to accurately represent key features of the motion of a system over long periods of time and are therefore commonly used in physics and engineering applications. For instance, any variational time integration scheme derived from an approximate discrete Lagrangian, as described in Section 2.4 is simplectic. The general proof of this statement is beyond the scope of this text. However, the next section will explore in detail the particular case of the mid-point integration scheme.

2.6.4 Discrete Hamiltonian Maps and Simplectic Time Integrators

The discrete time integration of Equation (2.121) is represented graphically in Figure 2.7. At each time step the integration scheme provides a means of obtaining $\boldsymbol{z}_{n+1}$ explicitly or implicitly from $\boldsymbol{z}_n$. This establishes a discrete Hamiltonian map $\widehat{\phi}_{\mathcal{H}}$ from n to $n+1$ as

$$\boldsymbol{z}_{n+1} = \widehat{\phi}_{\mathcal{H}}(\boldsymbol{z}_n, \Delta t). \tag{2.135}$$

For instance, the mid-point rule in the context of the Hamiltonian notation is established as

$$\boldsymbol{z}_{n+1} = \boldsymbol{z}_n + \Delta t \dot{\boldsymbol{z}}_{n+1/2}; \qquad \boldsymbol{z} = \begin{bmatrix} \mathbf{x} \\ \mathbf{p} \end{bmatrix}, \tag{2.136a,b}$$

which, after substitution for $\dot{\boldsymbol{z}}_{n+1/2}$ from Equation (2.121), becomes

$$\boldsymbol{z}_{n+1} = \boldsymbol{z}_n + \Delta t \boldsymbol{J} \boldsymbol{\nabla}_{\boldsymbol{z}} \mathcal{H}(\boldsymbol{z}_{n+1/2}) \; ; \quad \boldsymbol{z}_{n+1/2} = \frac{1}{2}(\boldsymbol{z}_n + \boldsymbol{z}_{n+1}). \tag{2.137a,b}$$

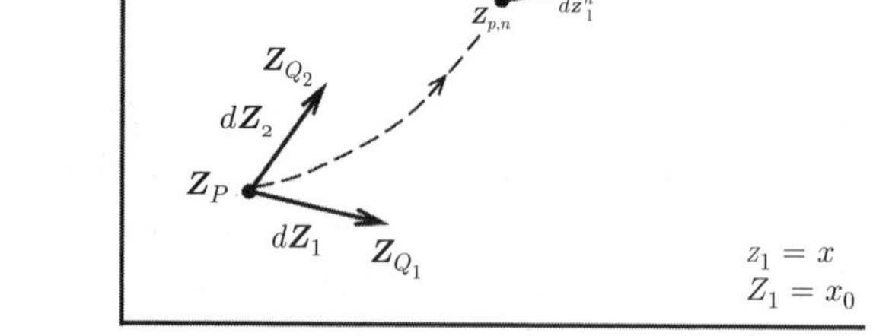

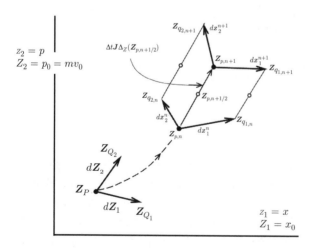

FIGURE 2.7 Discrete Hamiltonian map showing step n to $n+1$.

The incremental deformation gradient that maps the elemental vectors $d\boldsymbol{z}_1^n$ into $d\boldsymbol{z}_1^{n+1}$ and $d\boldsymbol{z}_2^n$ into $d\boldsymbol{z}_2^{n+1}$ is obtained by differentiating Equation (2.137a,b) (see also Equation (2.133)) to give

$$\boldsymbol{F}_{\boldsymbol{Z},\Delta t} = \frac{\partial \boldsymbol{z}_{n+1}}{\partial \boldsymbol{z}_n} = \boldsymbol{I} + \Delta t \boldsymbol{J}\, \frac{\partial^2 \mathcal{H}}{\partial \boldsymbol{z} \partial \boldsymbol{z}}\bigg|_{\boldsymbol{Z}_{n+1/2}} \left[\frac{\partial \boldsymbol{z}_{n+1/2}}{\partial \boldsymbol{z}_n}\right]. \tag{2.138}$$

The derivative of $\boldsymbol{z}_{n+1/2}$ with respect to $\boldsymbol{z}_n$ can be found from Equation (2.137a,b)$_b$ to give

$$\begin{aligned}\frac{\partial \boldsymbol{z}_{n+1/2}}{\partial \boldsymbol{z}_n} &= \frac{1}{2}\boldsymbol{I} + \frac{1}{2}\frac{\partial \boldsymbol{z}_{n+1}}{\partial \boldsymbol{z}_n}\\ &= \frac{1}{2}\big(\boldsymbol{I} + \boldsymbol{F}_{\boldsymbol{Z},\Delta t}\big).\end{aligned} \tag{2.139}$$

Substituting for this derivative into Equation (2.138) leads to an implicit expression with $\boldsymbol{F}_{\boldsymbol{Z},\Delta t}$ on both left and right sides of the equation as

$$\boldsymbol{F}_{\boldsymbol{Z},\Delta t} = \boldsymbol{I} + \frac{1}{2}\Delta t \boldsymbol{J}\, \frac{\partial^2 \mathcal{H}}{\partial \boldsymbol{z} \partial \boldsymbol{z}}\bigg|_{\boldsymbol{Z}_{n+1/2}} \big(\boldsymbol{I} + \boldsymbol{F}_{\boldsymbol{Z},\Delta t}\big). \tag{2.140}$$

Solving for $\boldsymbol{F}_{\boldsymbol{Z},\Delta t}$ gives, after some simple algebra and with $|_{\boldsymbol{Z}_{n+1/2}}$ omitted,

$$\boldsymbol{F}_{\boldsymbol{Z},\Delta t} = \left(\boldsymbol{I} - \frac{1}{2}\Delta t \boldsymbol{J}\frac{\partial^2 \mathcal{H}}{\partial \boldsymbol{z} \partial \boldsymbol{z}}\right)^{-1} \left(\boldsymbol{I} + \frac{1}{2}\Delta t \boldsymbol{J}\frac{\partial^2 \mathcal{H}}{\partial \boldsymbol{z} \partial \boldsymbol{z}}\right). \tag{2.141}$$

This operator maps elemental vectors from step n to step $n+1$ in the manner of an amplification matrix. In particular, following Equation (2.122),

$$d\boldsymbol{z}_i^{n+1} = \boldsymbol{F}_{\boldsymbol{Z},\Delta t} d\boldsymbol{z}_i^n; \quad i = 1, 2. \tag{2.142}$$

For simple linear systems, such as the spring-mass case shown in Example 2.6 (below), where $\boldsymbol{z}_{n+1}$ is directly related to $\boldsymbol{z}_n$ via a constant amplification matrix as $\boldsymbol{z}_{n+1} = \mathbf{A}\boldsymbol{z}_n$, $\boldsymbol{F}_{\boldsymbol{Z},\Delta t}$ is simply given by $\mathbf{A}$. However, more generally, the incremental deformation gradient $\boldsymbol{F}_{\boldsymbol{Z},\Delta t}$ extends the concept of the amplification matrix to the full nonlinear case.

The characteristics of a time integration scheme can be extracted from the properties of the incremental deformation gradient operator. For instance, stability requires the eigenvalues of $\boldsymbol{F}_{\boldsymbol{Z},\Delta t}$ to have a magnitude smaller than unity; see Example 2.7 (below) for the simple spring-mass system. Furthermore, a time integration scheme is simplectic if $\boldsymbol{F}_{\boldsymbol{Z},\Delta t}$ preserves the simplectic product, that is,

$$d\boldsymbol{z}_1^{n+1} \cdot \boldsymbol{J} d\boldsymbol{z}_2^{n+1} = d\boldsymbol{z}_1^n \cdot \boldsymbol{J} d\boldsymbol{z}_2^n, \tag{2.143}$$

which, after substitution for $d\boldsymbol{z}_i^{n+1}$ from Equation (2.142), gives a requirement for a simplectic incremental deformation gradient that ensures preservation of the simplectic product as

$$\boldsymbol{F}_{\boldsymbol{Z},\Delta t}^T \boldsymbol{J} \boldsymbol{F}_{\boldsymbol{Z},\Delta t} = \boldsymbol{J}. \tag{2.144}$$

In mapping terms this implies that $\boldsymbol{J}$ remains unchanged when pulled back from time step $n+1$ to n via the deformation gradient $\boldsymbol{F}_{\boldsymbol{Z},\Delta t}$. Additionally, taking the determinant on both sides of Equation (2.144) shows that if $\boldsymbol{F}_{\boldsymbol{Z},\Delta t}$ is simplectic its determinant must necessarily be equal to unity and consequently incrementally satisfy Liouville's theorem, that is, it conserves volume in the $6N$ phase space. Note, however, that the simplectic requirement enshrined in Equation (2.144) is far more onerous than the scalar constraint of volume conservation given by the determinant of $\boldsymbol{F}_{\boldsymbol{Z},\Delta t}$ (i.e. det $\boldsymbol{F}_{\boldsymbol{Z},\Delta t} = 1$).

The proof that the mid-point rule, characterized by the incremental deformation gradient given in Equation (2.141), satisfies the simplectic constraint Equation (2.144) is somewhat tedious and lengthy but is fundamentally based on simple matrix algebra manipulations using Equations (2.124a,b,c,d). Substituting Equation (2.141) into the simplectic condition Equation (2.144) gives

$$\boldsymbol{J} = \left[(\boldsymbol{I} + \boldsymbol{JS})^T (\boldsymbol{I} - \boldsymbol{JS})^{-T}\right] \boldsymbol{J} \left[(\boldsymbol{I} - \boldsymbol{JS})^{-1} (\boldsymbol{I} + \boldsymbol{JS})\right], \tag{2.145}$$

where $\boldsymbol{S}$ is the symmetric matrix contained in Equation (2.141) and given by

$$\boldsymbol{S} = \frac{1}{2} \Delta t \frac{\partial^2 \mathcal{H}}{\partial \boldsymbol{z} \partial \boldsymbol{z}}. \tag{2.146}$$

Consequently, proof of Equation (2.144) requires proof of Equation (2.145).

To begin, note that, since $\boldsymbol{J}^T = \boldsymbol{J}^{-1} = -\boldsymbol{J}$ and that the inverse product $(\boldsymbol{AB})^{-1} = \boldsymbol{B}^{-1}\boldsymbol{A}^{-1}$, then

$$\begin{aligned}
(\boldsymbol{I} + \boldsymbol{JS})^T (\boldsymbol{I} - \boldsymbol{JS})^{-T} \boldsymbol{J} &= -(\boldsymbol{I} - \boldsymbol{SJ})(\boldsymbol{I} + \boldsymbol{SJ})^{-1} \boldsymbol{J}^{-1} \\
&= -(\boldsymbol{I} - \boldsymbol{SJ})\left[\boldsymbol{J}(\boldsymbol{I} + \boldsymbol{SJ})\right]^{-1} \\
&= -(\boldsymbol{I} - \boldsymbol{SJ})\left[(\boldsymbol{J} + \boldsymbol{JSJ})\right]^{-1} \\
&= -(\boldsymbol{I} - \boldsymbol{SJ})\left[(\boldsymbol{I} + \boldsymbol{JS})\boldsymbol{J}\right]^{-1} \\
&= -(\boldsymbol{I} - \boldsymbol{SJ})\boldsymbol{J}^{-1}(\boldsymbol{I} + \boldsymbol{JS})^{-1} \\
&= (\boldsymbol{I} - \boldsymbol{SJ})\boldsymbol{J}(\boldsymbol{I} + \boldsymbol{JS})^{-1} \\
&= (\boldsymbol{J} + \boldsymbol{S})(\boldsymbol{I} + \boldsymbol{JS})^{-1} \\
&= (\boldsymbol{J} - \boldsymbol{JJS})(\boldsymbol{I} + \boldsymbol{JS})^{-1} \\
&= \boldsymbol{J}(\boldsymbol{I} - \boldsymbol{JS})(\boldsymbol{I} + \boldsymbol{JS})^{-1}.
\end{aligned} \tag{2.147}$$

Substituting this identity into Equation (2.145) gives the simplectic condition as

$$\boldsymbol{J} = \boldsymbol{J}(\boldsymbol{I} - \boldsymbol{JS})(\boldsymbol{I} + \boldsymbol{JS})^{-1}(\boldsymbol{I} - \boldsymbol{JS})^{-1}(\boldsymbol{I} + \boldsymbol{JS}). \tag{2.148}$$

The product of the two central terms on the right side of the above expression can be commuted since

$$\begin{aligned}(\boldsymbol{I} + \boldsymbol{JS})^{-1}(\boldsymbol{I} - \boldsymbol{JS})^{-1} &= \left[(\boldsymbol{I} - \boldsymbol{JS})(\boldsymbol{I} + \boldsymbol{JS})\right]^{-1} \\ &= \left[(\boldsymbol{I} - \boldsymbol{JSJS})\right]^{-1} \\ &= \left[(\boldsymbol{I} + \boldsymbol{JS})(\boldsymbol{I} - \boldsymbol{JS})\right]^{-1} \\ &= (\boldsymbol{I} - \boldsymbol{JS})^{-1}(\boldsymbol{I} + \boldsymbol{JS})^{-1}.\end{aligned} \tag{2.149}$$

Substituting Equation (2.149) into Equation (2.148) reveals the mid-point rule to be simplectic since

$$\begin{aligned}\boldsymbol{J} &= \boldsymbol{J}(\boldsymbol{I} - \boldsymbol{JS})(\boldsymbol{I} + \boldsymbol{JS})^{-1}(\boldsymbol{I} - \boldsymbol{JS})^{-1}(\boldsymbol{I} + \boldsymbol{JS}) \\ &= \boldsymbol{J}(\boldsymbol{I} - \boldsymbol{JS})(\boldsymbol{I} - \boldsymbol{JS})^{-1}(\boldsymbol{I} + \boldsymbol{JS})^{-1}(\boldsymbol{I} + \boldsymbol{JS}) \\ &= \boldsymbol{JII} \\ &= \boldsymbol{J}.\end{aligned} \tag{2.150}$$

Observe that this expression is satisfied for any symmetric matrix $\boldsymbol{S}$, which, as shown in Equation (2.146), will be the case for the mid-point rule irrespective of the value of the time step.

As a final observation, it is worth asking whether the mid-point rule in particular or simplectic time integrators in general satisfy the conservation of total energy of the system. Unfortunately the answer in most cases is negative, as already seen in Section 2.5.3 for the case of the mid-point rule. In order to show this, note first that the change in total energy from step n to $n+1$ is given by the change in the Hamiltonian, which can be approximated as

$$\mathcal{H}(\boldsymbol{z}_{n+1}) - \mathcal{H}(\boldsymbol{z}_n) \approx \left.\frac{\partial \mathcal{H}}{\partial \boldsymbol{z}}\right|_{\boldsymbol{z}_{n+1/2}} \cdot (\boldsymbol{z}_{n+1} - \boldsymbol{z}_n). \tag{2.151}$$

Substituting the increment in $\boldsymbol{z}$ from Equation (2.137a,b) for the mid-point rule, and noting that the simplectic product of a vector by itself is zero (see Equation (2.126)), gives the approximate energy change as

$$\begin{aligned}\mathcal{H}(\boldsymbol{z}_{n+1}) - \mathcal{H}(\boldsymbol{z}_n) &\approx \left.\frac{\partial \mathcal{H}}{\partial \boldsymbol{z}}\right|_{\boldsymbol{z}_{n+1/2}} \cdot (\boldsymbol{z}_{n+1} - \boldsymbol{z}_n) \\ &= \nabla_{\boldsymbol{z}}\mathcal{H}(\boldsymbol{z}_{n+1/2}) \cdot \Delta t \boldsymbol{J} \nabla_{\boldsymbol{z}}\mathcal{H}(\boldsymbol{z}_{n+1/2}) \\ &= 0.\end{aligned} \tag{2.152}$$

The equality above shows that the energy is approximately maintained from step n to $n+1$. But unfortunately the approximation only becomes exact for functional dependencies of $\mathcal{H}$ on $\boldsymbol{z}$ that are no more complex than quadratic. Only in these circumstances is the approximation in Equation (2.151) exact and energy is conserved. In more general cases this is not the situation. However, it is possible to remedy this by modifying the mid-point rule integration scheme by introducing a "corrected" gradient in Equation (2.137a,b) so that the step increment is

$$\boldsymbol{z}_{n+1} = \boldsymbol{z}_n + \Delta t \boldsymbol{J} \boldsymbol{\mathcal{G}}(\boldsymbol{z}_n, \boldsymbol{z}_{n+1}), \tag{2.153}$$

where $\boldsymbol{\mathcal{G}}$ is defined in such a manner that it approximates $\nabla_{\boldsymbol{z}} \mathcal{H}(\boldsymbol{z}_{n+1/2})$ but exactly meets the secant energy change condition given as

$$\boldsymbol{\mathcal{G}}(\boldsymbol{z}_n, \boldsymbol{z}_{n+1}) \cdot (\boldsymbol{z}_{n+1} - \boldsymbol{z}_n) = \mathcal{H}(\boldsymbol{z}_{n+1}) - \mathcal{H}(\boldsymbol{z}_n). \tag{2.154}$$

However, by so doing the incremental deformation gradient $\boldsymbol{F}_{\boldsymbol{z},\Delta t}$ is changed and will no longer satisfy the simplectic condition given by Equation (2.144). There is an extensive body of literature and debate about the merits of time integration schemes that satisfy energy and momentum conservation versus those that are simplectic.

In summary, simplectic schemes automatically satisfy momentum conservation but not energy conservation. The preservation of the simplectic product for any arbitrary pair of phase space vectors $d\boldsymbol{z}_1$ and $d\boldsymbol{z}_2$ places a very stringent set of restrictions on an adopted time integration scheme.

In effect, the area elements on the phase space plane of each degree of freedom are conserved through the analytical and discrete integration. This is equivalent to one scalar constraint per degree of freedom, whereas conservation of energy is a single overall scalar constraint. Moreover, it is also possible to prove that if a time-stepping scheme is simplectic, it is the exact integration of the flow of a specific Hamiltonian which differs from the exact one by an amount that converges to zero as $\Delta t \to 0$.** It would therefore appear that simplectic-based schemes are always preferable to schemes that only conserve energy and momentum in an overall integral sense. Nevertheless, both numerical evidence and theoretical analysis have shown that for very stiff systems energy and momentum conserving schemes can offer better stability characteristics for large time steps and are consequently often preferred. This is particularly the case when using energy-momentum methods that impose exact secant conditions, not just globally, but at element level as described in Section 4.8.

To conclude, both energy-momentum and simplectic methods have a role to play in the numerical integration of dynamic systems, and the optimum choice will inevitably depend on the problem being addressed.

** If a time-stepping scheme is proved to be simplectic, then if that scheme is used to solve the Newtonian equations of motion, simplecticity is maintained.

EXAMPLE 2.6: Hamiltonian formulation of a simple spring-mass system

Consider the one-dimensional spring-mass system shown below having a spring constant k, a mass m, and no external forces. The velocity v, momentum p, and phase vector $\boldsymbol{z}$ are given by

$$v = \dot{x}(t) \; ; \; p = mv \; ; \; \boldsymbol{z} = \begin{bmatrix} x \\ p \end{bmatrix}.$$

Simple one-degree-of-freedom spring-mass system.

The Hamiltonian and associated gradient are

$$\mathcal{H}(x,p) = \frac{1}{2m}p^2 + \frac{1}{2}kx^2,$$

$$\boldsymbol{\nabla}_{\boldsymbol{Z}}\mathcal{H} = \begin{bmatrix} \partial\mathcal{H}/\partial x \\ \partial\mathcal{H}/\partial p \end{bmatrix} = \begin{bmatrix} kx \\ p/m \end{bmatrix} = \begin{bmatrix} k & 0 \\ 0 & 1/m \end{bmatrix} \boldsymbol{z}.$$

From Equation (2.121) the equation of motion is

$$\dot{\boldsymbol{z}} = \boldsymbol{J}\boldsymbol{\nabla}_{\boldsymbol{Z}}\mathcal{H}(\boldsymbol{z})$$

$$= \begin{bmatrix} 0 & 1 \\ -1 & 0 \end{bmatrix} \begin{bmatrix} k & 0 \\ 0 & 1/m \end{bmatrix} \boldsymbol{z} = \begin{bmatrix} 0 & 1/m \\ -k & 0 \end{bmatrix} \boldsymbol{z}.$$

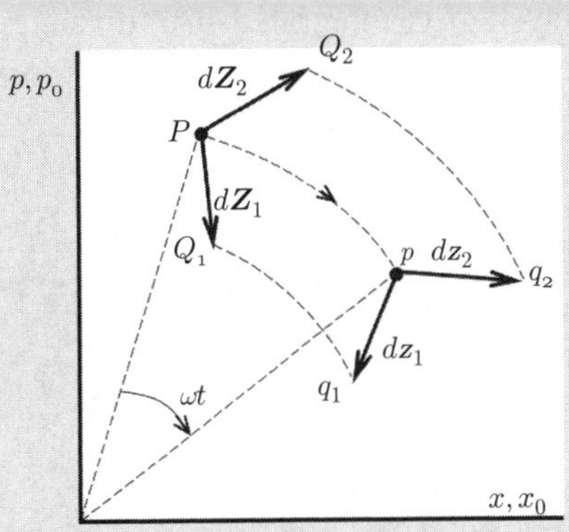

(a) One-degree-of-freedom system – phase space for $m\omega = 1$.

(continued)

Example 2.6: *(cont.)*

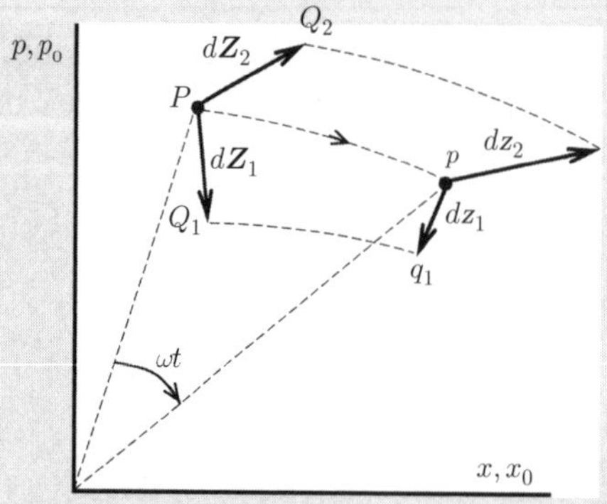

(b) One-degree-of-freedom system – phase space for $m\omega \neq 1$.

The above rate equation can be integrated analytically to give the Hamiltonian map $\boldsymbol{z} = \phi_{\mathcal{H}}(\boldsymbol{Z}, t)$ as

$$\boldsymbol{z}(t) = \begin{bmatrix} \cos(\omega t) & (\sin(\omega t))/m\omega \\ -m\omega \sin(\omega t) & \cos(\omega t) \end{bmatrix} \boldsymbol{Z}\,; \quad \omega = \sqrt{\frac{k}{m}}\,; \quad \boldsymbol{Z} = \begin{bmatrix} x_0 \\ p_0 \end{bmatrix}.$$

For the particular case where $m\omega = 1$, this represents a clockwise rotation in the phase space; see Figure (a) below. However, for general values when $m\omega \neq 1$, the map follows elliptical paths as shown in Figure (b).

EXAMPLE 2.7: Mid-point integration of the simple spring-mass Hamiltonian system

Using the mid-point rule, and in particular Equations (2.136a,b) or (2.137a,b), the Hamiltonian spring-mass system is integrated in a discrete manner as

$$\begin{aligned} \boldsymbol{z}_{n+1} &= \boldsymbol{z}_n + \Delta t \begin{bmatrix} 0 & 1/m \\ -k & 0 \end{bmatrix} \boldsymbol{z}_{n+1/2} \\ &= \boldsymbol{z}_n + \frac{1}{2}\Delta t \begin{bmatrix} 0 & 1/m \\ -k & 0 \end{bmatrix} (\boldsymbol{z}_{n+1} + \boldsymbol{z}_n). \end{aligned}$$

Solving for $\boldsymbol{z}_{n+1}$ from this implicit equation gives

$$\boldsymbol{z}_{n+1} = \begin{bmatrix} 1 & -\Delta t/2m \\ \Delta t/2k & 1 \end{bmatrix}^{-1} \begin{bmatrix} 1 & \Delta t/2m \\ -\Delta t/2k & 1 \end{bmatrix} \boldsymbol{z}_n.$$

(continued)

Example 2.7: *(cont.)*

In this simple case the discrete Hamiltonian map is given by a constant amplification matrix $\boldsymbol{A}$, which coincides with the incremental deformation gradient $\boldsymbol{F}_{\boldsymbol{Z},\Delta t}$ as

$$\boldsymbol{A} = \boldsymbol{F}_{\boldsymbol{Z},\Delta t} = \begin{bmatrix} 1 & -\Delta t/2m \\ \Delta t/2k & 1 \end{bmatrix}^{-1} \begin{bmatrix} 1 & \Delta t/2m \\ -\Delta t/2k & 1 \end{bmatrix}.$$

It is an easy exercise to check that this matrix has a determinant equal to unity, that is, it is isochoric (only distortional, without area change), and that its eigenvalues are

$$\lambda = \Pi_A \pm i\sqrt{1 - \Pi_A^2}\ ; \quad \Pi_A = \frac{4 - \omega^2 \Delta t^2}{4 + \omega^2 \Delta t^2} \leq 1.$$

The above complex conjugate eigenvalues have a magnitude of unity for any time-step size and hence the scheme is linearly unconditionally stable.

2.7 EXAMPLES

Two examples illustrate the nonlinear dynamic behavior that can be simulated using the formulations presented above. The first is a "rope" represented as a two-dimensional truss in order to include some bending behavior. The example clearly demonstrates the large deformation, large rotation action of a rope hinged at one end and allowed to swing freely under the action of gravity. The second example is a simple three-dimensional truss structure representing the well-known example proposed by Simo and Tarnow and employed since by many researchers.[††] Known as the L-shaped block, this benchmark example will be repeated using solid elements in Chapter 7.

2.7.1 Swinging Rope

The rope is configured in the $x - z$ plane, is initially horizontal, and is hinged at the origin. The length is 120 and the "thickness" in the $x - z$ plane is unity with a mesh comprising 24 nodes and 55 truss elements. At time $t = 0$, the nodal velocities are zero and the gravity force is ramped up linearly over a period of 0.5 s, equal, in this case, to the first time step. Using Newmark time integration, the control data for the FLagSHyP program is shown in Figure 2.8 together with the configuration at various times. The swinging behavior of the rope is satisfyingly evident

[††] Simo and Tarnow (1994), pp. 2527–2549.

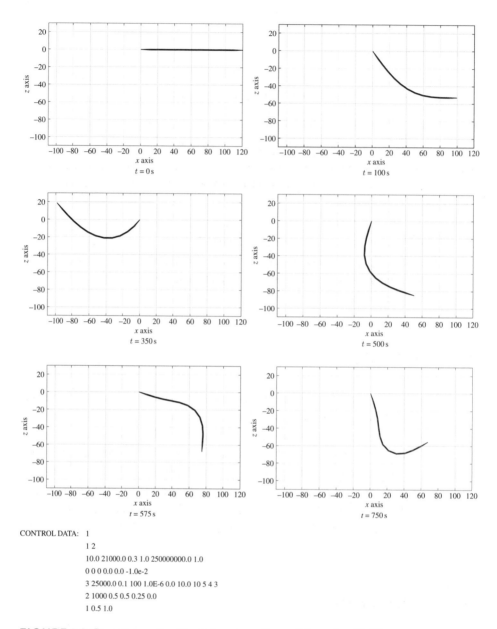

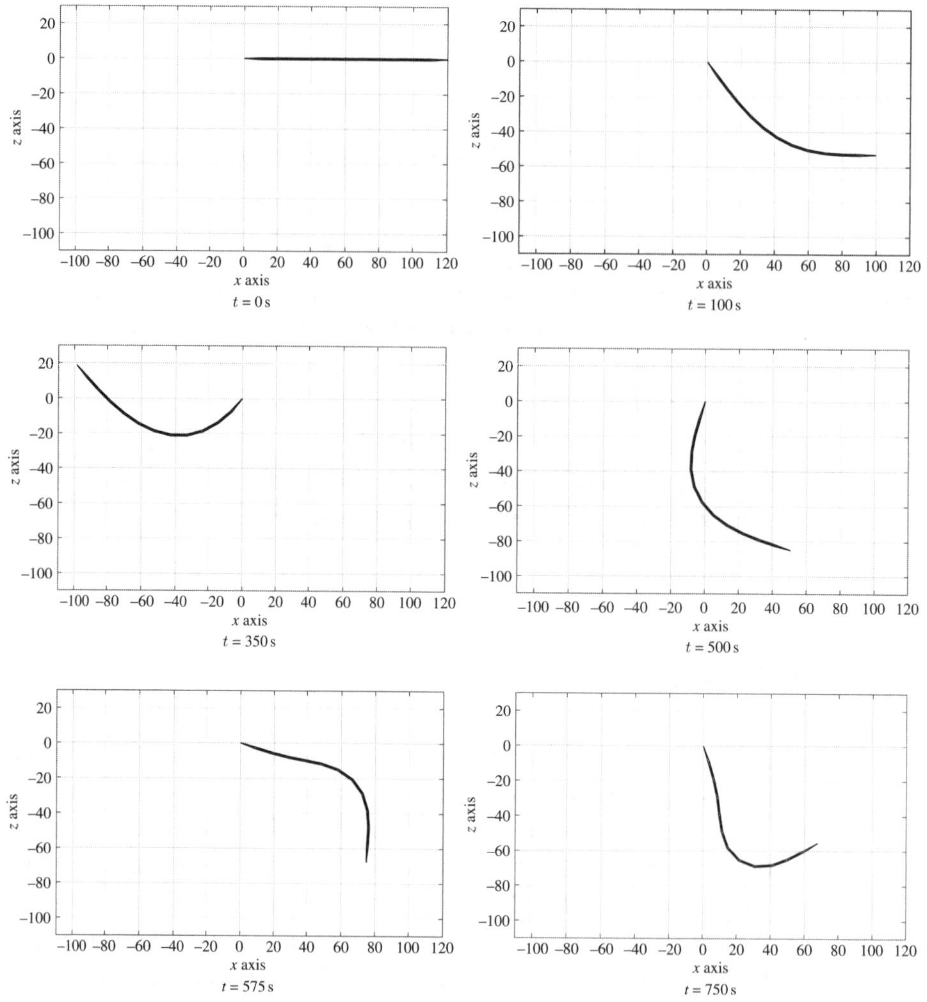

CONTROL DATA: 1

```
1 2
10.0 21000.0 0.3 1.0 250000000.0 1.0
0 0 0 0.0 0.0 -1.0e-2
3 25000.0 0.1 100 1.0E-6 0.0 10.0 10 5 4 3
2 1000 0.5 0.5 0.25 0.0
1 0.5 1.0
```

FIGURE 2.8 Rope truss: Position at various times (hinged at [0,0]).

in a video of the complete motion (see `www.flagshyp.com`).‡‡ Figure 2.9 shows the development of the various energy measures and the conservation of the total energy.

‡‡ Or `www.cambridge.org/9781107115620`.

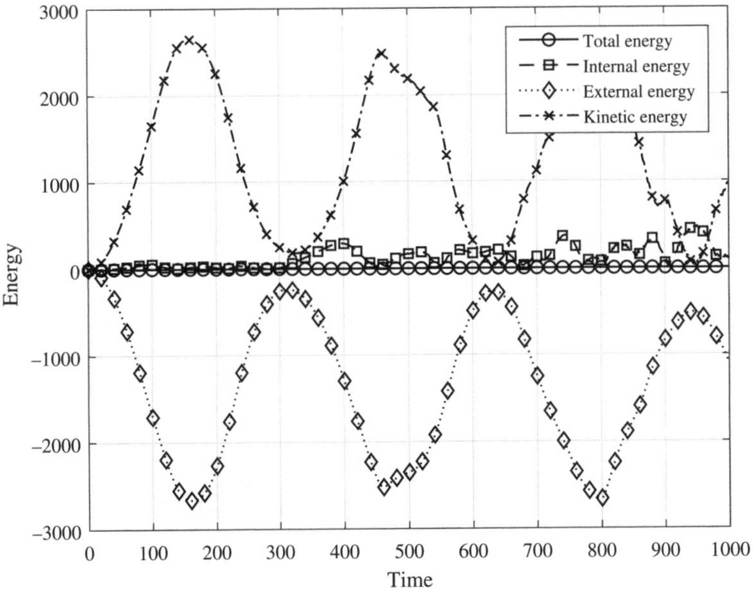

FIGURE 2.9 Rope truss: Conservation of total energy.

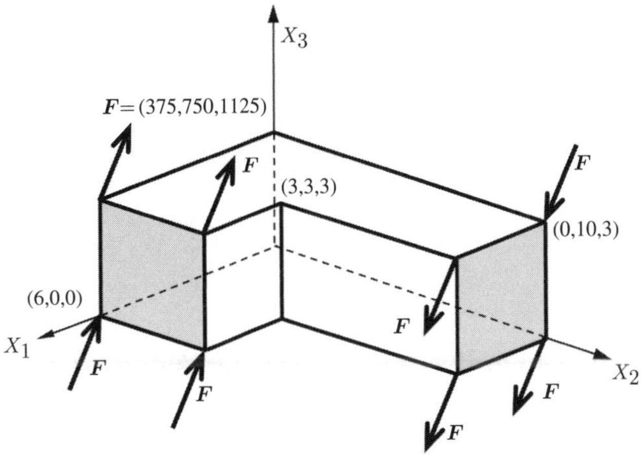

FIGURE 2.10 L block truss: Configuration.

2.7.2 L-Shaped Truss

The configuration of the L-block-shaped truss is given in Figure 2.10, where the loading is applied to achieve a spin in a zero gravity situation. The forces ramp up linearly to a peak value of 10 times the force at 0.05 s and return linearly to zero at 0.1 s, and thereafter the truss spins freely. The mesh comprises 20 nodes and 57 truss elements. Using a leap-frog time integration scheme, Figure 2.11 shows

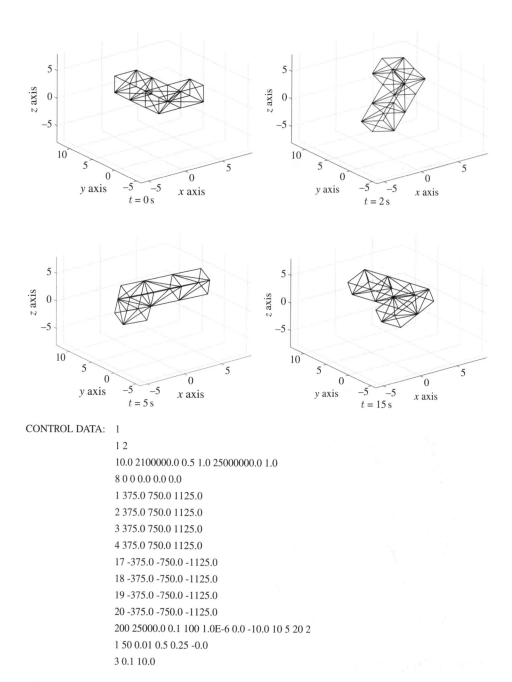

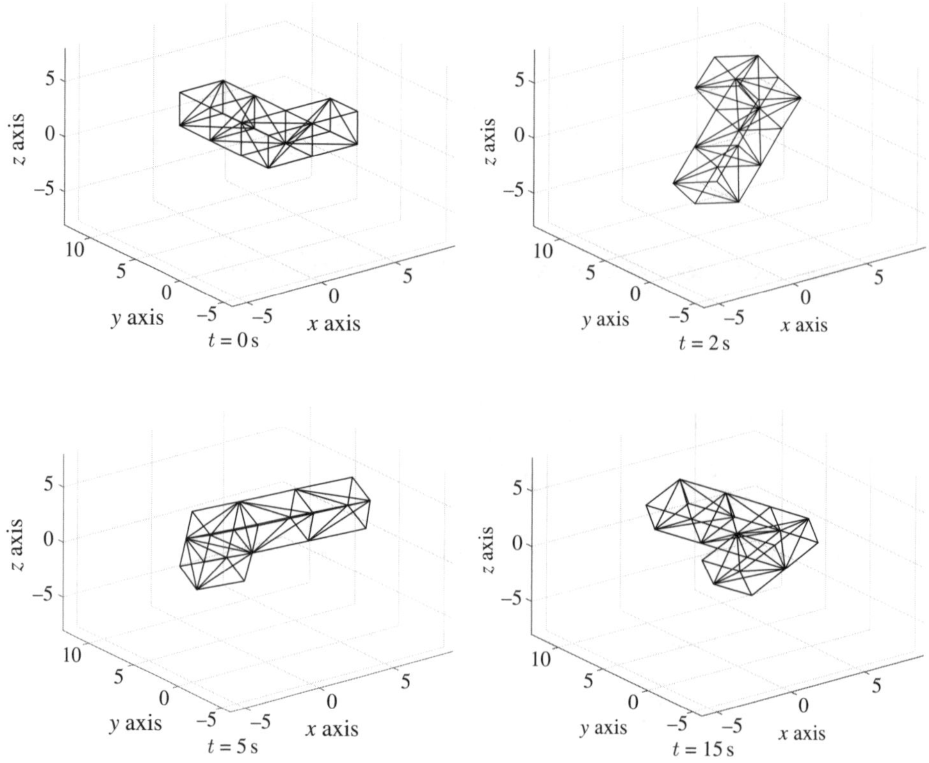


```
CONTROL DATA: 1
              1 2
              10.0 2100000.0 0.5 1.0 25000000.0 1.0
              8 0 0 0.0 0.0 0.0
              1 375.0 750.0 1125.0
              2 375.0 750.0 1125.0
              3 375.0 750.0 1125.0
              4 375.0 750.0 1125.0
              17 -375.0 -750.0 -1125.0
              18 -375.0 -750.0 -1125.0
              19 -375.0 -750.0 -1125.0
              20 -375.0 -750.0 -1125.0
              200 25000.0 0.1 100 1.0E-6 0.0 -10.0 10 5 20 2
              1 50 0.01 0.5 0.25 -0.0
              3 0.1 10.0
```

FIGURE 2.11 L block truss: Position at various times.

positions at various times together with the control data necessary to impose the force conditions.

Exercises

1. Show that for the one-degree-of-freedom example in Chapter 1 the Lagrangian can be written as

$$\mathcal{L}(\theta, \dot{\theta}) = \frac{1}{2} m L^2 \dot{\theta}^2 - \left(\frac{1}{2} k(\theta - \theta_0)^2 + mgL\cos\theta \right),$$

and derive the dynamic equilibrium for $\theta(t)$ using the equation of Lagrange.

2. For the same one-degree-of-freedom problem given in the previous exercise, show that the momentum variable p is $mL^2\dot{\theta}$ and that the Hamiltonian is

$$\mathcal{H}(\theta, p) = \frac{1}{2}\frac{1}{mL^2} p^2 + \frac{1}{2} k(\theta - \theta_0)^2 + mgL\cos\theta.$$

Derive the resulting equations of Hamilton for $z = (\theta, p)^T$; see Equation (2.117a,b).

3. Show that for the leap-frog time integration scheme the following expression for the angular momentum is conserved:

$$\boldsymbol{A}_{n+1} = \sum_a \boldsymbol{x}_{a,n+1} \times m_a \boldsymbol{v}_{a,n+1/2}.$$

4. The leap-frog scheme can be expressed in the so-called Verlet form as

$$\mathbf{v}_{n+1/2} = \mathbf{v}_n + \frac{\Delta t}{2} \mathbf{a}_n,$$

$$\mathbf{x}_{n+1} = \mathbf{x}_n + \Delta t \mathbf{v}_{n+1/2},$$

$$\mathbf{v}_{n+1} = \mathbf{v}_{n+1/2} + \frac{\Delta t}{2} \mathbf{a}_{n+1}.$$

Show that each of the three steps above can be interpreted as the exact (and therefore simplectic) integration of each of the components of the Hamiltonian broken down into three terms as $\mathcal{H} = \mathcal{H}_1 + \mathcal{H}_2 + \mathcal{H}_3$, where

$$\mathcal{H}_1 = \frac{1}{2}\Big(\Pi_{\text{int}}(\mathbf{x}) + \Pi_{\text{ext}}(\mathbf{x})\Big),$$

$$\mathcal{H}_2 = K(\mathbf{p}),$$

$$\mathcal{H}_3 = \frac{1}{2}\Big(\Pi_{\text{int}}(\mathbf{x}) + \Pi_{\text{ext}}(\mathbf{x})\Big).$$

5. (a) Modify the leap-frog program given in Box 1.3 to simulate the motion of the one-degree-of-freedom column example given in Figure 1.1, for which an example is shown in Figure 1.3.
   (b) Adapt the program to demonstrate that the simpletic product given by Equation (2.129) remains constant. Suggested problem data: $L = 2, \theta_0 = 1$, $k = 100$, $c = 0$, $g = 9.81$, $m = 1$, $dt = 0.0001$, and $t_{max} =$

0.2. Introduce an outer loop to enable the following initial data to run prior to the calculation of the simplectic product: $\theta = [1, 1.001, 1]$, $\dot{\theta} = [1, 1, 1.005]$.

6. Finally, for fun, modify the MATLAB program, available from `www.flagshyp.com`, and use the leap-frog time-stepping scheme to solve the "three body problem" that is, the motion of three bodies orbiting around each other under the influence of Newton's gravitational law $T = Gm_a m_b/l^2$, where G is the gravitational constant (assume $G = 1$ and simulate in two dimensions for simplicity). By fixing one of the masses (the sun), it is possible to plot the orbit of the moon around the earth. If the mid-point rule is used, then the tangent matrix has to be reformulated.
7. Revisit the swinging rope problem given in Section 2.7.1, using FlagShyp and the data given in `www.flagshyp.com` with zero gravity and a linear distribution of initial velocity $\boldsymbol{v}(\boldsymbol{X})|_{t=0} = [0, 0, -v_0(X_1/L)]^T$, where v_0 is any given value and L is the length of the rope.

CHAPTER THREE

DYNAMIC EQUILIBRIUM OF DEFORMABLE SOLIDS

3.1 INTRODUCTION

The previous two chapters have dealt with the motion, deformation, and equilibrium of simple examples such as the elastic pendulum and truss systems made up of assemblies of rods. This chapter will attempt to transfer the concepts introduced in these chapters to the motion of deformable solids. Consequently the main aim of the chapter is to derive the necessary equilibrium equations in a form suitable for the finite element discretization to be formulated later on in the text. Equilibrium will be introduced as a balance between the rate of change of linear momentum and the resultant and external forces, from which a dynamic statement of the principle of virtual work will be derived. This represents the generalization of the principle of virtual work used to describe static equilibrium by the addition of an inertial force term. As in the static case described in the NL-Statics volume, Section 5.4, the principle of virtual work can be presented in the reference or in the current configurations using the first Piola–Kirchhoff or Cauchy stress tensors respectively.

Before embarking on the description of dynamic equilibrium it will be necessary to review very briefly some basic kinematic concepts concerning the motion of deformable solids. These have been described and explained in greater detail in Chapter 4 of the NL-Statics volume. Similarly, the concepts of stress tensors in the reference and current configurations will be presented and used here, but again the reader may wish to consult NL-Statics, Section 5.5, for a more in-depth description.

The chapter will conclude by describing the equilibrium of a deformable solid as a more complex version of the variational principle presented in Chapter 2 for truss systems. This will include the definition and discussion of the Lagrangian for continuous systems and how, by using the action integral and invoking Hamilton's principle, it is possible to demonstrate that this principle is equivalent to the statement of equilibrium. However, it transpires that employing the action integral in the dynamic context has considerable advantages in that it enables spatial and time discretization to be considered in a holistic manner. Such an approach leads to discretizations that implicitly satisfy physical conservation requirements, as already presented in Chapter 2 and which will be elaborated on in detail in Chapter 4.

3.2 DYNAMIC EQUILIBRIUM

3.2.1 Motion, Velocity, and Acceleration

Consider the motion of a deformable solid from an initial reference configuration to a time-dependent current state, as shown in Figure 3.1. Each material particle can be defined by its original coordinates $\boldsymbol{X}$ in the reference configuration and its motion described by giving the current position $\boldsymbol{x}$ at time t via a mapping $\boldsymbol{\phi}$ as

$$\boldsymbol{x} = \boldsymbol{\phi}(\boldsymbol{X}, t). \tag{3.1}$$

The velocity and acceleration of the particle with reference coordinates $\boldsymbol{X}$ are obtained by time differentiation of the mapping function to give

$$\boldsymbol{v}(\boldsymbol{X}, t) = \frac{\partial \boldsymbol{x}}{\partial t} = \frac{\partial \boldsymbol{\phi}(\boldsymbol{X}, t)}{\partial t}, \tag{3.2a}$$

$$\boldsymbol{a}(\boldsymbol{X}, t) = \frac{\partial \boldsymbol{v}}{\partial t} = \frac{\partial^2 \boldsymbol{\phi}(\boldsymbol{X}, t)}{\partial t^2}. \tag{3.2b}$$

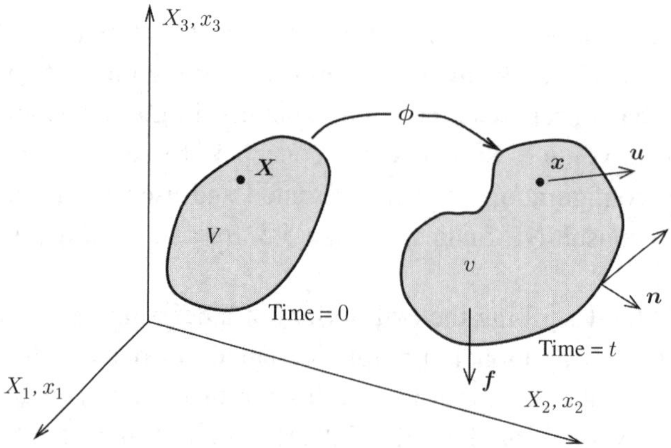

FIGURE 3.1 Motion of a deformable solid from reference configuration to current configuration at time t.

The deformation of the solid as it moves from the reference to the current configuration is described by the deformation gradient tensor $\boldsymbol{F}$ defined by the partial derivatives as

$$\boldsymbol{F} = \frac{\partial \boldsymbol{x}}{\partial \boldsymbol{X}} = \frac{\partial \boldsymbol{\phi}(\boldsymbol{X}, t)}{\partial \boldsymbol{X}}. \tag{3.3}$$

The tensor $\boldsymbol{F}$ contains information that describes how the body stretches and distorts through the motion. For instance, the determinant of $\boldsymbol{F}$, usually denoted by the letter J for Jacobian, measures the ratio between the current and reference elements of the volume as

$$J = \det \boldsymbol{F} = \frac{dv}{dV}. \tag{3.4}$$

This allows the reference and current densities, ρ_R and ρ, respectively, to be related as

$$\rho_R = \frac{dm}{dV} = J\frac{dm}{dv} = J\rho. \tag{3.5}$$

3.2.2 External Forces, Inertial Forces, and Global Dynamic Equilibrium in Current and Reference Configurations

The motion of a deformable solid such as that shown in Figure 3.1 is the consequence of the action of external forces. Such forces can either act on the surface ∂v of the solid as traction forces $\boldsymbol{t}$ per unit area or as body forces distributed throughout the internal volume v. A common example of a traction force would be the wind pressure on a building, whereas the most common example of a body force is that due to gravity or self weight. Body forces can be expressed per unit mass, in which case they are typically denoted by the letter $\boldsymbol{b}$.* In the case of self weight, $\boldsymbol{b}$ is simply given by the gravity acceleration vector $\boldsymbol{g}$ (since $\boldsymbol{b}$ is per unit mass). Alternatively, body forces can be expressed per unit current or undeformed volume as $\boldsymbol{f}$ or $\boldsymbol{f}_R$ respectively. The relationships between the reference and current density shown in Equation (3.5) enable these three representations of the same body force to be related as

$$\boldsymbol{b}\, dm = \boldsymbol{f} dv = \boldsymbol{f}_R\, dV; \quad \boldsymbol{f} = \rho \boldsymbol{b}; \quad \boldsymbol{f}_R = \rho_R \boldsymbol{b} = J\boldsymbol{f}. \tag{3.6a,b,c}$$

The total resultant external force, $\boldsymbol{E}$,† acting on a moving volume of deformable solid can now be calculated by integrating the body forces $\boldsymbol{f}$ over the current

* Although the context will be obvious, the body force $\boldsymbol{b}$ is not to be confused with $\boldsymbol{b} = \boldsymbol{F}\boldsymbol{F}^T$ which is the left Cauchy–Green tensor; see NL-Statics, Section 4.5.

† The external force $\boldsymbol{E}$ is not to be confused with the Lagrangian or Green's strain tensor $\boldsymbol{E} = (\boldsymbol{C} - \boldsymbol{I})/2$; see NL-Statics, Section 4.5.

volume and adding traction forces $\boldsymbol{t}$ integrated over that part of the current surface area upon which $\boldsymbol{t}$ acts to give

$$\boldsymbol{E} = \int_v \boldsymbol{f}\, dv + \int_{\partial v} \boldsymbol{t}\, da. \tag{3.7}$$

In the case of static equilibrium, this force must vanish. In contrast, in the dynamic case, the resultant external force leads to a change in linear momentum of the body following Newton's second law. The total linear momentum of a body is the integral of the momentum of each element of mass, that is,

$$\boldsymbol{L} = \int_v \boldsymbol{v}\, dm. \tag{3.8}$$

Applying Newton's second law, dynamic equilibrium can therefore be expressed as the balance between the rate of change of linear momentum and the resultant external force as

$$\frac{d}{dt}\int_v \boldsymbol{v}\, dm = \int_v \boldsymbol{f} dv + \int_{\partial v} \boldsymbol{t}\, da. \tag{3.9}$$

Alternatively, since the element of mass, dm, is not time dependent, the left-hand side of Equation (3.9) representing the inertial forces can be variously expressed as

$$\boldsymbol{I} = \frac{d}{dt}\int_v \boldsymbol{v}(\boldsymbol{x},t)\, dm = \frac{d}{dt}\int_V \boldsymbol{v}(\boldsymbol{X},t)\, dm \tag{3.10a}$$

$$= \int_V \boldsymbol{a}(\boldsymbol{X},t)\, dm = \int_v \boldsymbol{a}(\boldsymbol{x},t)\, dm \tag{3.10b}$$

$$= \int_v \rho\, \boldsymbol{a}(\boldsymbol{x},t)\, dv = \int_V \rho_R\, \boldsymbol{a}(\boldsymbol{X},t)\, dV. \tag{3.10c}$$

Note that in these relationships v denotes the volume currently occupied by the material initially occupying volume V. Observe also that the velocity $\boldsymbol{v}(\boldsymbol{X},t)$ of a particle located at $\boldsymbol{X}$ in the reference configuration is identical to the velocity of the same particle that has coordinates $\boldsymbol{x}(\boldsymbol{X},t)$ in the current configuration, that is, $\boldsymbol{v}(\boldsymbol{X},t) = \boldsymbol{v}(\boldsymbol{x}(\boldsymbol{X},t),t)$. Similarly for the acceleration $\boldsymbol{a}$. This distinction will not be repeatedly stated in an explicit manner given that the infinitesimal quantities dv or da and dV or dA already indicate whether the integrand is with respect to the current or reference configurations. See NL-Statics, Section 4.3, for more details.

Equilibrium can now be established as the balance between inertial and external forces in the current configuration as

$$\int_v \rho\boldsymbol{a}\, dv = \int_v \boldsymbol{f}\, dv + \int_{\partial v} \boldsymbol{t}\, da. \tag{3.11}$$

Alternatively, the same global equilibrium statement can be expressed in the reference volume as

$$\int_V \rho_R \boldsymbol{a} \, dV = \int_V \boldsymbol{f}_R \, dV + \int_{\partial V} \boldsymbol{t}_R \, dA, \tag{3.12}$$

where $\boldsymbol{t}_R$, see Figure 3.2, denotes the same traction vector $\boldsymbol{t}$ but is now expressed per unit undeformed area, that is,

$$\boldsymbol{t}_R \, dA = \boldsymbol{t} \, da; \quad \boldsymbol{t}_R = \boldsymbol{t} \frac{da}{dA}. \tag{3.13}$$

3.2.3 Local Dynamic Equilibrium

Equations (3.11) and (3.12) represent the global dynamic equilibrium of a volume v or V as the balance between total external and inertial forces. In order to derive a local equilibrium equation at a material point, note first that the same integral laws apply to any arbitrary embedded volume v^* and V^*, as shown in Figure 3.2. The balance of external and inertial forces for any arbitrary volume v^* gives

$$\int_{v^*} \rho \boldsymbol{a} \, dv = \int_{v^*} \boldsymbol{f} \, dv + \int_{\partial v^*} \boldsymbol{t} \, da, \tag{3.14}$$

or, equivalently, for any corresponding arbitrary volume V^*,

$$\int_{V^*} \rho_R \boldsymbol{a} \, dV = \int_{V^*} \boldsymbol{f}_R \, dV + \int_{\partial V^*} \boldsymbol{t}_R \, dA. \tag{3.15}$$

Furthermore, the traction vector $\boldsymbol{t}$ at an arbitrary material point on the boundary ∂v^* can be related to the Cauchy stress tensor $\boldsymbol{\sigma}$ by the surface unit normal $\boldsymbol{n}$ as

$$\boldsymbol{t} = \boldsymbol{\sigma} \boldsymbol{n}. \tag{3.16}$$

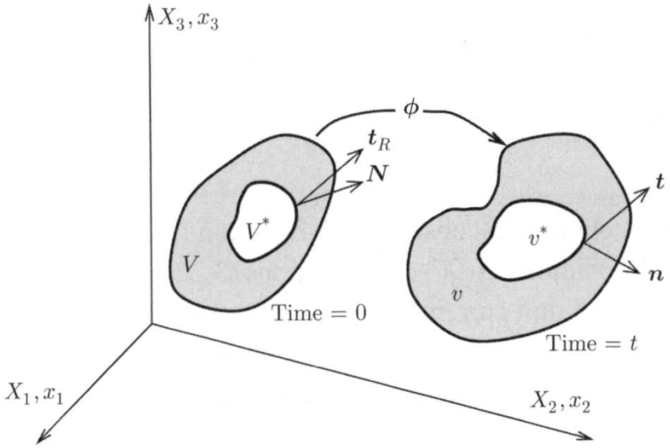

FIGURE 3.2 Embedded arbitrary volume V^* deforming to v^*.

In the reference configuration, the same corresponding relationship between surface normal $\boldsymbol{N}$ and traction vector $\boldsymbol{t}_R$ is furnished by the first Piola–Kirchhoff stress tensor $\boldsymbol{P}$ as

$$\boldsymbol{t}_R = \boldsymbol{P}\boldsymbol{N}, \tag{3.17}$$

where $\boldsymbol{\sigma}$ and $\boldsymbol{P}$ are related via $\boldsymbol{P} = J\boldsymbol{\sigma}\boldsymbol{F}^{-T}$; see NL-Statics, Section 5.5. Using the divergence integral theorem, it is now possible to convert area integrals over the external volume surfaces ∂v^* and ∂V^* into volume integrals as

$$\int_{\partial v^*} \boldsymbol{t}\, da = \int_{\partial v^*} \boldsymbol{\sigma}\boldsymbol{n}\, da = \int_{v^*} (\operatorname{div}\boldsymbol{\sigma})\, dv \tag{3.18}$$

and

$$\int_{\partial V^*} \boldsymbol{t}_R\, dA = \int_{\partial V^*} \boldsymbol{P}\boldsymbol{N}\, dA = \int_{V^*} (\mathrm{DIV}\boldsymbol{P})\, dV, \tag{3.19}$$

where div and DIV denote the divergence operators performed with respect to the current or reference coordinates, respectively; that is, in indicial notation

$$(\operatorname{div}\boldsymbol{\sigma})_i = \sum_j \frac{\partial \sigma_{ij}}{\partial x_j}; \quad (\mathrm{DIV}\boldsymbol{P})_i = \sum_J \frac{\partial \boldsymbol{P}_{iJ}}{\partial X_J}. \tag{3.20a,b}$$

The equilibrium equation in the current and reference configurations given by Equations (3.14) and (3.15) can now be transformed using Equations (3.18) and (3.19) to give

$$\int_{v^*} (\rho\, \boldsymbol{a} - \operatorname{div}\boldsymbol{\sigma} - \boldsymbol{f})\, dv = 0, \tag{3.21a}$$

$$\int_{V^*} (\rho_R\, \boldsymbol{a} - \mathrm{DIV}\boldsymbol{P} - \boldsymbol{f}_R)\, dV = 0. \tag{3.21b}$$

Given that these two expressions are valid for any arbitrary volumes v^* and V^*, the terms in the brackets above must vanish, thus leading to the local (pointwise) dynamic equilibrium equations in the current and reference configurations:

$$\rho\, \boldsymbol{a} \;= \operatorname{div}\boldsymbol{\sigma} + \boldsymbol{f}, \tag{3.22a}$$

$$\rho_R\, \boldsymbol{a} = \mathrm{DIV}\boldsymbol{P} + \boldsymbol{f}_R. \tag{3.22b}$$

Recalling discussions from Section 3.2.2, observe that Equation (3.22a) implies $\boldsymbol{a}(\boldsymbol{x}, t)$, whereas Equation (3.22b) implies $\boldsymbol{a}(\boldsymbol{X}, t)$.‡ Equations (3.22a) and (3.22b) describe the dynamic equilibrium of unit current and undeformed volumes respectively. They are versions of Newton's second law, where the mass per unit volume times the acceleration is given by the resultant forces acting on the unit volume,

‡ Equation (3.22a) can be rearranged as $\boldsymbol{r} = \operatorname{div}\boldsymbol{\sigma} + \boldsymbol{f} - \rho\boldsymbol{a} = 0$; this is known as D'Alembert's principle, where the inertial *force* $\rho\boldsymbol{a}$ is treated as an external force in a static equilibrium equation.

which include external forces $\boldsymbol{f}$ (or $\boldsymbol{f}_R$) and the net force per unit volume due to the gradient in stress.

EXAMPLE 3.1: Dynamic equilibrium in linear elasticity

It is instructive to study the equilibrium Equation (3.22a) in the case of *linear* elasticity. In so doing, it is possible to identify useful physical phenomena such as wave propagation at speeds dictated by the type of wave and the constitutive properties of the material. In order to achieve this, note first that in linear elasticity the stress is related to the small strain tensor $\boldsymbol{\varepsilon}$ by

$$\boldsymbol{\sigma} = \lambda(\text{tr}\boldsymbol{\varepsilon})\boldsymbol{I} + 2\mu\boldsymbol{\varepsilon}\ ;\ \boldsymbol{\varepsilon} = \frac{1}{2}(\boldsymbol{\nabla u} + \boldsymbol{\nabla u}^T),$$

where λ and μ are the Lamé coefficients describing the elastic properties of the material (for relationships between λ, μ, and Young's modulus E and Poisson's ratio ν see Section 6.6.5 of NL-Statics). Substituting the above equation into $(3.22)_a$ gives, after some simple algebra,

$$\rho\frac{\partial^2 \boldsymbol{u}}{\partial t^2} = (\lambda + \mu)\boldsymbol{\nabla}(\text{div}\,\boldsymbol{u}) + \mu\nabla^2\boldsymbol{u} + \boldsymbol{f}, \tag{a}$$

where the acceleration $\boldsymbol{a}$ has been expressed as the second time derivative of the displacement function $\boldsymbol{u}(\boldsymbol{x}, t) = \boldsymbol{x} - \boldsymbol{X}$ and the Laplacian of $\boldsymbol{u}$ is

$$\nabla^2\boldsymbol{u} = \text{div}\,(\boldsymbol{\nabla u}) = \sum_{i,j}\frac{\partial^2 u_i}{\partial x_j \partial x_j}\boldsymbol{e}_i.$$

EXAMPLE 3.2: Plane elastic waves

The linear elastic dynamic Equation (a) described in Example 3.1 can admit some simple analytical solutions in the absence of external forces for correctly chosen initial conditions and boundary conditions. The simplest of these analytical solutions, known as plane waves, are given by vibrations $\boldsymbol{u}(\boldsymbol{x}, t)$, which are expressed in the form

$$\boldsymbol{u}(\boldsymbol{x}, t) = \boldsymbol{n}f(\boldsymbol{x}\cdot\boldsymbol{\nu} - ct) \tag{b}$$

for an arbitrary function f and unit vectors $\boldsymbol{n}$ and $\boldsymbol{\nu}$. This expression represents a vibration in the direction of $\boldsymbol{n}$ which moves, or propagates, along an infinite elastic medium in the direction $\boldsymbol{\nu}$ at speed c, as shown in the figure below.

(continued)

Example 3.2: *(cont.)*

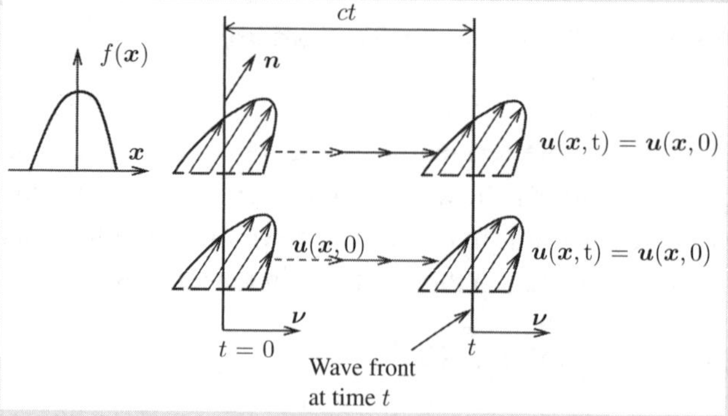

In order for Equation (b) to satisfy the linear elastic dynamic equilibrium Equation (a) in Example 3.1, $\boldsymbol{n}$ and $\boldsymbol{\nu}$ must be related to λ, μ, and ρ. To identify these dependencies simply substitute the vibration Equation (b) into the dynamic Equation (a) from Example 3.1, noting first that

$$\begin{aligned}
\frac{\partial^2 \boldsymbol{u}}{\partial t^2} &= c^2 \boldsymbol{n} f''(\boldsymbol{x} \cdot \boldsymbol{\nu} - ct), \\
\boldsymbol{\nabla u} &= (\boldsymbol{n} \otimes \boldsymbol{\nu}) f'(\boldsymbol{x} \cdot \boldsymbol{\nu} - ct), \\
\operatorname{div} \boldsymbol{u} &= (\boldsymbol{n} \cdot \boldsymbol{\nu}) f'(\boldsymbol{x} \cdot \boldsymbol{\nu} - ct), \\
\nabla^2 \boldsymbol{u} &= \boldsymbol{n} f''(\boldsymbol{x} \cdot \boldsymbol{\nu} - ct), \\
\boldsymbol{\nabla}(\operatorname{div} \boldsymbol{u}) &= (\boldsymbol{n} \cdot \boldsymbol{\nu}) \boldsymbol{\nu} f''(\boldsymbol{x} \cdot \boldsymbol{\nu} - ct),
\end{aligned}$$

to give, after division by f'',

$$\rho c^2 \boldsymbol{n} = (\lambda + \mu)(\boldsymbol{n} \cdot \boldsymbol{\nu})\boldsymbol{\nu} + \mu \boldsymbol{n}.$$

There are two possible solutions to this equation corresponding to the well-known cases of longitudinal and shear waves. In the first of these cases, $\boldsymbol{\nu} = \boldsymbol{n}$ and the speed of propagation is c_L, given by

$$c_L^2 = \frac{\lambda + 2\mu}{\rho}.$$

In the shear wave case, $\boldsymbol{n} \cdot \boldsymbol{\nu} = 0$ (the vibration is perpendicular to the direction of propagation) and the speed of propagation is c_S, given by

$$c_S^2 = \frac{\mu}{\rho}.$$

3.3 PRINCIPLE OF VIRTUAL WORK AND CONSERVATION OF GLOBAL VARIABLES

The local dynamic equilibrium equations derived in the previous section can be presented in an integral form more suitable for direct finite element discretization by employing the principle of virtual work (or the principle of virtual power as it is more accurately called whenever velocities are used in the formulation). This is carried out below in both the reference and current configurations. In order to continue, it is important to first introduce the concepts of virtual displacement and virtual velocity.§ These are simply arbitrary fields of displacements or velocities that are zero at points where the displacement of the solid is constrained by boundary conditions. For instance, in Figure 3.3 virtual fields would be arbitrary everywhere except for the part of the boundary $\partial_u v$ where the the motion of the solid is prescribed, where they would vanish.

At this stage, the distinction between virtual displacement or virtual velocity is somewhat superfluous as they are both arbitrary fields of any magnitude. The use of virtual velocities is sometimes preferred over the use of virtual displacements as velocities avoid any possible confusion over the misplaced need for virtual displacements to be designated *small* in magnitude. However, later in the chapter it will be necessary to use both virtual displacements and virtual velocities in a connected manner in order to derive dynamic variational principles.

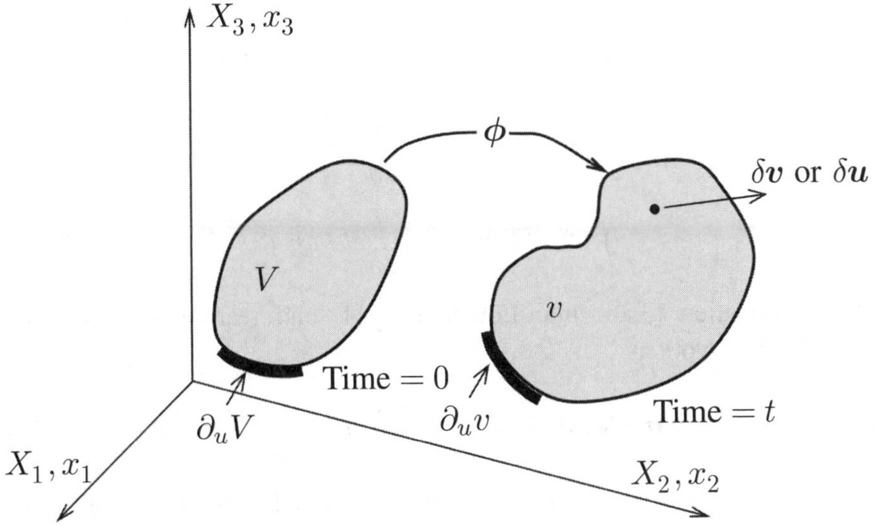

FIGURE 3.3 Virtual velocity.

§ Virtual – an imaginary convenience. Virtual quantities are introduced to facilitate a formulation, but such quantities do not exist in reality. It is the arbitrary nature of virtual quantities that enables real statements to be developed such as differential equilibrium equations or discretized equilibrium equations suitable for finite element implementation. See NL-Statics, Sections 5.4 and 9.3, respectively.

3.3.1 Principle of Virtual Work in the Current Configuration

Consider again the local equilibrium per unit current volume given in Equation (3.22a). Rearranging this equation and multiplying through by an arbitrary virtual velocity $\delta \boldsymbol{v}$ (or displacement $\delta \boldsymbol{u}$) gives

$$(\rho \boldsymbol{a} - \operatorname{div} \boldsymbol{\sigma} - \boldsymbol{f}) \cdot \delta \boldsymbol{v} = 0, \tag{3.23}$$

for all $\delta \boldsymbol{v}$ such that $\delta \boldsymbol{v} = \boldsymbol{0}$ on the boundary $\partial_u v$. Integration over the whole current volume v gives

$$\int_v \rho \boldsymbol{a} \cdot \delta \boldsymbol{v} \, dv - \int_v (\operatorname{div} \boldsymbol{\sigma}) \cdot \delta \boldsymbol{v} \, dv - \int_v \boldsymbol{f} \cdot \delta \boldsymbol{v} \, dv = 0. \tag{3.24}$$

The second term on the left can be transformed by first noting that

$$\operatorname{div} (\boldsymbol{\sigma} \delta \boldsymbol{v}) = (\operatorname{div} \boldsymbol{\sigma}) \cdot \delta \boldsymbol{v} + \boldsymbol{\sigma} : \boldsymbol{\nabla} \delta \boldsymbol{v}, \tag{3.25}$$

where $\boldsymbol{\nabla} \delta \boldsymbol{v}$ denotes the spatial gradient of $\delta \boldsymbol{v}$, that is, $\boldsymbol{\nabla} \delta \boldsymbol{v} = \partial \delta \boldsymbol{v} / \partial \boldsymbol{x}$. Now, with the help of the divergence theorem and noting the symmetry of the Cauchy stress $\boldsymbol{\sigma}$ so that $(\boldsymbol{\sigma a}) \cdot \boldsymbol{b} = (\boldsymbol{\sigma b}) \cdot \boldsymbol{a}$ for any two vectors $\boldsymbol{a}$ and $\boldsymbol{b}$, Equation (3.25) can be expanded to give

$$\begin{aligned}
\int_v (\operatorname{div} \boldsymbol{\sigma}) \cdot \delta \boldsymbol{v} \, dv &= \int_v \operatorname{div} (\boldsymbol{\sigma} \delta \boldsymbol{v}) \, dv - \int_v (\boldsymbol{\sigma} : \boldsymbol{\nabla} \delta \boldsymbol{v}) \, dv \\
&= -\int_v (\boldsymbol{\sigma} : \boldsymbol{\nabla} \delta \boldsymbol{v}) \, dv + \int_{\partial v} (\boldsymbol{\sigma} \delta \boldsymbol{v}) \cdot \boldsymbol{n} \, da \\
&= -\int_v (\boldsymbol{\sigma} : \boldsymbol{\nabla} \delta \boldsymbol{v}) \, dv + \int_{\partial v} (\boldsymbol{\sigma n}) \cdot \delta \boldsymbol{v} \, da \\
&= -\int_v (\boldsymbol{\sigma} : \boldsymbol{\nabla} \delta \boldsymbol{v}) \, dv + \int_{\partial v} \boldsymbol{t} \cdot \delta \boldsymbol{v} \, da.
\end{aligned} \tag{3.26}$$

Substituting Equation (3.26) into Equation (3.24) and rearranging gives the principle of virtual work as

$$\int_v \rho \boldsymbol{a} \cdot \delta \boldsymbol{v} \, dv + \int_v (\boldsymbol{\sigma} : \boldsymbol{\nabla} \delta \boldsymbol{v}) \, dv = \int_v \boldsymbol{f} \cdot \delta \boldsymbol{v} \, dv + \int_{\partial v} \boldsymbol{t} \cdot \delta \boldsymbol{v} \, da. \tag{3.27}$$

This simply states that the virtual work due to the external forces, namely the right-hand side of Equation (3.27), must match the virtual work done by the inertial and internal forces, namely the left-hand side of Equation (3.27). Finally, it is worth noting that the surface term in Equation (3.27) need only be integrated over that part of the boundary where the motion is not prescribed since $\delta \boldsymbol{v} = \boldsymbol{0}$ on $\partial_u v$. For future reference, Equation (3.27) can be rewritten in terms of the virtual rate of deformation tensor $\delta \boldsymbol{d}$ as

$$\int_v \rho \boldsymbol{a} \cdot \delta \boldsymbol{v} \, dv + \int_v \boldsymbol{\sigma} : \delta \boldsymbol{d} \, dv = \int_v \boldsymbol{f} \cdot \delta \boldsymbol{v} \, dv + \int_{\partial v} \boldsymbol{t} \cdot \delta \boldsymbol{v} \, da, \tag{3.28a}$$

where

$$\delta \boldsymbol{d} = \frac{1}{2}(\boldsymbol{\nabla} \delta \boldsymbol{v} + (\boldsymbol{\nabla} \delta \boldsymbol{v})^T). \tag{3.28b}$$

The virtual rate of deformation tensor $\delta \boldsymbol{d}$ is derived in NL-Statics, Section 5.4.

3.3.2 Principle of Virtual Work in the Reference Configuration

It is possible to repeat the derivations in Section 3.2.3 starting from Equation (3.22b) in order to directly formulate the principle of virtual work in the reference configuration. Alternatively, Equation (3.27) can be transformed using the relationships between reference and current density, body force per unit volume, and traction per unit area given in Sections 3.2.1 and 3.2.2. Note that a particle in the reference configuration retains the same virtual velocity associated with the same particle in the current configuration. To begin, using Equation (3.5) the inertial virtual force term is transformed as

$$\int_v \rho \boldsymbol{a} \cdot \delta \boldsymbol{v} \, dv = \int_V (\rho J) \boldsymbol{a} \cdot \delta \boldsymbol{v} \, dV = \int_V \rho_R \boldsymbol{a} \cdot \delta \boldsymbol{v} \, dV. \tag{3.29}$$

Similarly, the body force term can be expressed, using Equation $(3.6\text{a,b,c})_a$, as

$$\int_v \boldsymbol{f} \cdot \delta \boldsymbol{v} \, dv = \int_V J \boldsymbol{f} \cdot \delta \boldsymbol{v} \, dV = \int_V \boldsymbol{f}_R \cdot \delta \boldsymbol{v} \, dV, \tag{3.30}$$

and the traction term is transformed using Equation (3.13) into

$$\int_{\partial v} \boldsymbol{t} \cdot \delta \boldsymbol{v} \, da = \int_{\partial V} \boldsymbol{t}_R \cdot \delta \boldsymbol{v} \, dA. \tag{3.31}$$

The evaluation of the internal virtual work due to the stress field requires the use of the chain rule to relate the gradient of the virtual velocities in the current and reference configurations. This is achieved using the deformation gradient $\boldsymbol{F}$ as

$$\begin{aligned} \boldsymbol{\nabla}_0 \delta \boldsymbol{v} &= \frac{\partial \delta \boldsymbol{v}}{\partial \boldsymbol{X}} \\ &= \frac{\partial \delta \boldsymbol{v}}{\partial \boldsymbol{x}} \frac{\partial \boldsymbol{x}}{\partial \boldsymbol{X}} \\ &= (\boldsymbol{\nabla} \delta \boldsymbol{v}) \boldsymbol{F}. \end{aligned} \tag{3.32}$$

Consequently, using the properties of the trace, namely $\text{tr}(\boldsymbol{AB}) = \text{tr}(\boldsymbol{BA}) = \boldsymbol{A}^T : \boldsymbol{B}$, the internal virtual work term becomes

$$
\begin{aligned}
\int_v (\boldsymbol{\sigma} : \boldsymbol{\nabla}\delta \boldsymbol{v})\, dv &= \int_V \operatorname{tr}(\boldsymbol{\sigma}\boldsymbol{\nabla}\delta\boldsymbol{v}) J\, dV \\
&= \int_V \operatorname{tr}\left[\boldsymbol{\sigma}(\boldsymbol{\nabla}_0\delta\boldsymbol{v})\boldsymbol{F}^{-1}\right] J\, dV \\
&= \int_V \operatorname{tr}(\boldsymbol{F}^{-1}\boldsymbol{\sigma}\boldsymbol{\nabla}_0\delta\boldsymbol{v}) J\, dV \\
&= \int_V (J\boldsymbol{\sigma}\boldsymbol{F}^{-T}) : (\boldsymbol{\nabla}_0\delta\boldsymbol{v})\, dV.
\end{aligned} \tag{3.33}
$$

Observe that the first term in brackets in the final expression in Equation (3.33) is precisely the definition of the first Piola–Kirchhoff stress tensor $\boldsymbol{P}$. Hence the internal virtual work can now be expressed in the reference configuration as

$$
\int_v \boldsymbol{\sigma} : \boldsymbol{\nabla}\delta\boldsymbol{v}\, dv = \int_V \boldsymbol{P} : \boldsymbol{\nabla}_0\delta\boldsymbol{v}\, dV. \tag{3.34}
$$

On substituting Equations (3.29), (3.30), (3.31), and (3.34) into the principle of virtual work in the current configuration, Equation (3.27) gives the principle of virtual work in the reference configuration as

$$
\int_V \rho_R \boldsymbol{a}\cdot\delta\boldsymbol{v}\, dV + \int_V \boldsymbol{P} : \boldsymbol{\nabla}_0\delta\boldsymbol{v}\, dV = \int_V \boldsymbol{f}_R\cdot\delta\boldsymbol{v}\, dV + \int_{\partial V} \boldsymbol{t}_R\cdot\delta\boldsymbol{v}\, dA. \tag{3.35}
$$

3.3.3 Conservation of Global Momentum Quantities

The principle of virtual work derived in the previous sections can be used to demonstrate that global quantities such as linear momentum, angular momentum, and energy associated with the deforming solid are conserved under certain circumstances during deformation. The simplest case is the conservation of linear momentum in the absence of external forces. This is in fact a trivial consequence of the starting equation used to establish dynamic equilibrium, namely Equation (3.8). Clearly, in the case where external forces vanish, the total linear momentum $\boldsymbol{L}$ is constant, that is,

$$
\frac{d\boldsymbol{L}}{dt} = \frac{d}{dt}\int_v \boldsymbol{v}\, dm = 0. \tag{3.36}
$$

It is possible to derive this expression starting from the principle of virtual work by taking a uniform virtual velocity field $\delta\boldsymbol{v}(\boldsymbol{X}) = \delta\boldsymbol{v}_0$, that is, a pure translation where the velocity of every particle in the solid is the same. In this case the internal virtual work terms $\boldsymbol{\sigma} : \boldsymbol{\nabla}\delta\boldsymbol{v}$ and $\boldsymbol{P} : \boldsymbol{\nabla}_0\delta\boldsymbol{v}$ vanish, and the principle of virtual work becomes

$$
\int_v \rho\boldsymbol{a}\cdot\delta\boldsymbol{v}_0\, dv = \delta\boldsymbol{v}_0\cdot\int_v \boldsymbol{a}\, dm = \delta\boldsymbol{v}_0\cdot\left(\frac{d}{dt}\int_v \boldsymbol{v}\, dm\right) = 0. \tag{3.37}
$$

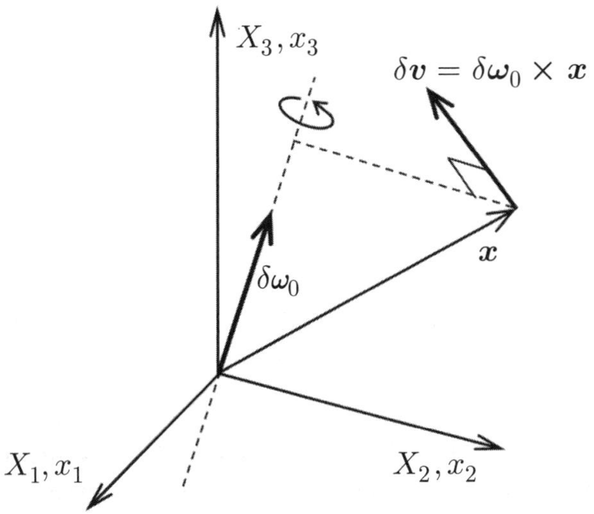

FIGURE 3.4 Virtual velocities due to rotation about an axis aligned with $\delta \boldsymbol{\omega}_0$ passing through the origin.

Given that this must be satisfied for any arbitrary uniform virtual velocity vector $\delta \boldsymbol{v}_0$, the statement of conservation of linear momentum in the absence of external forces given by Equation (3.36) is recovered.

A more insightful case can be revealed by taking a virtual velocity $\delta \boldsymbol{v}$, corresponding to an arbitrary rotation of the solid about the origin of the coordinate system around an axis $\delta \boldsymbol{\omega}_0$, as illustrated in Figure 3.4. Under such a virtual velocity, the inertial virtual work becomes

$$\begin{aligned}
\int_v \rho \boldsymbol{a} \cdot \delta \boldsymbol{v}\, dv &= \int_v \boldsymbol{a} \cdot (\delta \boldsymbol{\omega}_0 \times \boldsymbol{x})\, dm \\
&= \int_v \delta \boldsymbol{\omega}_0 \cdot (\boldsymbol{x} \times \boldsymbol{a})\, dm \\
&= \delta \boldsymbol{\omega}_0 \cdot \int_v (\boldsymbol{x} \times \boldsymbol{a})\, dm \\
&= \delta \boldsymbol{\omega}_0 \cdot \left[\int_v \frac{\partial}{\partial t} (\boldsymbol{x} \times \boldsymbol{v}\, dm) - \int_v \boldsymbol{v} \times \boldsymbol{v}\, dm \right] \\
&= \delta \boldsymbol{\omega}_0 \cdot \frac{d}{dt} \boldsymbol{A},
\end{aligned} \tag{3.38}$$

where $\boldsymbol{A} = \int_v \boldsymbol{x} \times \boldsymbol{v}\, dm$ denotes the total angular momentum of the solid about the origin. From Equation (3.27), the principle of virtual work in the absence of external forces then becomes

$$\delta\boldsymbol{\omega}_0 \cdot \frac{d}{dt}\boldsymbol{A} + \int_v \boldsymbol{\sigma} : \boldsymbol{\nabla}(\delta\boldsymbol{\omega}_0 \times \boldsymbol{x})\, dv = 0. \tag{3.39}$$

The vector $(\delta\boldsymbol{\omega}_0 \times \boldsymbol{x})$ is clearly linear in $\boldsymbol{x}$ and therefore its gradient is a constant second-order tensor denoted as $\delta\boldsymbol{W}_0$ and given by

$$\delta\boldsymbol{W}_0 = \boldsymbol{\nabla}(\delta\omega_0 \times \boldsymbol{x}) = \begin{bmatrix} 0 & -\delta\omega_3 & \delta\omega_2 \\ \delta\omega_3 & 0 & -\delta\omega_1 \\ -\delta\omega_2 & \delta\omega_1 & 0 \end{bmatrix}, \tag{3.40}$$

where $\delta\omega_1$, $\delta\omega_2$, and $\delta\omega_3$ are the components of the axis of rotation vector $\delta\boldsymbol{\omega}_0 = [\delta\omega_1, \delta\omega_2, \delta\omega_3]^T$. Since $\delta\boldsymbol{W}_0$ is skew symmetric and the Cauchy stress $\boldsymbol{\sigma}$ is symmetric, the contraction $\boldsymbol{\sigma} : \delta\boldsymbol{W}_0$ in Equation (3.39) vanishes, revealing that the total angular momentum of a deformable solid in the absence of external forces remains constant, that is,

$$\frac{d}{dt}\int_v \boldsymbol{x} \times \boldsymbol{v}\, dm = 0. \tag{3.41}$$

In the presence of external forces, the conservation of linear momentum implies Equation (3.9), that is, the balance between the change of momentum and the resultant external force. In the case of angular momentum, the rate of change of angular momentum is given by the resultant moment of the external forces, as described in Example 3.3. A final global conserved quantity, namely the energy, can be derived by taking a virtual velocity field $\delta\boldsymbol{v}$ equal to the actual velocity $\boldsymbol{v}$. This is permissible since the real velocity $\boldsymbol{v}$ is a valid virtual velocity candidate as it fulfils the required boundary conditions. This will lead to the conservation of energy equation considered in the next section.

EXAMPLE 3.3: Resultant moment of external forces

The conservation of angular momentum derived in Section 3.3.3 from the principle of virtual work can be extended to include the presence of external forces. This is achieved by introducing the resultant moment of the external forces via the principle of virtual work, whereby a virtual velocity field corresponds to an arbitrary rotation $\delta\boldsymbol{\omega}_0$ about an axis, as shown in Figure 3.4 and given by

$$\delta\boldsymbol{v} = \delta\boldsymbol{\omega}_0 \times \boldsymbol{x}.$$

(continued)

Example 3.3: *(cont.)*

Introducing this term into the principle of virtual work given in the current configuration by Equation (3.27) gives

$$\int_v \rho \boldsymbol{a} \cdot (\delta \boldsymbol{\omega}_0 \times \boldsymbol{x})\, dv + \int_v \boldsymbol{\sigma} : \nabla(\delta \boldsymbol{\omega}_0 \times \boldsymbol{x})\, dv$$

$$= \int_v \boldsymbol{f} \cdot (\delta \boldsymbol{\omega}_0 \times \boldsymbol{x})\, dv + \int_{\partial v} \boldsymbol{t} \cdot (\delta \boldsymbol{\omega}_0 \times \boldsymbol{x})\, da. \qquad \text{(a)}$$

The terms on the left-hand side of this equation have been shown in Section 3.3.3 to give the rate of change of the angular momentum as

$$\int_v \rho \boldsymbol{a} \cdot (\delta \boldsymbol{\omega}_0 \times \boldsymbol{x})\, dv + \int_v \boldsymbol{\sigma} : \nabla(\delta \boldsymbol{\omega}_0 \times \boldsymbol{x})\, dv = \delta \boldsymbol{\omega}_0 \cdot \frac{d}{dt}\boldsymbol{A}, \qquad \text{(b)}$$

where $\boldsymbol{A} = \int_v \boldsymbol{x} \times \boldsymbol{v}\, dm$. The right-hand side of Equation (a) can be re-expressed as

$$\int_v \boldsymbol{f} \cdot (\delta \boldsymbol{\omega}_0 \times \boldsymbol{x})\, dv + \int_{\partial v} \boldsymbol{t} \cdot (\delta \boldsymbol{\omega}_0 \times \boldsymbol{x})\, da$$

$$= \delta \boldsymbol{\omega}_0 \cdot \int_v \boldsymbol{x} \times \boldsymbol{f}\, dv + \delta \boldsymbol{\omega}_0 \cdot \int_{\partial v} \boldsymbol{x} \times \boldsymbol{t}\, da = \delta \boldsymbol{\omega}_0 \cdot \boldsymbol{T}, \qquad \text{(c)}$$

where $\boldsymbol{T}$ is the resultant moment of the external forces defined at each instant in time as

$$\boldsymbol{T} = \int_v \boldsymbol{x} \times \boldsymbol{f}\, dv + \int_{\partial v} \boldsymbol{x} \times \boldsymbol{t}\, da.$$

Finally, substituting Equations (b) and (c) in to (a) and noting that $\delta \boldsymbol{\omega}_0$ is arbitrary gives the conservation of global angular momentum in the presence of external forces as

$$\frac{d}{dt}\boldsymbol{A} = \boldsymbol{T}.$$

This implies that the rate of change of angular momentum is given by the resultant moment of the external forces.

3.3.4 Conservation of Energy

On substituting the virtual velocity $\delta \boldsymbol{v}$ by the real velocity $\boldsymbol{v}$ in Equation (3.35), the principle of virtual work expressed in the reference configuration gives, in the absence of external forces,

$$\int_V \rho_R \boldsymbol{a} \cdot \boldsymbol{v}\, dV + \int_V \boldsymbol{P} : \boldsymbol{\nabla}_0 \boldsymbol{v}\, dV = 0. \qquad (3.42)$$

Now consider the first inertial term above and note that simple algebra yields

$$\begin{aligned}\int_V \rho_R \boldsymbol{a}\cdot\boldsymbol{v}\,dV &= \int_V \rho_R \frac{\partial}{\partial t}\left(\tfrac{1}{2}\boldsymbol{v}\cdot\boldsymbol{v}\right)dV \\ &= \frac{d}{dt}\int_V \tfrac{1}{2}\rho_R \boldsymbol{v}\cdot\boldsymbol{v}\,dV \\ &= \frac{dK}{dt},\end{aligned} \tag{3.43}$$

where

$$K = \int_V \tfrac{1}{2}\rho_R \boldsymbol{v}\cdot\boldsymbol{v}\,dV \tag{3.44}$$

is the total kinetic energy of the solid. The internal energy term in Equation (3.42) can also be transformed by introducing the rate of the deformation gradient $\dot{\boldsymbol{F}}$ to give

$$\begin{aligned}\int_V \boldsymbol{P} : \boldsymbol{\nabla}_0\boldsymbol{v}\,dV &= \int_V \boldsymbol{P} : \frac{\partial}{\partial \boldsymbol{X}}\frac{\partial}{\partial t}\phi(\boldsymbol{X},t)\,dV \\ &= \int_V \boldsymbol{P} : \frac{\partial}{\partial t}\frac{\partial}{\partial \boldsymbol{X}}\phi(\boldsymbol{X},t)\,dV \\ &= \int_V \boldsymbol{P} : \dot{\boldsymbol{F}}\,dV.\end{aligned} \tag{3.45}$$

For the case where the material follows a hyperelastic constitutive model, the first Piola–Kirchhoff stress tensor is given by the derivative of the elastic energy Ψ with respect to the deformation gradient $\boldsymbol{F}$ as (see Chapter 6 of NL-Statics)

$$\boldsymbol{P} = \frac{\partial \Psi}{\partial \boldsymbol{F}}. \tag{3.46}$$

Consequently, the internal virtual work with $\delta\boldsymbol{v} = \boldsymbol{v}$ becomes

$$\begin{aligned}\int_V \boldsymbol{P} : \boldsymbol{\nabla}_0\boldsymbol{v}\,dV &= \int_V \frac{\partial \Psi}{\partial \boldsymbol{F}} : \dot{\boldsymbol{F}}\,dV = \int_V \frac{\partial}{\partial t}\Psi\big(\boldsymbol{F}(\boldsymbol{X},t)\big)\,dV \\ &= \frac{d}{dt}\int_V \Psi(\boldsymbol{F})\,dV = \frac{d}{dt}\Pi_{\text{int}},\end{aligned} \tag{3.47}$$

where

$$\Pi_{\text{int}} = \int_V \Psi(\boldsymbol{F})\,dV \tag{3.48}$$

represents the total internal elastic energy in the material. Finally, substituting Equations (3.43) and (3.47) into Equation (3.42) gives, in the absence of external forces and with $\delta \boldsymbol{v} = \boldsymbol{v}$, the conservation principle for total energy of the solid as

$$\frac{d}{dt}(K + \Pi_{\text{int}}) = 0. \tag{3.49}$$

Clearly this shows that, in the absence of external forces acting on the body, the motion is such that there is an energy balance between kinetic and elastic energy, that is, a gain in one occasions a loss in the other.

In order to extend this energy conservation principle to the more general case involving external forces, note first that in general the principle of virtual work applied to a field of virtual velocity that coincides with the real velocity of the solid gives (see Equation (3.35))

$$\int_V \rho_R \boldsymbol{a} \cdot \boldsymbol{v}\, dV + \int_V \boldsymbol{P} : \boldsymbol{\nabla}_0 \boldsymbol{v}\, dV = \int_V \boldsymbol{f}_R \cdot \boldsymbol{v}\, dV + \int_{\partial V} \boldsymbol{t}_R \cdot \boldsymbol{v}\, dA. \tag{3.50}$$

The first two terms in this equation have been shown above to measure the rate of change of kinetic and internal elastic energy of the body, hence

$$\frac{d}{dt}(K + \Pi_{\text{int}}) = \int_V \boldsymbol{f}_R \cdot \boldsymbol{v}\, dV + \int_{\partial V} \boldsymbol{t}_R \cdot \boldsymbol{v}\, dA. \tag{3.51}$$

The right-hand side of the above expression represents the rate of work done by the external forces, $\dot{W}_{\text{ext}}$, thus allowing the energy balance equation to be written as

$$\frac{d}{dt}(K + \Pi_{\text{int}}) = \dot{W}_{\text{ext}}\,; \; \dot{W}_{\text{ext}} = \int_V \boldsymbol{f}_R \cdot \boldsymbol{v}\, dV + \int_{\partial V} \boldsymbol{t}_R \cdot \boldsymbol{v}\, dA. \tag{3.52a,b}$$

This implies that the rate of change of the internal and kinetic energy is the result of the work done by the external forces.

For certain cases of external loading, $\boldsymbol{f}_R$ and $\boldsymbol{t}_R$, it is possible to define an external energy, Π_{ext}, such that

$$\frac{d}{dt}\Pi_{\text{ext}} = -\dot{W}_{\text{ext}}, \tag{3.53}$$

which enables the energy conservation equation to be rewritten in terms of a total energy $(K + \Pi_{\text{int}} + \Pi_{\text{ext}})$ as

$$\frac{d}{dt}(K + \Pi_{\text{int}} + \Pi_{\text{ext}}) = 0. \tag{3.54}$$

Such cases of loading are known as *conservative*, insofar as they allow the definition of the total energy that is conserved during the motion. The simplest of these cases arises when $\boldsymbol{f}_R$ and $\boldsymbol{t}_R$ are constant in time and therefore independent of the motion. In this case Π_{ext} is simply

$$\Pi_{\text{ext}} = -\int_V \boldsymbol{f}_R \cdot \boldsymbol{\phi}\, dV - \int_{\partial V} \boldsymbol{t}_R \cdot \boldsymbol{\phi}\, dA; \tag{3.55}$$

consequently, Equation (3.52a,b) is trivially satisfied. The cases where $\boldsymbol{f}_R$ and $\boldsymbol{t}_R$ are dependent upon the motion are known as follower forces and, with notable exceptions, tend to be nonconservative, that is, an external energy that satisfies Equation (3.54) does not exist.

EXAMPLE 3.4: Gravitational potential energy

A simple case of an external force for which it is possible to derive a potential is the case of self weight. In this case the external body force per unit mass is gravity $\boldsymbol{b} = \boldsymbol{g} = [0, 0, -g]^T$ and the corresponding body force vector per unit reference volume is

$$\boldsymbol{f}_R = \rho_R \boldsymbol{b} = \begin{bmatrix} 0 \\ 0 \\ -\rho_R g \end{bmatrix},$$

where it has been assumed that gravity acts vertically downwards in the z direction. During the motion of a body, the rate of work done by gravity can be evaluated as

$$\dot{W}_g = \int_V \boldsymbol{f}_R \cdot \boldsymbol{v} \, dV = -\int_V \rho_R g v_z \, dV,$$

where v_z denotes the vertical component of the velocity vector $\boldsymbol{v}$. Recalling Equation (3.53), it is now possible to obtain a gravitational external potential as

$$\Pi_g = \int_V \rho_R g z \, dV,$$

where z is the vertical component of the position vector $\boldsymbol{x}$. Note that the above equation corresponds to the standard definition of gravitational potential energy of a system.

3.4 THE ACTION INTEGRAL AND HAMILTON'S PRINCIPLE

Chapter 2, concerning truss systems, showed how the dynamic equilibrium of a system of masses connected by rods can be expressed in the form of a variational statement as the stationary point of a specific functional given by the action integral over a discrete time period of the Lagrangian function; see Equation (2.31). Such a formulation enables the development of time integration schemes that satisfy, in a discrete manner, the continuous time global conservation properties. This section will show that the same variational treatment is possible for the case of a deforming solid. The resulting variational principle, known as Hamilton's principle, is equivalent to the total energy minimization principle employed in static solutions. For simplicity, only external forces that are conservative and constant will be considered. Recall, as explained in the Introduction to this chapter and in Chapter 2 for trusses, Hamilton's principle endows the resulting discretization with considerable robustness.

3.4.1 The Lagrangian

The kinetic energy, the elastic energy of the solid, and the external potential energies have been defined in the previous section as integrals over the reference volume as

$$K(\boldsymbol{\phi}(\boldsymbol{X},t)) = \int_V \tfrac{1}{2}\rho_R \frac{\partial \boldsymbol{\phi}}{\partial t} \cdot \frac{\partial \boldsymbol{\phi}}{\partial t}\, dV, \tag{3.56a}$$

$$\Pi_{\text{int}}(\boldsymbol{\phi}(\boldsymbol{X},t)) = \int_V \Psi\left(\frac{\partial \boldsymbol{\phi}}{\partial \boldsymbol{X}}\right) dV, \tag{3.56b}$$

$$\Pi_{\text{ext}}(\boldsymbol{\phi}(\boldsymbol{X},t)) = -\int_V \boldsymbol{f}_R \cdot \boldsymbol{\phi}\, dV - \int_{\partial V} \boldsymbol{t}_R \cdot \boldsymbol{\phi}\, dA. \tag{3.56c}$$

Note that in these expressions the functional dependency of K and Π with respect to the mapping function $\boldsymbol{\phi}(\boldsymbol{X},t)$ has been explicitly stated, rather than being hidden by replacing $\partial\boldsymbol{\phi}/\partial t$ by the velocity $\boldsymbol{v}$ and $\partial\boldsymbol{\phi}/\partial\boldsymbol{X}$ by the deformation gradient $\boldsymbol{F}$. This explicit introduction of the mapping function makes it clear that in Equation (3.56a) the kinetic energy is an integral function of the time derivative of the motion mapping $\boldsymbol{\phi}(\boldsymbol{X},t)$. Similarly, the internal elastic energy Π_{int} is an integral function of the material gradient of the mapping function. It is now possible to define the *Lagrangian functional* as

$$\begin{aligned}
\mathcal{L}(\boldsymbol{\phi}(\boldsymbol{X},t)) &= K(\boldsymbol{\phi}(\boldsymbol{X},t)) - \Pi_{\text{int}}(\boldsymbol{\phi}(\boldsymbol{X},t)) - \Pi_{\text{ext}}(\boldsymbol{\phi}(\boldsymbol{X},t)) \\
&= \int_V \left[\tfrac{1}{2}\rho_R \frac{\partial \boldsymbol{\phi}}{\partial t} \cdot \frac{\partial \boldsymbol{\phi}}{\partial t} - \Psi\left(\frac{\partial \boldsymbol{\phi}}{\partial \boldsymbol{X}}\right) + \boldsymbol{f}_R \cdot \boldsymbol{\phi}\right] dV \\
&\quad + \int_{\partial V} \boldsymbol{t}_R \cdot \boldsymbol{\phi}\, dA.
\end{aligned} \tag{3.57}$$

Observe that, like K and Π, $\mathcal{L}$ being a function of the function $\boldsymbol{\phi}(\boldsymbol{X},t)$ is called a *functional* of $\boldsymbol{\phi}(\boldsymbol{X},t)$, hence the nomenclature Lagrangian functional.

3.4.2 The Action Integral and Hamilton's Principle

Consider a deformable solid moving between two positions from time t_1 to time t_2 as shown in Figure 3.5. Here t_1 and t_2 represent two arbitrary time values; they

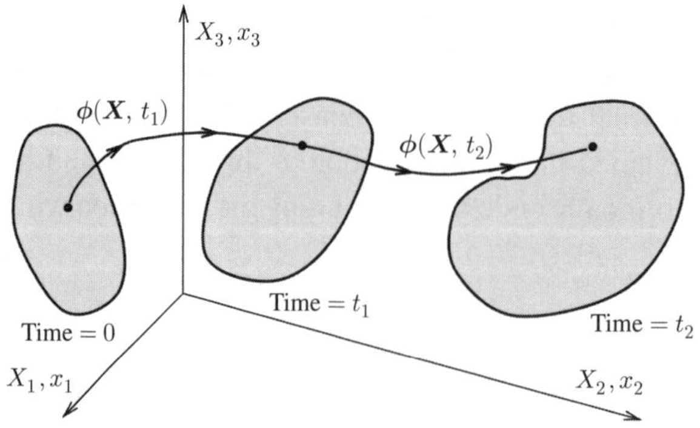

FIGURE 3.5 Motion of a deformable solid from times t_1 to t_2.

could equally well be from time 0 to a final time t_F, or a discrete time step Δt such that $t_2 = t_1 + \Delta t$. The action integral between times t_1 and t_2 is defined as the integral of the Lagrangian in this time interval, given as

$$\begin{aligned}\mathcal{A}(\boldsymbol{\phi}(\boldsymbol{X},t)) &= \int_{t_1}^{t_2} \mathcal{L}(\boldsymbol{\phi}(\boldsymbol{X},t))\,dt \\ &= \int_{t_1}^{t_2}\int_V \left[\tfrac{1}{2}\rho_R \frac{\partial \boldsymbol{\phi}}{\partial t}\cdot\frac{\partial \boldsymbol{\phi}}{\partial t} - \Psi\left(\frac{\partial \boldsymbol{\phi}}{\partial \boldsymbol{X}}\right) + \boldsymbol{f}_R\cdot\boldsymbol{\phi}\right] dV dt \\ &\quad + \int_{t_1}^{t_2}\int_{\partial V} \boldsymbol{t}_R\cdot\boldsymbol{\phi}\, dA dt. \end{aligned} \tag{3.58}$$

Hamilton's principle states that paths $\boldsymbol{\phi}(\boldsymbol{X},t)$ that satisfy dynamic equilibrium throughout the time interval $t = t_1$ to $t = t_2$ are such that they render the above action integral functional stationary.

This is similar to the statement that static equilibrium solutions $\boldsymbol{\phi}(\boldsymbol{X})$ minimize the total energy of an elastic system.

In order to prove the equivalence of dynamic equilibrium with paths which make the action integral stationary, as stated by Hamilton's principle, take first the directional derivative of $\mathcal{A}$ with respect to a variation of $\boldsymbol{\phi}$ (see NL-Statics, Section 2.3, for an in-depth discussion of the concept of the directional derivative) to give

$$D\mathcal{A}(\boldsymbol{\phi}(\boldsymbol{X},t))\left[\delta\boldsymbol{\phi}(\boldsymbol{X},t)\right] = \frac{d}{d\epsilon}\bigg|_{\epsilon=0} \mathcal{A}(\boldsymbol{\phi}(\boldsymbol{X},t) + \epsilon\delta\boldsymbol{\phi}(\boldsymbol{X},t)). \tag{3.59}$$

In this expression, $\delta\boldsymbol{\phi}(\boldsymbol{X},t)$ denotes an arbitrary variation of the path between the positions of the solid at times t_1 and t_2, as shown in Figure 3.6. Observe that $\delta\boldsymbol{\phi}(\boldsymbol{X},t_1) = \delta\boldsymbol{\phi}(\boldsymbol{X},t_2) = \boldsymbol{0}$ as there are no implied changes in the starting and final positions, that is, $\delta\boldsymbol{\phi}$ simply describes alternative paths from the two fixed positions. Using Equation (3.58), the above directional derivative can

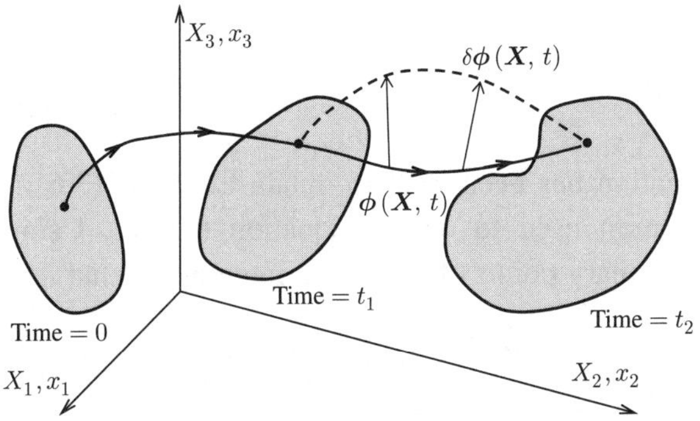

FIGURE 3.6 Variation in the deformation path from times t_1 to t_2 described by $\phi(\boldsymbol{X}, t)$.

be formulated as

$$
\begin{aligned}
D\mathcal{A}\big(\phi(\boldsymbol{X},t)\big)[\delta\phi(\boldsymbol{X},t)] = \frac{d}{d\epsilon}\bigg|_{\epsilon=0} \int_{t_1}^{t_2}\int_V \bigg[&\tfrac{1}{2}\rho_R \frac{\partial}{\partial t}(\boldsymbol{\phi}+\epsilon\delta\boldsymbol{\phi})\cdot\frac{\partial}{\partial t}(\boldsymbol{\phi}+\epsilon\delta\boldsymbol{\phi}) \\
&- \Psi\bigg(\frac{\partial}{\partial \boldsymbol{X}}(\boldsymbol{\phi}+\epsilon\delta\boldsymbol{\phi})\bigg) + \boldsymbol{f}_R\cdot(\boldsymbol{\phi}+\epsilon\delta\boldsymbol{\phi})\bigg]\, dV\,dt \\
&+ \frac{d}{d\epsilon}\bigg|_{\epsilon=0}\int_{t_1}^{t_2}\int_{\partial V} \boldsymbol{t}_R\cdot(\boldsymbol{\phi}+\epsilon\delta\boldsymbol{\phi})\, dA dt, \qquad (3.60)
\end{aligned}
$$

where for simplicity the dependency of ϕ and $\delta\phi$ upon $\boldsymbol{X}$ and t has not been explicitly stated. The derivatives with respect to ϵ in the two terms in the above integral can be taken separately to give

$$
\begin{aligned}
&\frac{d}{d\epsilon}\bigg|_{\epsilon=0}\int_{t_1}^{t_2}\int_V \bigg[\tfrac{1}{2}\rho_R\bigg(\frac{\partial\boldsymbol{\phi}}{\partial t}+\epsilon\frac{\partial\delta\boldsymbol{\phi}}{\partial t}\bigg)\cdot\bigg(\frac{\partial\boldsymbol{\phi}}{\partial t}+\epsilon\frac{\partial\delta\boldsymbol{\phi}}{\partial t}\bigg) dV\,dt, \\
&\quad = \int_{t_1}^{t_2}\int_V \rho_R \boldsymbol{v}\cdot\frac{\partial\delta\boldsymbol{\phi}}{\partial t}\, dV\,dt, \qquad (3.61a)
\end{aligned}
$$

$$
\frac{d}{d\epsilon}\bigg|_{\epsilon=0}\int_{t_1}^{t_2}\int_V \Psi\bigg(\frac{\partial\boldsymbol{\phi}}{\partial\boldsymbol{X}}+\epsilon\frac{\partial\delta\boldsymbol{\phi}}{\partial\boldsymbol{X}}\bigg)\, dV\,dt = \int_{t_1}^{t_2}\int_V \boldsymbol{P} : \frac{\partial\delta\boldsymbol{\phi}}{\partial\boldsymbol{X}}\, dV\,dt, \qquad (3.61b)
$$

$$
\frac{d}{d\epsilon}\bigg|_{\epsilon=0}\int_{t_1}^{t_2}\int_V \big(\boldsymbol{f}_R\cdot(\boldsymbol{\phi}+\epsilon\delta\boldsymbol{\phi})\big)\, dV\,dt = \int_{t_1}^{t_2}\int_V \boldsymbol{f}_R\cdot\delta\boldsymbol{\phi}\, dV\,dt, \qquad (3.61c)
$$

$$
\frac{d}{d\epsilon}\bigg|_{\epsilon=0}\int_{t_1}^{t_2}\int_{\partial V} \boldsymbol{t}_R\cdot(\boldsymbol{\phi}+\epsilon\delta\boldsymbol{\phi})\, dA dt = \int_{t_1}^{t_2}\int_{\partial V} \boldsymbol{t}_R\cdot\delta\boldsymbol{\phi}\, dA dt, \qquad (3.61d)
$$

where as usual

$$\boldsymbol{v} = \frac{\partial \boldsymbol{\phi}}{\partial t}; \; \boldsymbol{P} = \frac{\partial \Psi}{\partial \boldsymbol{F}}; \; \boldsymbol{F} = \frac{\partial \boldsymbol{\phi}}{\partial \boldsymbol{X}}, \tag{3.61e,f,g}$$

and the product rule of differentiation has been used to obtain Equation (3.61a) and the chain rule of differentiation used to obtain Equation (3.61b). Using Equations (3.61a)–(3.61d), the stationary points of the action integral $\mathcal{A}$ satisfy

$$D\mathcal{A}(\boldsymbol{\phi}(\boldsymbol{X},t))\left[\delta\boldsymbol{\phi}(\boldsymbol{X},t)\right] = \int_{t_1}^{t_2}\int_V \left[\rho_R \boldsymbol{v}\cdot\frac{\partial \delta\boldsymbol{\phi}}{\partial t} - \boldsymbol{P} : \frac{\partial \delta\boldsymbol{\phi}}{\partial \boldsymbol{X}} + \boldsymbol{f}_R\cdot\delta\boldsymbol{\phi}\right] dV\,dt$$

$$+ \int_{t_1}^{t_2}\int_{\partial V} \boldsymbol{t}_R\cdot\delta\boldsymbol{\phi}\,dA\,dt = 0. \tag{3.62}$$

It is now a simple algebraic exercise to show that Equation (3.62) is equivalent to dynamic equilibrium either expressed through the local differential Equation (3.22b) or via the principle of virtual work given by Equation (3.35). To accomplish this, integrate by parts in time the first term in the integral in Equation (3.62) to give

$$\int_{t_1}^{t_2} \rho_R \boldsymbol{v}\cdot\frac{\partial \delta\boldsymbol{\phi}}{\partial t}\,dt = \rho_R \boldsymbol{v}\cdot\delta\boldsymbol{\phi}\Big|_{t_1}^{t_2} - \int_{t_1}^{t_2} \rho_R \frac{\partial \boldsymbol{v}}{\partial t}\cdot\delta\boldsymbol{\phi}\,dt \tag{3.63a}$$

$$= -\int_{t_1}^{t_2} \rho_R \boldsymbol{a}\cdot\delta\boldsymbol{\phi}\,dt. \tag{3.63b}$$

Observe that the first term on the right-hand side of Equation (3.63a) vanishes since $\delta\boldsymbol{\phi}(\boldsymbol{X},t_1) = \delta\boldsymbol{\phi}(\boldsymbol{X},t_2) = \boldsymbol{0}$.

Similarly the second term in Equation (3.62) can be transformed using the divergence integral theorem to give

$$\begin{aligned}\int_V \boldsymbol{P} : \frac{\partial \delta\boldsymbol{\phi}}{\partial \boldsymbol{X}}\,dV &= \int_V \mathrm{DIV}(\boldsymbol{P}^T\delta\boldsymbol{\phi})\,dV - \int_V (\mathrm{DIV}\boldsymbol{P})\cdot\delta\boldsymbol{\phi}\,dV \\ &= \int_{\partial V} \delta\boldsymbol{\phi}\cdot\boldsymbol{PN}\,dA - \int_V (\mathrm{DIV}\boldsymbol{P})\cdot\delta\boldsymbol{\phi}\,dV \\ &= \int_{\partial V} \boldsymbol{t}_R\cdot\delta\boldsymbol{\phi}\,dA - \int_V (\mathrm{DIV}\boldsymbol{P})\cdot\delta\boldsymbol{\phi}\,dV.\end{aligned} \tag{3.64}$$

Substituting Equations (3.63) and (3.64) into Equation (3.62) yields

$$\int_{t_1}^{t_2}\int_V (\mathrm{DIV}\boldsymbol{P} + \boldsymbol{f}_R - \rho_R\boldsymbol{a})\cdot\delta\boldsymbol{\phi}\;dV\,dt = 0. \tag{3.65}$$

Since this integral needs to vanish for all $\delta\boldsymbol{\phi}$, the term in brackets must satisfy the local dynamic equilibrium Equation (3.22b) for conservative external forces, that is,

$$\rho_R\,\boldsymbol{a} = \mathrm{DIV}\boldsymbol{P} + \boldsymbol{f}_R, \tag{3.66}$$

which is again the dynamic equilibrium equation expressed in the reference configuration.

In order to arrive at the principle of virtual work directly from Equation (3.62), it is necessary to decompose the term $\delta\boldsymbol{\phi}(\boldsymbol{x},t)$ into time and spatial variation components. Since $\delta\boldsymbol{\phi}(\boldsymbol{x},t)$ is arbitrary, except for the constraint that $\delta\boldsymbol{\phi}(\boldsymbol{x},t_1) = \delta\boldsymbol{\phi}(\boldsymbol{x},t_2) = 0$, it is possible to choose $\delta\boldsymbol{\phi}(\boldsymbol{x},t)$ as

$$\delta\boldsymbol{\phi}(\boldsymbol{x},t) = \delta f(t)\,\delta\boldsymbol{u}(\boldsymbol{X});\; \delta f(t_1) = \delta f(t_2) = 0, \tag{3.67a,b}$$

where $\delta\boldsymbol{u}(\boldsymbol{X})$ denotes an arbitrary virtual displacement field that is a function of $\boldsymbol{X}$ but is constant in time with the time dependency being contained in $\delta f(t)$. Substituting this decomposition into Equation (3.62) gives

$$\begin{aligned}
D\mathcal{A}(\boldsymbol{\phi}(\boldsymbol{X},t))\,[\delta\boldsymbol{\phi}(\boldsymbol{X},t)] &= \int_V\int_{t_1}^{t_2}\left[\rho_R\boldsymbol{v}\cdot\delta\boldsymbol{u}\frac{d\delta f}{dt} - (\boldsymbol{P}:\boldsymbol{\nabla}_0\delta\boldsymbol{u})\delta f\right]dtdV \\
&\quad + \int_V\int_{t_1}^{t_2}(\boldsymbol{f}_R\cdot\delta\boldsymbol{u}\delta f)\,dtdV + \int_{t_1}^{t_2}\int_{\partial V}(\boldsymbol{t}_R\cdot\delta\boldsymbol{u}\delta f)\,dAdt \\
&= \int_V\left[\rho_R\delta\boldsymbol{u}\cdot\left(\int_{t_1}^{t_2}\boldsymbol{v}\frac{d\delta f}{dt}\,dt\right) - \int_{t_1}^{t_2}(\boldsymbol{P}:\boldsymbol{\nabla}_0\delta\boldsymbol{u} - \boldsymbol{f}_R\cdot\delta\boldsymbol{u})\delta f\,dt\right]dV \\
&\quad + \int_{t_1}^{t_2}\int_{\partial V}(\boldsymbol{t}_R\cdot\delta\boldsymbol{u})\delta f\,dAdt \\
&= \int_V\left[\rho_R\delta\boldsymbol{u}\cdot\left(\boldsymbol{v}\delta f\Big|_{t_1}^{t_2} - \int_{t_1}^{t_2}\boldsymbol{a}\delta f\,dt\right)\ldots\right. \\
&\quad \left. - \int_{t_1}^{t_2}(\boldsymbol{P}:\boldsymbol{\nabla}_0\delta\boldsymbol{u} - \boldsymbol{f}_R\cdot\delta\boldsymbol{u})\delta f\,dt\right]dV + \int_{t_1}^{t_2}\int_{\partial V}(\boldsymbol{t}_R\cdot\delta\boldsymbol{u})\delta f\,dAdt \\
&= -\int_{t_1}^{t_2}\left[\int_V\left(\rho_R\boldsymbol{a}\cdot\delta\boldsymbol{u} + \boldsymbol{P}:\boldsymbol{\nabla}_0\delta\boldsymbol{u} - \boldsymbol{f}_R\cdot\delta\boldsymbol{u}\right)dV\ldots\right. \\
&\quad \left. - \int_{\partial V}\boldsymbol{t}_R\cdot\delta\boldsymbol{u}\,dA\right]\delta f\,dt.
\end{aligned} \tag{3.68}$$

Since the stationary conditions of $\mathcal{A}$ imply that the above expression needs to vanish for all $\delta f(t)$, it follows that

$$\int_V \rho_R\boldsymbol{a}\cdot\delta\boldsymbol{u}\,dV + \int_V \boldsymbol{P}:\boldsymbol{\nabla}_0\delta\boldsymbol{u}\,dV = \int_V \boldsymbol{f}_R\cdot\delta\boldsymbol{u}\,dV + \int_{\partial V}\boldsymbol{t}_R\cdot\delta\boldsymbol{u}\,dA, \tag{3.69}$$

for any arbitrary $\delta\boldsymbol{u}(\boldsymbol{X})$, which is simply the principle of virtual work statement in the reference configuration, now expressed in terms of $\delta\boldsymbol{u}$ instead of $\delta\boldsymbol{v}$.

It is worth emphasizing that the use of Hamilton's principle in a time and space discretization context ensures that the resulting numerical procedure is endowed with the same conservation properties that are present in the continuum formulation, as explained in Chapter 2.

Exercises

1. Show that if the mass center $\boldsymbol{x}_c$ of a solid is defined by

$$\boldsymbol{x}_c = \frac{\int_V \rho_R \boldsymbol{x}\, dV}{\int_V \rho_R\, dV},$$

then its movement satisfies the equation

$$M\boldsymbol{a}_c = \boldsymbol{E},$$

where the total mass M, total external force $\boldsymbol{E}$, and mass center acceleration $\boldsymbol{a}_c$ are

$$M = \int_V \rho_R\, dV; \ \boldsymbol{E} = \int_V \boldsymbol{f}_R\, dV + \int_{\partial V} \boldsymbol{t}_R\, dA; \ \boldsymbol{a}_c = \frac{\partial^2 \boldsymbol{x}_c}{\partial t^2}.$$

2. Show that the kinetic energy of a deformable solid can be expressed as

$$K = \frac{1}{2} M \boldsymbol{v}_c \cdot \boldsymbol{v}_c + \int_V \frac{1}{2} \rho_R \boldsymbol{v}_r \cdot \boldsymbol{v}_r\, dV,$$

where M is the total mass and $\boldsymbol{v}_r = \boldsymbol{v} - \boldsymbol{v}_c$ is the velocity of a particle relative to the velocity of the mass center $\boldsymbol{v}_c = d\boldsymbol{x}_c/dt$.

3. Show that the kinetic energy of a rigid body with a mass center velocity $\boldsymbol{v}_c$ rotating about its mass center with angular velocity $\boldsymbol{\omega}$ is given by

$$K = \frac{1}{2} M \boldsymbol{v}_c \cdot \boldsymbol{v}_c + \frac{1}{2} \boldsymbol{\omega} \cdot \mathbf{D} \boldsymbol{\omega}\, dV,$$

where M is the total mass and $\mathbf{D}$ is the inertia tensor given as

$$\mathbf{D} = \int_v \rho \big[(\boldsymbol{x}_r \cdot \boldsymbol{x}_r) \boldsymbol{I} - \boldsymbol{x}_r \otimes \boldsymbol{x}_r \big]\, dv \ ; \ \boldsymbol{x}_r = \boldsymbol{x} - \boldsymbol{x}_c.$$

(Hint: note that the velocity vector at each particle is $\boldsymbol{v} = \boldsymbol{v}_c + \boldsymbol{\omega} \times \boldsymbol{x}_r$.)

4. Show that the equilibrium equation and the principle of virtual work can be expressed in terms of the second Piola–Kirchhoff stress tensor $\boldsymbol{S}$ as

$$\rho_R \boldsymbol{a} = \text{DIV}(\boldsymbol{F}\boldsymbol{S}) + \boldsymbol{f}_R,$$

$$\int_V \rho_R \boldsymbol{a} \cdot \delta\boldsymbol{v}\, dV + \int_V \boldsymbol{S} : \delta\dot{\boldsymbol{E}}\, dV = \int_V \boldsymbol{f}_R \cdot \delta\boldsymbol{v}\, dV + \int_{\partial V} \boldsymbol{t}_R \cdot \delta\boldsymbol{v}\, dA,$$

where $\delta\dot{\boldsymbol{E}}$ is the virtual Lagrangian rate of deformation tensor (see Section 5.5.3 of NL-Statics).

5. Show that in one-dimensional linear elasticity, where $\sigma = E\varepsilon$, E being Young's modulus and ε the linear strain $\varepsilon = \partial u/\partial x$, the equilibrium equation, in the absence of external forces, becomes the so-called one-dimensional second-order wave equation for $u(x,t)$ as

$$\frac{\partial^2 u}{\partial t^2} = c^2 \frac{\partial^2 u}{\partial x^2} \; ; \; c^2 = \frac{E}{\rho}.$$

Derive the equation directly from the pointwise Equation (3.22a) or from the Lagrangian

$$\mathcal{L}\big(u(x,t)\big) = \int_L \frac{1}{2}\left(\frac{\partial u}{\partial t}\right)^2 dx - \int_L \frac{1}{2}E\left(\frac{\partial u}{\partial x}\right)^2 dx,$$

where L is an arbitrary length in a current configuration.

6. For the one-dimensional linear elastic dynamic equation shown in Exercise 4, the analytical solution for an infinite medium is given by D'Alembert's expression as

$$u(x,t) = f(x+ct) + g(x-ct).$$

Prove that this satisfies the equilibrium equation for any f and g functions and determine the value of f and g for the case where the initial conditions are

$$u(x,0) = 2\cos\left(\frac{\pi x}{2}\right) \; ;\text{for } -1 < x < 1 \; ; \text{ elsewhere } u = 0,$$

$$\frac{\partial u}{\partial t}(x,0) = 0.$$

Plot the resulting solution at various times, taking, for instance, $c = 10$.

7. The solution for the one-dimensional linear elastic dynamic equation in the presence of boundary conditions leads to stationary waves. For instance, if $u(0,t) = u(L,t) = 0$, then $u(x,t)$ can be expressed as a harmonic series:

$$u(x,t) = \sum_{n=1}^{\infty}\left(A_n \sin(\omega_n t)\sin\left(\frac{n\pi x}{L}\right) + B_n \cos(\omega_n t)\sin\left(\frac{n\pi x}{L}\right)\right).$$

Determine the values of ω_n by substituting the above expression into the equilibrium equation, and also find the values of A_n and B_n using the initial conditions

$$u(x,0) = 2\sin\left(\frac{\pi x}{L}\right),$$

$$\frac{\partial u}{\partial t}(x,0) = 0.$$

CHAPTER FOUR

DISCRETIZATION AND SOLUTION

4.1 INTRODUCTION

The dynamic equilibrium equations established in Chapter 3 were expressed, via the principle of virtual work, in either a material or spatial setting involving integrals over the initial volume V or the current spatial volume v respectively. Either description could provide a basis for discretization; however, in the following development, the spatial configuration will be adopted for consistency with the presentation in Chapter 9 of NL-Statics. With the exception of inertia terms involving the mass, the discretization has been formulated in detail in Chapter 9 of NL-Statics. Nevertheless, for completeness, the key equations will be summarized herein. The discretized equations derived in this chapter have been coded by "extending" the static program FlagShyp originally developed for the NL-Statics volume. The resulting code is presented in Chapter 8 with a set of simple illustrative examples.

4.2 DISCRETIZED KINEMATICS

In the process of discretizing the dynamic equilibrium equations, it is necessary to discretize the fundamental kinematics measure known as the deformation gradient tensor $\boldsymbol{F}$. Consequently it will be necessary to consider the discretization of a number of kinematic quantities. To begin, the initial configuration is represented using an isoparametric element interpolation which defines the initial position $\boldsymbol{X}$ within an element in terms of the element nodal positions $\boldsymbol{X}_a$ as

$$\boldsymbol{X} = \sum_{a=1}^{n} N_a(\xi_1, \xi_2, \xi_3)\boldsymbol{X}_a, \tag{4.1}$$

where $N_a(\xi_1, \xi_2, \xi_3)$ is the standard shape function corresponding to node a, with n denoting the number of nodes per element. Similarly, positions in the current

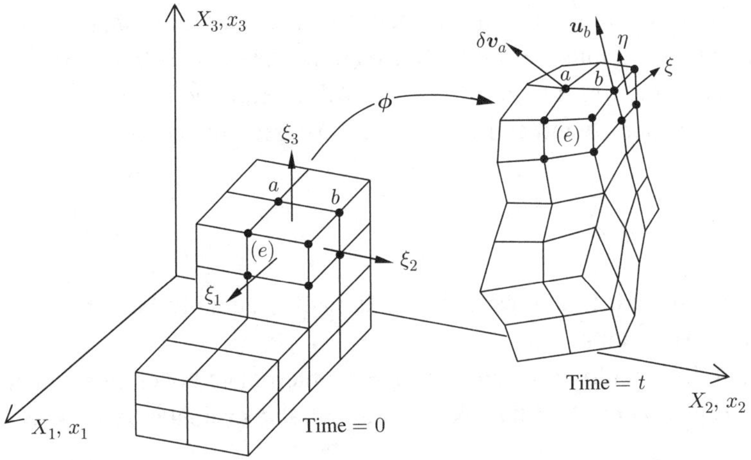

FIGURE 4.1 Discretization of a moving solid using isoparametric finite elements.

configuration are described in terms of the time-dependent current nodal particle positions $\boldsymbol{x}_a(t)$ (see Figure 4.1) as follows:

$$\boldsymbol{x} = \sum_{a=1}^{n} N_a(\xi_1, \xi_2, \xi_3)\boldsymbol{x}_a(t). \tag{4.2}$$

Equations (4.1) and (4.2) establish an indirect relationship between $\boldsymbol{x}$ and $\boldsymbol{X}$ through the nondimensional isoparametric coordinates ξ_1, ξ_2, ξ_3 particular to each element. In contrast to the static situation, where time is somewhat of a notional concept, time is now clearly "real" in a dynamic case.

Observe that in Equation (4.1), although $N_a(\xi_1, \xi_2, \xi_3)$ depends upon the coordinates ξ_i, the nodal coordinates $\boldsymbol{x}_a(t)$ depend on the time t. In this sense the interpolation can be considered as a separation of space- and time-dependent terms exemplified by $N_a(\xi_1, \xi_2, \xi_3)$ and $\boldsymbol{x}_a(t)$. As a consequence of the time differentiation of Equation (4.2), the velocity $\boldsymbol{v}$ and acceleration $\boldsymbol{a}$ can be interpolated in the same manner as

$$\boldsymbol{v} = \sum_{a=1}^{n} N_a \boldsymbol{v}_a(t); \quad \boldsymbol{a} = \sum_{a=1}^{n} N_a \boldsymbol{a}_a(t), \tag{4.3a,b}$$

$$\boldsymbol{v}_a(t) = \frac{d\boldsymbol{x}_a}{dt}; \quad \boldsymbol{a}_a(t) = \frac{d\boldsymbol{v}_a}{dt}, \tag{4.3c,d}$$

where $\boldsymbol{v}_a$ and $\boldsymbol{a}_a$ are the nodal velocities and accelerations respectively and the dependency of N_a on ξ_1, ξ_2, and ξ_3 has been assumed implicitly rather than repeated explicitly in order to simplify the notation. Note that in order to ensure consistency with the differential relationships between $\boldsymbol{x}$, $\boldsymbol{v}$, and $\boldsymbol{a}$ the shape functions employed in either of Equations (4.3c,d) need to be the same as those used in Equation (4.2). Finally, the formulations to follow will require the introduction of an incremental displacement $\boldsymbol{u}$ and a virtual velocity $\delta\boldsymbol{v}$ interpolated as

$$\boldsymbol{u} = \sum_{a=1}^{n} N_a \boldsymbol{u}_a; \quad \delta\boldsymbol{v} = \sum_{a=1}^{n} N_a \delta\boldsymbol{v}_a, \tag{4.4a,b}$$

where neither $\boldsymbol{u}$ nor $\delta\boldsymbol{v}$ are functions of time but will eventually be subject to boundary condition restrictions.

The fundamental deformation gradient tensor $\boldsymbol{F}$ is interpolated over an element by differentiating Equation (4.2) with respect to the initial coordinates $\boldsymbol{X}$ to give,*

$$\boldsymbol{F} = \sum_{a=1}^{n} \boldsymbol{x}_a \otimes \boldsymbol{\nabla}_0 N_a, \tag{4.5}$$

where $\boldsymbol{\nabla}_0 N_a = \partial N_a / \partial \boldsymbol{X}$ is formulated in the reference configuration using the standard chain rule in terms of the initial nodal coordinates $\boldsymbol{X}_a$ and the isoparametric gradient $\boldsymbol{\nabla}_\xi$ as

$$\boldsymbol{\nabla}_0 N_a = \left(\frac{\partial \boldsymbol{X}}{\partial \boldsymbol{\xi}}\right)^{-T} \boldsymbol{\nabla}_\xi N_a; \quad \frac{\partial \boldsymbol{X}}{\partial \boldsymbol{\xi}} = \sum_{a=1}^{n} \boldsymbol{X}_a \otimes \boldsymbol{\nabla}_\xi N_a. \tag{4.6a,b}$$

In indicial notation (see NL-Statics, Section 9.2),

$$\frac{\partial X_I}{\partial \xi_\alpha} = \sum_{a=1}^{n} X_{a,I} \frac{\partial N_a}{\partial \xi_\alpha}; \quad F_{iJ} = \sum_{a=1}^{n} x_{a,i} \frac{\partial N_a}{\partial X_J}; \quad i, J = 1, 2, 3. \tag{4.7a,b}$$

As an example of the importance of the deformation gradient, recall that the left Cauchy–Green tensor $\boldsymbol{b}$ (see NL-Statics, Sections 4.5 and 6.4.3) used in the stress evaluation for neo-Hookean materials is $\boldsymbol{b} = \boldsymbol{F}\boldsymbol{F}^T$. Alternatively, the right Cauchy–Green tensor $\boldsymbol{C} = \boldsymbol{F}^T\boldsymbol{F}$ is used in the sections below and in discretized form it is given by Equation (9.9a,b) of NL-Statics as

$$\begin{aligned}\boldsymbol{C} &= \left(\sum_{a=1}^{n} \boldsymbol{x}_a \otimes \boldsymbol{\nabla}_0 N_a\right)^T \left(\sum_{b=1}^{n} \boldsymbol{x}_b \otimes \boldsymbol{\nabla}_0 N_b\right) \\ &= \sum_{a,b=1}^{n} (\boldsymbol{x}_a \cdot \boldsymbol{x}_b) \boldsymbol{\nabla}_0 N_a \otimes \boldsymbol{\nabla}_0 N_b. \end{aligned} \tag{4.8}$$

* Note: $\boldsymbol{\nabla}(f\boldsymbol{v}) = f\boldsymbol{\nabla}\boldsymbol{v} + \boldsymbol{v} \otimes \boldsymbol{\nabla} f$.

Finally the discretized virtual rate of deformation tensor $\delta\boldsymbol{d}$ required for the virtual work equation, see Equation (3.28b), is given as

$$\delta\boldsymbol{d} = \frac{1}{2}\sum_{a=1}^{n}(\delta\boldsymbol{v}_a \otimes \boldsymbol{\nabla} N_a + \boldsymbol{\nabla} N_a \otimes \delta\boldsymbol{v}_a), \tag{4.9}$$

where $\boldsymbol{\nabla} N_a$ is found by replacing $\boldsymbol{X}$ by $\boldsymbol{x}$ and $\boldsymbol{\nabla}_0$ by $\boldsymbol{\nabla}$ in Equation (4.6a,b).

4.3 MASS MATRIX

This section will concentrate on the formulation of the mass matrix. The mass matrix derives from the discretization of the inertia term in the virtual work expression of dynamic equilibrium. The discretization of the remaining internal and external virtual work terms is explained in detail in NL-Statics; consequently, only final results will be included in what follows. The mass matrix, while containing many zero entries, is nevertheless a full matrix, which implies a computational penalty. In order to mitigate this problem, the matrix can be systematically diagonalized, and by doing so efficient explicit time-stepping schemes can be introduced.

To begin, the virtual work expression of dynamic equilibrium given by Equation (3.28a) is expressed in separate internal (int), external (ext), and mass inertia (m) terms as

$$\delta W(\boldsymbol{\phi}, \delta\boldsymbol{v}) = \delta W_{\text{int}}(\boldsymbol{\phi}, \delta\boldsymbol{v}) - \delta W_{\text{ext}}(\boldsymbol{\phi}, \delta\boldsymbol{v}) + \delta W_m(\boldsymbol{\phi}, \delta\boldsymbol{v}), \tag{4.10a}$$

where

$$\delta W_{\text{int}}(\boldsymbol{\phi}, \delta\boldsymbol{v}) = \int_v \boldsymbol{\sigma} : \delta\boldsymbol{d}\, dv, \tag{4.10b}$$

$$\delta W_{\text{ext}}(\boldsymbol{\phi}, \delta\boldsymbol{v}) = \int_v \boldsymbol{f} \cdot \delta\boldsymbol{v}\, dv + \int_{\partial v} \boldsymbol{t} \cdot \delta\boldsymbol{v}\, da, \tag{4.10c}$$

$$\delta W_m(\boldsymbol{\phi}, \delta\boldsymbol{v}) = \int_v \rho\boldsymbol{a} \cdot \delta\boldsymbol{v}\, dv. \tag{4.10d}$$

The discretized internal and external virtual work terms per element (e) per node a can be expressed in terms of the internal and external equivalent nodal forces $\boldsymbol{T}_a^{(e)}$ and $\boldsymbol{F}_a^{(e)}$ (see NL-Statics, Section 9.3), as

$$\delta W_{\text{int}}^{(e)}(\boldsymbol{\phi}, N_a\delta\boldsymbol{v}_a) + \delta W_{\text{ext}}^{(e)}(\boldsymbol{\phi}, N_a\delta\boldsymbol{v}_a) = \delta\boldsymbol{v}_a \cdot (\boldsymbol{T}_a^{(e)} - \boldsymbol{F}_a^{(e)}), \tag{4.11}$$

where

$$\boldsymbol{T}_a^{(e)} = \int_{v^{(e)}} \boldsymbol{\sigma}\boldsymbol{\nabla} N_a\, dv; \quad \boldsymbol{F}_a^{(e)} = \int_{v^{(e)}} N_a \boldsymbol{f}\, dv + \int_{\partial v^{(e)}} N_a \boldsymbol{t}\, da. \tag{4.12a,b}$$

The development of the mass matrix begins by considering the contribution to the inertia term $\delta W_m(\boldsymbol{\phi}, \delta \boldsymbol{v})$ caused by a single virtual velocity $\boldsymbol{v}_a$ occurring at node a of element (e). Introducing the interpolation for $\delta \boldsymbol{v}$ given by Equation (4.4a,b)$_b$ into Equation (4.10d) gives

$$\begin{aligned} \delta W_m^{(e)}(\boldsymbol{\phi}, N_a \delta \boldsymbol{v}_a) &= \int_{v^{(e)}} \rho \boldsymbol{a} \cdot (N_a \delta \boldsymbol{v}_a)\, dv \\ &= \delta \boldsymbol{v}_a \cdot \int_{v^{(e)}} N_a \rho \boldsymbol{a}\, dv. \end{aligned} \tag{4.13}$$

The above expression can be further expanded by substituting the interpolation for the acceleration $\boldsymbol{a}$ given by Equation (4.3a,b)$_b$ to yield

$$\begin{aligned} \delta W_m^{(e)}(\boldsymbol{\phi}, N_a \delta \boldsymbol{v}_a) &= \delta \boldsymbol{v}_a \cdot \int_{v^{(e)}} \rho N_a \left(\sum_{b=1}^{n} N_b \boldsymbol{a}_b \right) dv \\ &= \delta \boldsymbol{v}_a \cdot \sum_{b=1}^{n} \left(\int_{v^{(e)}} \rho N_a N_b \, dv \right) \boldsymbol{a}_b \\ &= \delta \boldsymbol{v}_a \cdot \sum_{b=1}^{n} \boldsymbol{M}_{ab}^{(e)} \boldsymbol{a}_b, \end{aligned} \tag{4.14}$$

where n is the number of nodes in the element and the element mass matrix $\boldsymbol{M}_{ab}^{(e)}$ is expressed in terms of the (3×3) identity matrix $\boldsymbol{I}$ as

$$\boldsymbol{M}_{ab}^{(e)} = m_{ab}^{(e)} \boldsymbol{I}; \quad m_{ab}^{(e)} = \int_{v^{(e)}} \rho N_a N_b \, dv. \tag{4.15a,b}$$

The above expression is worth some elaboration. First, observe that the inertial virtual work $\delta W_m^{(e)}(\boldsymbol{\phi}, N_a \delta \boldsymbol{v}_a)$ is a function of the *virtual* velocity $\delta \boldsymbol{v}_a$ at node a and the *real* mapping $\boldsymbol{\phi}(t)$. The real mapping, being a function of time, is responsible for the acceleration $\boldsymbol{a}_b$ at node b. Consequently, for element (e), $\boldsymbol{M}_{ab}^{(e)}$ represents the contribution to the inertial virtual work term due to a virtual velocity at node a acting in conjunction with the inertia force due to the acceleration at node b. Finally Equation (4.14) represents the contribution to node a from all nodes in the element.

Equation (4.15a,b)$_b$ can also easily be computed in the initial undeformed configuration by using the conservation of mass equation, $\rho_R = \rho J$, which gives

$$m_{ab}^{(e)} = \int_{V^{(e)}} \rho_R N_a N_b \, dV; \quad dv = J dV. \tag{4.16a,b}$$

Note that the above equation need only be computed once (and stored) as it is independent of any configuration update. This is an inevitable consequence of mass conservation.

In order to assemble the complete inertial virtual work term, the following sequence of sub-assemblies is developed. The contribution to $\delta W_m^{(e)}(\boldsymbol{\phi}, N_a \delta \boldsymbol{v}_a)$ from all elements (e) (1 to M_a) containing node a ($e \ni a$) is

$$
\begin{aligned}
\delta W_m(\boldsymbol{\phi}, N_a \delta \boldsymbol{v}_a) &= \sum_{\substack{e=1 \\ e \ni a}}^{M_a} \delta W_m^{(e)}(\boldsymbol{\phi}, N_a \delta \boldsymbol{v}_a) \\
&= \delta \boldsymbol{v}_a \cdot \left(\sum_{\substack{e=1 \\ e \ni a}}^{M_a} \sum_{b=1}^{n} \boldsymbol{M}_{ab}^{(e)} \boldsymbol{a}_b \right).
\end{aligned} \tag{4.17}
$$

The above equation can be rewritten by swapping the symbols for summation and by varying b from 1 to the number of nodes n_a connected to node b ($b \leftrightarrow a$) via any given element (e) as

$$
\delta W_m(\boldsymbol{\phi}, N_a \delta \boldsymbol{v}_a) = \delta \boldsymbol{v}_a \cdot \left(\sum_{\substack{b=1 \\ b \leftrightarrow a}}^{n_a} \boldsymbol{M}_{ab} \, \boldsymbol{a}_b \right), \tag{4.18}
$$

where $\boldsymbol{M}_{ab}$ is obtained by adding contributions from all elements that connect nodes a and b (1 to $M_{a,b}$) as

$$
\boldsymbol{M}_{ab} = m_{ab} \boldsymbol{I} \quad \text{and} \quad m_{ab} = \left(\sum_{\substack{e=1 \\ e \ni a,b}}^{M_{a,b}} m_{ab}^{(e)} \right). \tag{4.19}
$$

Finally, noting that the term m_{ab}, and therefore M_{ab}, will vanish for pairs of nodes not connected via an element (e), the complete contribution to $\delta W_m(\boldsymbol{\phi}, \delta \boldsymbol{v})$ from the finite element mesh Equation (4.18) is summed over all nodes N to give

$$
\begin{aligned}
\delta W_m(\boldsymbol{\phi}, \delta \boldsymbol{v}) &= \sum_{a=1}^{N} \delta W_m(\boldsymbol{\phi}, N_a \delta \boldsymbol{v}_a) \\
&= \sum_{a=1}^{N} \delta \boldsymbol{v}_a \cdot \left(\sum_{b=1}^{N} \boldsymbol{M}_{ab} \, \boldsymbol{a}_b \right).
\end{aligned} \tag{4.20}
$$

This standard finite element assembly procedure can also be expressed using the complete virtual velocity vector $\delta \mathbf{v}$ together with the complete acceleration vector $\mathbf{a}$ as

$$
\delta W_m(\boldsymbol{\phi}, \delta \boldsymbol{v}) = \delta \mathbf{v}^T \mathbf{M} \mathbf{a}, \tag{4.21}
$$

where

$$\delta \mathbf{v} = \begin{bmatrix} \delta \boldsymbol{v}_1 \\ \delta \boldsymbol{v}_2 \\ \vdots \\ \delta \boldsymbol{v}_N \end{bmatrix}; \qquad \mathbf{a} = \begin{bmatrix} \boldsymbol{a}_1(t) \\ \boldsymbol{a}_2(t) \\ \vdots \\ \boldsymbol{a}_N(t) \end{bmatrix}, \tag{4.22a,b}$$

and where the assembled consistent mass matrix $\mathbf{M}$ is defined by assembling the nodal components as

$$\mathbf{M} = \begin{bmatrix} m_{11}\boldsymbol{I} & m_{12}\boldsymbol{I} & \cdots & m_{1N}\boldsymbol{I} \\ m_{21}\boldsymbol{I} & m_{22}\boldsymbol{I} & \cdots & m_{2N}\boldsymbol{I} \\ \vdots & \vdots & & \vdots \\ m_{N1}\boldsymbol{I} & m_{N2}\boldsymbol{I} & \cdots & m_{NN}\boldsymbol{I} \end{bmatrix}. \tag{4.23}$$

It is easy to see that the above matrix is symmetric since $m_{ab} = m_{ba}$ for any pair $a \neq b$ and, as explained above, $m_{ab} = 0$ if a is not connected to b via a finite element.

EXAMPLE 4.1: Mass matrix for a two-node element

This simple example illustrates the derivation of the mass matrix for a two-node truss element.

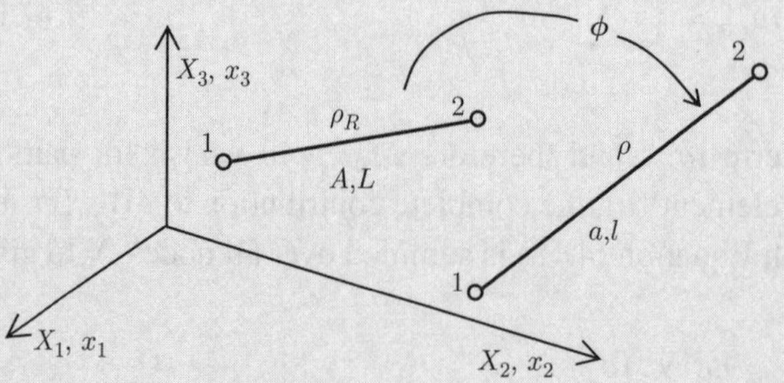

Two-node truss element.

The two-node element shown has an initial configuration with a cross-sectional area of A, length L, and density ρ_R, while the corresponding items in the current configuration at time t are a, l, and ρ, respectively. Over the interval $\xi = [0, 1]$ the shape functions are

$$N_1(\xi) = 1 - \xi; \quad N_2(\xi) = \xi.$$

(continued)

Example 4.1: *(cont.)*

As an example these shape functions would enable the current coordinates given by Equation (4.2) to be represented as

$$\boldsymbol{x} = (1-\xi)\boldsymbol{x}_1 + \xi\boldsymbol{x}_2; \quad \boldsymbol{x} = [x_1, x_2, x_3]^T.$$

From Equations (4.15a,b) and (4.16a,b) the mass matrix coefficient is

$$m_{ab}^{(e)} = m\int_0^1 N_a(\xi)N_b(\xi)\,d\xi; \quad m = \rho_R AL.$$

Substituting the shape functions N_1 and N_2 given above yields

$$\begin{aligned} m_{11}^{(e)} &= m\int_0^1 (1-\xi)^2\,d\xi = \frac{m}{3}, \\ m_{22}^{(e)} &= m\int_0^1 \xi^2\,d\xi = \frac{m}{3}, \\ m_{12}^{(e)} &= m_{21}^{(e)} = m\int_0^1 (1-\xi)\xi\,d\xi = \frac{m}{6}. \end{aligned}$$

Finally the assembled consistent element mass matrix given by Equation (4.23) becomes

$$\mathbf{M} = \rho_R AL \begin{bmatrix} \frac{1}{3}\boldsymbol{I} & \frac{1}{6}\boldsymbol{I} \\ \frac{1}{6}\boldsymbol{I} & \frac{1}{3}\boldsymbol{I} \end{bmatrix}.$$

4.3.1 Lumped Mass Matrix

To obtain the full advantage of an explicit time integration scheme, that is, avoiding the solution of a set of linear equations, it is advantageous to replace the consistent matrix with an alternative, the so-called "lumped" diagonal matrix approximation, $\mathbf{M}^L$. There are several alternatives for the construction of $\mathbf{M}^L$. A simple "row sum" technique consists of replacing $\mathbf{M}$ in Equation (4.23) with $\mathbf{M}^L$, defined as

$$\mathbf{M}^L = \begin{bmatrix} m_{11}^L\boldsymbol{I} & 0 & \cdots & 0 \\ 0 & m_{22}^L\boldsymbol{I} & \cdots & 0 \\ \vdots & \vdots & & \vdots \\ 0 & 0 & \cdots & m_{NN}^L\boldsymbol{I} \end{bmatrix}; \quad m_{aa}^L = \sum_{b=1}^{N} m_{ab}. \tag{4.24a,b}$$

The nature of the finite element interpolation is such that most terms m_{ab} in Equation $(4.16\text{a,b})_b$ will be zero except where nodes a and b are connected via an element (e). Substituting Equation $(4.16\text{a,b})_a$ into Equation (4.18) enables m_{aa}^L given by Equation $(4.24\text{a,b})_b$ to be rewritten more precisely as

$$
\begin{aligned}
m^L_{aa} &= \sum_{\substack{b=1 \\ b\leftrightarrow a}}^{n_a} \sum_{\substack{e=1 \\ e\ni a}}^{Ma} \int_{V^{(e)}} \rho_R N_a N_b \, dV \\
&= \sum_{\substack{e=1 \\ e\ni a}}^{Ma} \int_{V^{(e)}} \rho_R N_a \left(\sum_{\substack{b=1 \\ b\leftrightarrow a}}^{n_a} N_b \right) dV \\
&= \sum_{\substack{e=1 \\ e\ni a}}^{Ma} \int_{V^{(e)}} \rho_R N_a \, dV \\
&= \sum_{\substack{e=1 \\ e\ni a}}^{Ma} m^{L(e)}_{aa},
\end{aligned}
\tag{4.25}
$$

where the "partition of unity" property of the shape functions N_a; that is, the sum in the term in parenthesis containing N_b above is equal to one, has been invoked. The above equation shows that m^L_{aa} comprises the elementary $m^{L(e)}_{aa}$, which are the "tributary" masses accumulating from all elements containing node a.

EXAMPLE 4.2: Lumped mass matrix for a two-node element

The lumped mass matrix for the two-node element is easily constructed as

$$
\begin{aligned}
\mathbf{M}^L &= \rho_R A L \begin{bmatrix} (\frac{1}{3}+\frac{1}{6})\boldsymbol{I} & \mathbf{0} \\ \mathbf{0} & (\frac{1}{3}+\frac{1}{6})\boldsymbol{I} \end{bmatrix} \\
&= \rho_R A L \begin{bmatrix} \frac{1}{2}\boldsymbol{I} & \mathbf{0} \\ \mathbf{0} & \frac{1}{2}\boldsymbol{I} \end{bmatrix},
\end{aligned}
$$

which is what would be intuitively expected with such a simple element, that is, half the mass of the element is attached to each node. This is the intuitive process used in Chapter 2 in order to obtain the mass matrix of a truss.

4.4 DISCRETIZED DYNAMIC EQUILIBRIUM EQUATIONS

Having considered the development of the mass matrix in the preceding section and the discretized kinematics in Section 4.2, the virtual work expression given by Equation (4.10a) can be expressed in discrete form by combining Equations (4.11) and (4.20) to give

$$\delta W(\phi, \delta \boldsymbol{v}) = \sum_{a=1}^{N} \delta \boldsymbol{v}_a \cdot \left[\left(\sum_{b=1}^{N} \boldsymbol{M}_{ab}\, \boldsymbol{a}_b \right) + \boldsymbol{T}_a - \boldsymbol{F}_a \right] = 0. \tag{4.26}$$

Since the virtual work equation must be satisfied for any arbitrary nodal velocity, $\delta \boldsymbol{v}_a$, the discretized dynamic equilibrium equations can be expressed in terms of the nodal residual for each node a as

$$\boldsymbol{R}_a = \sum_{b=1}^{N} \boldsymbol{M}_{ab}\, \boldsymbol{a}_b + \boldsymbol{T}_a - \boldsymbol{F}_a = \boldsymbol{0}; \quad a = 1, \ldots, N. \tag{4.27}$$

An alternative view of the above equation is gained by recalling Newton's second law of motion and introducing an equivalent nodal inertia force $\boldsymbol{F}_{m,a}$ defined as

$$\boldsymbol{F}_{m,a} = \sum_{b=1}^{N} \boldsymbol{M}_{ab}\, \boldsymbol{a}_b, \tag{4.28}$$

which enables Equation (4.27) to be rewritten as

$$\boldsymbol{R}_a = \boldsymbol{F}_{m,a} + \boldsymbol{T}_a - \boldsymbol{F}_a = \boldsymbol{0}; \quad a = 1, \ldots, N. \tag{4.29}$$

The above equation represents the dynamic equilibrium between equivalent inertial, internal, and external forces at each of the nodes $a = 1, \ldots, N$. In other words, the equivalent external forces must be balanced by the inertial and internal forces.

Two particular cases of Equation (4.29) are worth consideration, namely rigid body motion and quasistatic loading. In the first case, the equivalent internal force vector $\boldsymbol{T}_a$ reduces to zero and the external forces can only be absorbed by inertial forces. In the second case, the acceleration vector $\boldsymbol{a}_b$ can be neglected and then $\boldsymbol{F}_{m,a}$ reduces to zero, yielding the discretized static equilibrium equations; see NL-Statics, Section 9.3. Other scenarios in which the inertial forces can be neglected are those where the density is small, or just simply the vector $\boldsymbol{F}_{m,a}$ is very small, in comparison with the other contributing forces $\boldsymbol{T}_a$ and $\boldsymbol{F}_a$. An example would be slow metal forming process such as creep forming or superplastic forming.

For convenience, all the nodal equivalent forces are assembled into single arrays to define the complete global internal and external forces $\mathbf{T}$ and $\mathbf{F}$ respectively, as well as the complete residual force $\mathbf{R}$, as

$$\mathbf{T} = \begin{bmatrix} \boldsymbol{T}_1 \\ \boldsymbol{T}_2 \\ \vdots \\ \boldsymbol{T}_N \end{bmatrix}; \quad \mathbf{F} = \begin{bmatrix} \boldsymbol{F}_1 \\ \boldsymbol{F}_2 \\ \vdots \\ \boldsymbol{F}_N \end{bmatrix}; \quad \mathbf{R} = \begin{bmatrix} \boldsymbol{R}_1 \\ \boldsymbol{R}_2 \\ \vdots \\ \boldsymbol{R}_N \end{bmatrix}. \tag{4.30a,b,c}$$

These definitions enable the discretized virtual work Equation (4.26) to be rewritten as

$$\delta W(\phi, \delta \boldsymbol{v}) = \delta \mathbf{v}^T \mathbf{R} = \delta \mathbf{v}^T (\mathbf{M}\mathbf{a} + \mathbf{T} - \mathbf{F}) = 0, \tag{4.31}$$

where the virtual velocity vector $\delta\mathbf{v}$ and acceleration $\mathbf{a}$ are given by Equation (4.22a,b) respectively.

By defining the complete global list of nodal positions $\boldsymbol{x}_a$ and velocities $\boldsymbol{v}_a$ as

$$\mathbf{x} = \begin{bmatrix} \boldsymbol{x}_1(t) \\ \boldsymbol{x}_2(t) \\ \vdots \\ \boldsymbol{x}_N(t) \end{bmatrix}; \qquad \mathbf{v} = \begin{bmatrix} \boldsymbol{v}_1(t) \\ \boldsymbol{v}_2(t) \\ \vdots \\ \boldsymbol{v}_N(t) \end{bmatrix}, \tag{4.32a,b}$$

and recalling that the internal equivalent forces are nonlinear functions of the nodal positions, enables the complete nonlinear dynamic equilibrium equations to be symbolically assembled as[†]

$$\mathbf{R}(\mathbf{x}, \mathbf{a}) = \mathbf{M}\mathbf{a} + \mathbf{T}(\mathbf{x}) - \mathbf{F}(\mathbf{x}, t) = \mathbf{0}. \tag{4.33}$$

The above equation can easily be expanded to include viscous damping, which is a function of the nodal velocities $\mathbf{v}$, to give

$$\mathbf{R}(\mathbf{x}, \mathbf{v}, \mathbf{a}) = \mathbf{M}\mathbf{a} + \mathbf{C}\mathbf{v} + \mathbf{T}(\mathbf{x}) - \mathbf{F}(\mathbf{x}, t) = \mathbf{0}, \tag{4.34}$$

where the global velocity vector $\mathbf{v}$ is given by Equation (4.32a,b)$_b$ and $\mathbf{C}$ is the so-called viscous damping matrix. In what follows this matrix will be assumed to be constant, independent of the deformation and known *ab initio*.

EXAMPLE 4.3: Velocity-dependent forces

In some engineering applications the internal force vector $\mathbf{T}$ can be a function of both position $\mathbf{x}$ and velocity $\mathbf{v}$, where $\mathbf{x}$ and $\mathbf{v}$ are defined in Equation (4.32a,b). For instance, a typical case occurs in material forming problems where the constitutive law defining the behavior of the material is not only hyperelastic, but also incorporates a viscoplastic component dependent upon the velocity field in a nonlinear fashion. In such a case the internal force is additively decomposed into elastic (e) and viscous (v) components as

$$\mathbf{T}(\mathbf{x}, \mathbf{v}) = \mathbf{T}_e(\mathbf{x}) + \mathbf{T}_v(\mathbf{x}, \mathbf{v}).$$

Another typical example is the inclusion of a damping mechanism in structures subjected to dynamic loading. The simplest case is linear damping, where the internal force vector is expressed as

$$\mathbf{T}(\mathbf{x}, \mathbf{v}) = \mathbf{T}_e(\mathbf{x}) + \mathbf{C}\mathbf{v}.$$

(continued)

[†] The external forces may well be functions of time and also nonlinear functions of the nodal positions when considering follower forces; for example, a pressure loading on a component undergoing large deformation.

Example 4.3: *(cont.)*

Here, $\mathbf{C}$ is the so-called damping matrix, which tends to be obtained in an ad hoc manner with numerous procedures being employed. A classical method is "Rayleigh" damping, whereby $\mathbf{C}$ is computed as a linear combination of mass and stiffness matrices as

$$\mathbf{C} = \alpha \mathbf{M} + \beta \mathbf{K},$$

where the parameters α and β are deduced from experimental data.

4.5 KINETIC ENERGY, POTENTIAL ENERGY, AND THE LAGRANGIAN

In the context of trusses, Chapter 2, Section 2.3.1 provides a gentle introduction to the topics contained in this section, which deals with the energy terms for a solid discretized via finite elements.

4.5.1 Kinetic energy

Expanding the truss definition of kinetic energy to the continuum case gives the kinetic energy as

$$K(\boldsymbol{v}) = \int_v \frac{1}{2}\rho \boldsymbol{v} \cdot \boldsymbol{v} \, dv = \int_V \frac{1}{2}\rho_R \boldsymbol{v} \cdot \boldsymbol{v} \, dV, \tag{4.35}$$

where the differential implies the functionality of the velocity vector, that is, either $\boldsymbol{v}(\boldsymbol{x})$ or $\boldsymbol{v}(\boldsymbol{X})$. The velocity field $\boldsymbol{v}$ over an element (e) is interpolated in the usual manner (see Equation (4.3a,b)$_a$), enabling the kinetic energy of the continuum to be written in terms of nodal velocities $\boldsymbol{v}_a$ as

$$\begin{aligned} K^{(e)}(\boldsymbol{v}) &= \int_{V^{(e)}} \frac{1}{2}\rho_R \left(\sum_{a=1}^{n} N_a \boldsymbol{v}_a\right) \cdot \left(\sum_{b=1}^{n} N_b \boldsymbol{v}_a\right) dV \\ &= \frac{1}{2}\sum_{a=1}^{n}\sum_{b=1}^{n} m_{ab}^{(e)} \boldsymbol{v}_a \cdot \boldsymbol{v}_b, \end{aligned} \tag{4.36}$$

where $m_{ab}^{(e)}$ is the elemental mass contribution corresponding to nodes a and b; see Equation (4.16a,b). Note that the spatial discretization of the velocity field $\boldsymbol{v}$ in terms of nodal velocities $\boldsymbol{v}_a$ and shape functions N_a renders the kinetic energy in Equation (4.36) as a function of the discrete nodal velocities.

Assembling all elemental contributions enables the total kinetic energy for the continuum to be written in vector/matrix form as

$$K(\mathbf{v}) = \sum_{e=1}^{M} K^{(e)} = \frac{1}{2}\mathbf{v}^T \mathbf{M}\mathbf{v}, \tag{4.37}$$

where $\mathbf{M}$ is the assembled consistent mass matrix and M is the total number of elements in the mesh. For an explicit time-stepping scheme, an approximation to $K(\mathbf{v})$ can be achieved by replacing the consistent mass matrix $\mathbf{M}$ by its lumped mass counterpart $\mathbf{M}^L$. This is equivalent to concentrating the mass of the body at the nodes of the finite element mesh.

4.5.2 Total Energy and the Lagrangian

The total energy of the dynamic system is defined as the sum of the kinetic energy, the elastic internal (int) energy, and the external (ext) potential energy, and is given by Equation (2.23) as

$$\Pi(\mathbf{x}, \mathbf{v}) = K(\mathbf{v}) + \Pi_{\text{int}}(\mathbf{x}) + \Pi_{\text{ext}}(\mathbf{x}). \tag{4.38}$$

The internal energy contribution from element (e) is a function of the positions $\boldsymbol{x}_a$ of the nodes in the element and a strain energy function per unit initial volume $\Psi(\boldsymbol{F})$, giving (see NL-Statics, Section 6.2)

$$\Pi_{\text{int}}^{(e)}(\boldsymbol{x}_a) = \int_{V^{(e)}} \Psi(\boldsymbol{F})\, dV. \tag{4.39}$$

The total internal energy accumulated over all elements M is obtained by adding elemental contributions to give

$$\Pi_{\text{int}}(\mathbf{x}) = \sum_{e=1}^{M} \Pi_{\text{int}}^{(e)}(\boldsymbol{x}_a). \tag{4.40}$$

In the presence of constant external forces, such as self weight (gravity force), the elemental contribution to the external potential energy is obtained as

$$\begin{aligned}
\Pi_{\text{ext}}^{(e)}(\boldsymbol{x}_a) &= \int_{V^{(e)}} -\boldsymbol{f}_R \cdot \left(\sum_{a=1}^{n} N_a \boldsymbol{x}_a\right) dV \\
&= \sum_{a=1}^{n} \left(-\int_{V^{(e)}} \boldsymbol{f}_R N_a\, dV\right) \cdot \boldsymbol{x}_a \\
&= \sum_{a=1}^{n} -\boldsymbol{F}_a^{(e)} \cdot \boldsymbol{x}_a,
\end{aligned} \tag{4.41}$$

where $\boldsymbol{f}_R$ is the body force vector per unit undeformed volume and $\boldsymbol{F}_a^{(e)}$ is the nodal equivalent body force at node a for element (e). The contribution to the external potential energy from all elements connecting to node a is given as

$$\Pi_{\text{ext}}(\boldsymbol{x}_a) = -\boldsymbol{F}_a \cdot \boldsymbol{x}_a \; ; \; \boldsymbol{F}_a = \sum_{\substack{e=1 \\ e \ni a}}^{Ma} \boldsymbol{F}_a^{(e)} \; ; \; \boldsymbol{F}_a^{(e)} = \int_{V^{(e}} \boldsymbol{f}_R N_a \, dV, \qquad (4.42\text{a,b,c})$$

where $\boldsymbol{F}_a$ is the assembled equivalent nodal force at node a. Listing the equivalent nodal forces and corresponding nodal positions (see Equations (4.30a,b,c)$_b$ and (4.32a,b)$_a$) enables the total contribution to the external potential energy to be conveniently expressed in vector form as

$$\Pi_{\text{ext}}(\mathbf{x}) = -\mathbf{F}^T\mathbf{x}. \qquad (4.43)$$

Finally, substituting Equations (4.37), (4.40), and (4.43) into Equation (4.38) gives the total energy of the body as an explicit representation on the discrete nodal positions $\mathbf{x}$ and velocities $\mathbf{v}$.

Similarly to the total energy function $\Pi(\mathbf{x}, \mathbf{v})$, the Lagrangian function $\mathcal{L}$ can now be introduced as the kinetic energy minus the sum of the internal and external energy:

$$\mathcal{L}(\mathbf{x}, \mathbf{v}) = K(\mathbf{v}) - \Pi_{\text{int}}(\mathbf{x}) - \Pi_{\text{ext}}(\mathbf{x}), \qquad (4.44)$$

where clearly the Lagrangian again depends explicitly on the discrete nodal positions $\mathbf{x}$ and velocities $\mathbf{v} = \dot{\mathbf{x}}$. This expression will be used in Section 4.7 as the starting point to derive variational time integrators in a manner similar to that employed in Chapter 2 for trusses.

4.6 NEWMARK TIME INTEGRATION SCHEME

In this section consideration is given to the numerical solution of the dynamic equilibrium equation represented in discrete form by Equation (4.33) subject to initial conditions defined at time $t = t_0$, that is

$$\mathbf{R}(\mathbf{x}, \mathbf{a}) = \mathbf{0} \; ; \quad \mathbf{x}(t_0) = \mathbf{x}_0 \; ; \; \mathbf{v}(t_0) = \mathbf{v}_0. \qquad (4.45\text{a,b,c})$$

Given these initial conditions, a solution for $\mathbf{x}(t)$ and $\mathbf{v}(t)$ is required during the time interval $t \in [t_0, T]$. Note that in general $\mathbf{x}_0$ need not coincide with the undeformed configuration of the body since initial displacements could have been imposed.

The basis for the schemes presented below is the so-called Newmark family of time integrators. However, it will be shown that, by choosing certain parameters, both the leap-frog and the trapezoidal methods presented in Chapters 1 and 2 can be retrieved.

At this stage the reader may wish to check back to Chapter 2, where the leap-frog and trapezoidal methods were introduced in the context of simple truss analysis.

Equation (4.45a,b,c) is an initial value problem (in time), involving, through the acceleration $\mathbf{a}$, a system of second-order ordinary differential equations. With a numerical time integration strategy in mind, the time domain, as expected, is discretized in a series of N_t time steps as

$$t_{n+1} = t_n + \Delta t \; ; \; n = 0, 1, \ldots, N_t, \tag{4.46}$$

such that $t_{N_t} = T$. For simplicity in what follows, it will be assumed that Δt is constant during the chosen time interval. The problem now becomes one whereby a time-discrete solution is found for the vectors $\mathbf{x}_{n+1}, \mathbf{a}_{n+1}$, and, if viscous forces are present, $\mathbf{v}_{n+1}$, satisfying

$$\mathbf{R}(\mathbf{x}_{n+1}, \mathbf{a}_{n+1}) = \mathbf{0} \; ; \; n = 0, 1, \ldots, N_t - 1, \tag{4.47}$$

and the initial conditions. This equation can be expressed in more detail using Equation (4.33) as‡

$$\mathbf{M}\mathbf{a}_{n+1} + \mathbf{T}(\mathbf{x}_{n+1}) - \mathbf{F}(t_{n+1}) = \mathbf{0} \; ; \; n = 0, 1, \ldots, N_t - 1. \tag{4.48}$$

This equation represents a system of nonlinear algebraic equations which are functions of $\mathbf{x}_{n+1}$ and $\mathbf{a}_{n+1}$. To solve these equations a kinematic relationship is required between $\mathbf{x}$ and $\mathbf{a}$ so that Equation (4.48) becomes a function solely of either $\mathbf{a}$ or $\mathbf{x}$. This is furnished by expanding the position $\mathbf{x}$ as a third-order Taylor series with respect to time and approximating the third-order derivative of $\mathbf{x}$ by assuming a linear variation of acceleration over time; see Example 4.4. Such a procedure results in the Newmark family of time-stepping schemes, giving the following kinematic relationships:

$$\mathbf{x}_{n+1} = \mathbf{x}_n + \Delta t \mathbf{v}_n + \frac{\Delta t^2}{2}\Big[(1 - 2\beta)\mathbf{a}_n + 2\beta\mathbf{a}_{n+1}\Big], \tag{4.49a}$$

$$\mathbf{v}_{n+1} = \mathbf{v}_n + \Delta t\big[(1 - \gamma)\mathbf{a}_n + \gamma\mathbf{a}_{n+1}\big], \tag{4.49b}$$

where γ and β are parameters which can be arbitrarily chosen (with some restrictions) to yield different time integration schemes. Alternative choices for γ and β determine the stability and accuracy of the time integration scheme. Equations (4.49) establish relationships between $\mathbf{x}_{n+1}$, $\mathbf{a}_{n+1}$, and, if necessary, $\mathbf{v}_{n+1}$, which can now be used in conjunction with Equation (4.48) in order to advance the solution in time, that is, the triad $\{\mathbf{x}, \mathbf{v}, \mathbf{a}\}$ from time t_n to time t_{n+1}.

‡ The dependence of $\mathbf{F}$ upon the position $\mathbf{x}$ will be ignored here; see further comments in Section 4.6.1 on Newmark implementation.

Two specific choices of the Newmark family are of interest, namely the average acceleration and leap-frog (alternatively called central difference) schemes. For the average acceleration scheme, the parameters are chosen as $\beta = 1/4$ and $\gamma = 1/2$, which when substituted into Equations (4.49) yield

$$\mathbf{x}_{n+1} = \mathbf{x}_n + \Delta t \mathbf{v}_n + \frac{\Delta t^2}{4}(\mathbf{a}_n + \mathbf{a}_{n+1}), \tag{4.50a}$$

$$\mathbf{v}_{n+1} = \mathbf{v}_n + \frac{\Delta t}{2}(\mathbf{a}_n + \mathbf{a}_{n+1}), \tag{4.50b}$$

where the use of the average acceleration between time step n and $n+1$ is clearly evident in advancing the velocity from $\mathbf{v}_n$ to $\mathbf{v}_{n+1}$ and the positions from $\mathbf{x}_n$ to $\mathbf{x}_{n+1}$. The scheme can also be interpreted as a trapezoidal integration of the acceleration to give the appropriate increments in velocity and position. The average acceleration scheme is known to be unconditionally stable and hence is traditionally preferred in many numerical simulations, particularly those involving linear problems.

The alternative leap-frog (central difference) scheme sets $\beta = 0$ and $\gamma = 1/2$, the resulting scheme emerging as

$$\mathbf{x}_{n+1} = \mathbf{x}_n + \Delta t \mathbf{v}_n + \frac{\Delta t^2}{2}\mathbf{a}_n, \tag{4.51a}$$

$$\mathbf{v}_{n+1} = \mathbf{v}_n + \frac{\Delta t}{2}(\mathbf{a}_n + \mathbf{a}_{n+1}), \tag{4.51b}$$

The leap-frog scheme is conditionally stable and requires smaller time steps than the average acceleration method.

An important difference is observed between Equations (4.50) and (4.51) in that the average acceleration scheme is implicit whereas the central difference scheme is explicit. By choosing $\beta = 0$ in Equation (4.49a), the acceleration $\mathbf{a}_{n+1}$ is removed, which means that, provided $\{\mathbf{x}_n, \mathbf{v}_n, \mathbf{a}_n\}$ are known, then Equation (4.48) is explicit with respect to $\mathbf{a}_{n+1}$, thereby enabling $\mathbf{v}_{n+1}$ to be found from Equation (4.51b).

EXAMPLE 4.4: Derive the Newmark time-stepping equations

Expanding the position $\mathbf{x}$ and velocity $\mathbf{v}$ as a Taylor series about the time t gives

$$\mathbf{x}_{t+\Delta t} \approx \mathbf{x}_t + \Delta t \left.\frac{d\mathbf{x}}{dt}\right|_t + \frac{\Delta t^2}{2}\left.\frac{d^2\mathbf{x}}{dt^2}\right|_t + \frac{\Delta t^3}{6}\left.\frac{d^3\mathbf{x}}{dt^3}\right|_t + \cdots$$

(continued)

Example 4.4: *(cont.)*

$$\left.\frac{d\mathbf{x}}{dt}\right|_{t+\Delta t} \approx \left.\frac{d\mathbf{x}}{dt}\right|_{t} + \Delta t\left.\frac{d^2\mathbf{x}}{dt^2}\right|_{t} + \frac{\Delta t^2}{2}\left.\frac{d^3\mathbf{x}}{dt^3}\right|_{t} + \cdots$$

The above equations can be truncated and rewritten in terms of velocity $\mathbf{v}$ and acceleration $\mathbf{a}$ with the substitution of, as yet unknown, parameters β and γ, which compensate for the truncation error to give

$$\mathbf{x}_{t+\Delta t} = \mathbf{x}_t + \mathbf{v}_t\Delta t + \frac{1}{2}\mathbf{a}_t\Delta t^2 + \beta\left.\frac{d^3\mathbf{x}}{dt^3}\right|_t \Delta t^3,$$

$$\mathbf{v}_{t+\Delta t} = \mathbf{v}_t + \mathbf{a}_t\Delta t + \gamma\left.\frac{d^3\mathbf{x}}{dt^3}\right|_t \Delta t^2.$$

The third-order derivative is now approximated by assuming a linear variation of the acceleration $\mathbf{a}$ over the time interval Δt, to give

$$\left.\frac{d^3\mathbf{x}}{dt^3}\right|_t = \left(\left.\frac{d^2\mathbf{x}}{dt^2}\right|_{t+\Delta t} - \left.\frac{d^2\mathbf{x}}{dt^2}\right|_t\right)\frac{1}{\Delta t}$$

$$= \left(\mathbf{a}_{t+\Delta t} - \mathbf{a}_t\right)\frac{1}{\Delta t}.$$

Substitution of the above approximation into the third and fourth equations above yields the following Newmark expressions in which the only unknown is the acceleration at time $t+\Delta t$:

$$\mathbf{x}_{t+\Delta t} = \mathbf{x}_t + \mathbf{v}_t\Delta t + \frac{1}{2}\left(1-2\beta\right)\mathbf{a}_t\Delta t^2 + \beta\mathbf{a}_{t+\Delta t}\Delta t^2,$$

$$\mathbf{v}_{t+\Delta t} = \mathbf{v}_t + \left(1-\gamma\right)\mathbf{a}_t\Delta t + \gamma\mathbf{a}_{t+\Delta t}\Delta t.$$

EXAMPLE 4.5: Demonstration that if $\beta = 0$ and $\gamma = 1/2$ then the Newmark and central difference schemes are identical

Recall the central difference expression given in Equation $(2.48\text{a,b})_b$ as

$$\mathbf{a}_n = \frac{\mathbf{x}_{n+1} - 2\mathbf{x}_n + \mathbf{x}_{n-1}}{\Delta t^2}.$$

Rearranging Equation (4.51a) for time step n gives

$$\mathbf{x}_{n+1} - \mathbf{x}_n = \Delta t\mathbf{v}_n + \frac{\Delta t^2}{2}\mathbf{a}_n,$$

(continued)

Example 4.5: *(cont.)*

and, for time step $n-1$,

$$\mathbf{x}_n - \mathbf{x}_{n-1} = \Delta t \mathbf{v}_{n-1} + \frac{\Delta t^2}{2}\mathbf{a}_{n-1}.$$

Subtracting these equations and dividing by Δt yields

$$\frac{\mathbf{x}_{n+1} - 2\mathbf{x}_n + \mathbf{x}_{n-1}}{\Delta t} = (\mathbf{v}_n - \mathbf{v}_{n-1}) + \frac{\Delta t}{2}(\mathbf{a}_n - \mathbf{a}_{n-1}).$$

Rearranging Equation (4.51b) to obtain $(\mathbf{v}_n - \mathbf{v}_{n-1})$ (make $n \to n-1$) and substituting into the above equation gives, upon further rearrangement,

$$\mathbf{a}_n = \frac{\mathbf{x}_{n+1} - 2\mathbf{x}_n + \mathbf{x}_{n-1}}{\Delta t^2},$$

which is the "standard" central difference expression for the acceleration over time steps $n-1$ to $n+1$.

4.6.1 Implementation of the Newmark Acceleration Corrector Method

Attention is now focused on the algorithm for the solution of the dynamic equilibrium Equation (4.48) using the Newmark kinematic relationships given by Equation (4.49). As previous implied, when $\beta \neq 0$ the resulting time integration algorithm is implicit and the system of nonlinear equations is typically solved by making use of a Newton–Raphson solution strategy to compute $\mathbf{a}_{n+1}$ iteratively. At each time step the triad $\{\mathbf{x}, \mathbf{v}, \mathbf{a}\}$ is predicted followed by repeated use of Equation (4.48) and (4.49) to correct $\{\mathbf{x}, \mathbf{v}, \mathbf{a}\}$ until dynamic equilibrium is satisfied (at least to some convergence criteria); this is known as a *predictor–corrector* procedure.[§]

At each time step $n+1$, an initial prediction of the acceleration is made as $\mathbf{a}_{n+1}^{(0)} = \mathbf{a}_n$ from the solution at the previous time step t_n, which establishes the predictor phase, formally from Equations (4.49), as

$$\mathbf{a}_{n+1}^{(0)} = \mathbf{a}_n, \tag{4.52a}$$

$$\mathbf{v}_{n+1}^{(0)} = \mathbf{v}_n + \Delta t\big[(1-\gamma)\mathbf{a}_n + \gamma \mathbf{a}_{n+1}^{(0)}\big], \tag{4.52b}$$

$$\mathbf{x}_{n+1}^{(0)} = \mathbf{x}_n + \Delta t \mathbf{v}_n + \frac{\Delta t^2}{2}\left[(1-2\beta)\mathbf{a}_n + 2\beta \mathbf{a}_{n+1}^{(0)}\right]. \tag{4.52c}$$

[§] In this section bracketed superscripts refer to Newton–Raphson iteration steps and subscripts refer to time steps.

Obviously, the predictor Equations (4.52b) and (4.52c) can be reformulated simply as

$$\mathbf{v}_{n+1}^{(0)} = \mathbf{v}_n + \Delta t \mathbf{a}_n, \tag{4.53a}$$

$$\mathbf{x}_{n+1}^{(0)} = \mathbf{x}_n + \Delta t \mathbf{v}_n + \frac{\Delta t^2}{2} \mathbf{a}_n. \tag{4.53b}$$

A residual, or out-of-balance, force $\mathbf{R}$ is now evaluated using either the above predictions or subsequent iterative values for $\{\mathbf{x}, \mathbf{v}, \mathbf{a}\}$ in Equation (4.48):

$$\mathbf{R}\big(\mathbf{x}_{n+1}^{(k)}, \mathbf{a}_{n+1}^{(k)}\big) = \mathbf{M}\mathbf{a}_{n+1}^{(k)} + \mathbf{T}\big(\mathbf{x}_{n+1}^{(k)}\big) - \mathbf{F}(t_{n+1}), \tag{4.54}$$

where k is the iteration number starting from the predicted values of $\{\mathbf{x}^{(0)}, \mathbf{v}^{(0)}, \mathbf{a}^{(0)}\}$ at $k = 0$, given by Equations (4.52). For $k > 0$ Equations (4.49) give iterative corrections as

$$\mathbf{a}_{n+1}^{(k+1)} = \mathbf{a}_{n+1}^{(k)} + \Delta \mathbf{a}_{n+1}, \tag{4.55a}$$

$$\mathbf{v}_{n+1}^{(k+1)} = \mathbf{v}_n + \Delta t \big[(1-\gamma)\mathbf{a}_n + \gamma \mathbf{a}_{n+1}^{(k+1)}\big], \tag{4.55b}$$

$$\mathbf{x}_{n+1}^{(k+1)} = \mathbf{x}_n + \Delta t \mathbf{v}_n + \frac{\Delta t^2}{2}\left[(1-2\beta)\mathbf{a}_n + 2\beta \mathbf{a}_{n+1}^{(k+1)}\right], \tag{4.55c}$$

where $\Delta \mathbf{a}_{n+1}$ is the update of the acceleration at iteration k computed from the Newton–Raphson solution to Equation (4.54). This corrector phase can be conveniently rewritten as

$$\mathbf{a}_{n+1}^{(k+1)} = \mathbf{a}_{n+1}^{(k)} + \Delta \mathbf{a}_{n+1}, \tag{4.56a}$$

$$\mathbf{v}_{n+1}^{(k+1)} = \mathbf{v}_{n+1}^{(k)} + \Delta \mathbf{v}_{n+1}\,; \quad \Delta \mathbf{v}_{n+1} = \gamma \Delta t \Delta \mathbf{a}_{n+1}, \tag{4.56b}$$

$$\mathbf{x}_{n+1}^{(k+1)} = \mathbf{x}_{n+1}^{(k)} + \Delta \mathbf{x}_{n+1}\,; \quad \Delta \mathbf{x}_{n+1} = \beta \Delta t^2 \Delta \mathbf{a}_{n+1}. \tag{4.56c}$$

The Newton–Raphson solution to Equation (4.54) is accomplished by linearizing the residual force $\mathbf{R}$ at iteration $k+1$. Noting that neither the mass matrix $\mathbf{M}$ nor the force $\mathbf{F}$ are functions of either $\mathbf{x}$ or $\mathbf{a}$, this gives

$$\begin{aligned}\mathbf{R}\big(\mathbf{x}_{n+1}^{(k+1)}, \mathbf{a}_{n+1}^{(k+1)}\big) &\approx \mathbf{R}\big(\mathbf{x}_{n+1}^{(k)}, \mathbf{a}_{n+1}^{(k)}\big) + \mathbf{M}\Delta \mathbf{a}_{n+1} \\ &\quad + D\mathbf{T}\big(\mathbf{x}_{n+1}^{(k)}\big)[\Delta \mathbf{x}_{n+1}] = \mathbf{0},\end{aligned} \tag{4.57}$$

where the linearization of the internal force term $\mathbf{T}(\mathbf{x})$ with respect to an increment in position $\Delta \mathbf{x}_{n+1}$ is given within a virtual work formulation in NL-Statics, Section 9.4 as¶

$$D\mathbf{T}\big(\mathbf{x}_{n+1}^{(k)}\big)[\Delta \mathbf{x}_{n+1}] = \mathbf{K}\big(\mathbf{x}_{n+1}^{(k)}\big)\Delta \mathbf{x}_{n+1}, \tag{4.58}$$

¶ In NL-Statics, $\mathbf{u}$ is used for $\Delta \mathbf{x}_{n+1}$.

where $\mathbf{K}(\mathbf{x}_{n+1}^{(k)})$ is the tangent matrix which results from the linearization of the internal force term $\mathbf{T}(\mathbf{x}_{n+1}^{(k)})$. This tangent matrix essentially comprises a constitutive component, which represents the change in internal forces due to the change in stresses, together with an initial stress component, which represents the change in internal forces caused by existing stresses and the change in geometry. Specifically, the linearization of the elemental (e) contribution of the internal force vector at node a, namely $\boldsymbol{T}_a^{(e)}$, with respect to an incremental displacement at node b given by $N_b\boldsymbol{u}_b$, can be obtained as follows (see Equation (9.32) in NL-Statics):

$$D\boldsymbol{T}_a^{(e)}[N_b\boldsymbol{u}_b] = \boldsymbol{K}_{ab}^{(e)}\boldsymbol{u}_b = \left(\boldsymbol{K}_{c,ab}^{(e)} + \boldsymbol{K}_{\sigma,ab}^{(e)}\right)\boldsymbol{u}_b. \tag{4.59}$$

Here, $\boldsymbol{K}_{c,ab}^{(e)}$ and $\boldsymbol{K}_{\sigma,ab}^{(e)}$ denote the constitutive and initial stress components relating node a to node b in element (e), respectively, with indicial representations as follows (see Equations (9.35) and (9.44c) in NL-Statics):

$$\left[\boldsymbol{K}_{c,ab}^{(e)}\right]_{ij} = \int_{v^{(e)}} \sum_{k,l=1}^{3} \frac{\partial N_a}{\partial x_k}\mathcal{c}_{ikjl}\frac{\partial N_b}{\partial x_l}\,dv; \qquad i,j = 1,2,3; \tag{4.60a}$$

$$\left[\boldsymbol{K}_{\sigma,ab}^{(e)}\right]_{ij} = \int_{v^{(e)}} \sum_{k,l=1}^{3} \frac{\partial N_a}{\partial x_k}\sigma_{kl}\frac{\partial N_b}{\partial x_l}\delta_{ij}\,dv; \qquad i,j = 1,2,3, \tag{4.60b}$$

where $\mathcal{c}_{ikjl}$ represents the spatial or Eulerian elasticity tensor (see Equation (6.14a,b)$_b$ in NL-Statics). For further information, the reader is referred to NL-Statics, Section 9.4, where all details can be found.

Note however that, in the case where external loads are a function of the geometry $\mathbf{x}$, known as *follower loads*, Equation (4.57) will need to be augmented to include the linearization of the follower loads. For the case of pressure loading, this is described in NL-Statics, Section 9.4.

Recall now that $\Delta\mathbf{x}_{n+1}$ is related through the Newmark expressions to $\Delta\mathbf{a}_{n+1}$ by Equation (4.56c), which can be substituted into Equation (4.57) to yield the corrector equation for the Newmark integration algorithm as

$$[\mathbf{K}_\alpha]\Delta\mathbf{a}_{n+1} = -\mathbf{R}\left(\mathbf{x}_{n+1}^{(k)}, \mathbf{a}_{n+1}^{(k)}\right);\ \mathbf{K}_\alpha = \left[\mathbf{M} + \beta\Delta t^2\mathbf{K}(\mathbf{x}_{n+1}^{(k)})\right]. \tag{4.61}$$

Observe that the above equation determines the iterative correction to the acceleration $\mathbf{a}$ necessary to drive the solution towards equilibrium; consequently, this is known as the Newmark acceleration corrector method, sometimes referred to as the Newmark a-method. It is equally possible to substitute $\Delta\mathbf{a}_{n+1} = \Delta\mathbf{x}_{n+1}/(\beta\Delta t^2)$ into Equation (4.57) and derive a corrector for $\mathbf{x}$ leading to a "displacement-corrector" method.

The Newmark a-method algorithm for time t_n to t_{n+1} can now be easily summarized as follows: (1) predict $\{\mathbf{a}, \mathbf{v}, \mathbf{x}\}$ using Equations (4.52); (2) compute iterative acceleration $\Delta\mathbf{a}_{n+1}$ using Equation (4.61); (3) update $\{\mathbf{a}, \mathbf{v}, \mathbf{x}\}$ using corrector Equations (4.56); repeat (2) and (3) until the residual force $\mathbf{R} \approx \mathbf{0}$.

Finally, for the algorithm to get started it is necessary to determine the acceleration $\mathbf{a}_0$ when $n = 0$ in Equation (4.52a). This can be accomplished by solving the dynamic equilibrium equations at t_0 for $\mathbf{a}_0$ as

$$\mathbf{M}\mathbf{a}_0 + \mathbf{T}(\mathbf{x}_0) - \mathbf{F}(t_0) = \mathbf{0}, \tag{4.62}$$

where, typically, for $t_0 = 0$, $\mathbf{x}_0 = \mathbf{X}$ and $\mathbf{F}(t_0) = \mathbf{0}$. Note that if $\mathbf{X}$ is the initial configuration then $\mathbf{a}_0 = \mathbf{0}$; however, $\mathbf{X}$ could be a set of prescribed positions different from the initial configuration, in which case the acceleration would be generated by internal forces.

The resulting acceleration based Newmark/Newton–Raphson algorithm is shown in Box 4.1.

BOX 4.1: Newmark acceleration corrector algorithm

- INPUT geometry, material properties, and solution parameters
- INITIALIZE $\mathbf{F} = 0$, $\mathbf{x} = \mathbf{X}$, $\mathbf{v} = \mathbf{v}_0$
- INITIALIZE $\mathbf{a} = \mathbf{a}_0$ (4.62)
- WHILE $t < T$ (time steps)
  - SET $\mathbf{F} = \mathbf{F}(t)$
  - PREDICT $\mathbf{a}_{n+1}, \mathbf{v}_{n+1}, \mathbf{x}_{n+1}$ (4.52)
  - FIND $\mathbf{R}$ (4.54)
  - FIND $\mathbf{K}_a$ (4.61)
  - WHILE ($\|\mathbf{R}\|/\|\mathbf{F}\| >$ tolerance)
    - SOLVE $\mathbf{K}_a\Delta\mathbf{a}_{n+1} = -\mathbf{R}$ (4.61)
    - CORRECT $\mathbf{a}_{n+1}, \mathbf{v}_{n+1}, \mathbf{x}_{n+1}$ (4.56)
    - FIND $\mathbf{R}$ (4.54)
    - FIND $\mathbf{K}_a$ (4.61)
  - ENDLOOP
- ENDLOOP

4.6.2 Alpha-Method Time Integration

The numerical solution of dynamic equilibrium problems in solid mechanics often leads to time-discrete solutions that contain high frequency oscillations of small amplitude. Although it is possible that these oscillations are also present in the real motion, it is more likely that they are the result of the approximations enshrined in the discrete time integration process. Such behavior can be damped out and the solution stabilized by introducing viscosity into the simulation through, for instance, a Rayleigh damping viscosity matrix. As a result of introducing physical viscosity, energy is not conserved, that is, it is *dissipated*. An alternative approach is to modify a time-stepping scheme to have the same effect. Such schemes are said to introduce *numerical dissipation* or artificial viscosity without sacrificing the overall accuracy of the time integration. One widely used scheme, called the α-method, that achieves this numerical dissipation due to Hilber, Hughes, and Taylor is discussed in this section.

In the α-method the Newmark kinematic relations, see Equations (4.49), are retained while the dynamic equilibrium equations are modified in the following way:

$$\mathbf{M}\mathbf{a}_{n+1} + \mathbf{T}(\mathbf{x}_{n+1+\alpha}) - \mathbf{F}(t_{n+1+\alpha}) = \mathbf{0}, \tag{4.63}$$

where‖

$$t_{n+1+\alpha} = (1+\alpha)t_{n+1} - \alpha t_n = t_{n+1} + \alpha\Delta t, \tag{4.64a}$$

$$\mathbf{x}_{n+1+\alpha} = (1+\alpha)\mathbf{x}_{n+1} - \alpha\mathbf{x}_n, \tag{4.64b}$$

$$\mathbf{v}_{n+1+\alpha} = (1+\alpha)\mathbf{v}_{n+1} - \alpha\mathbf{v}_n. \tag{4.64c}$$

Clearly the fundamental difference with respect to the Newmark method is in the evaluation of the external and internal force terms, $\mathbf{F}$ and $\mathbf{T}$ respectively, which, with $\alpha < 0$, are now evaluated at times and positions intermediate between time steps n and $n+1$.

Notice that if $\alpha = 0$ the method reverts to the Newmark method. If the parameters α, γ, and β are selected such that $\alpha \in [-1/3, 0]$ and

$$\gamma = (1-2\alpha)/2\,; \quad \beta = (1-\alpha)^2/4, \tag{4.65a,b}$$

this results in an unconditionally stable and second-order accurate scheme provided the time step remains constant. Alternatively, if $\alpha = 0$, $\gamma = 1/2$, and $\beta = 1/4$ the trapezoidal time integration scheme emerges.

From an implementation standpoint, only a few modifications are necessary with respect to the Newmark method. The first modification, required for the predictor phase, is the evaluation of the positions $\mathbf{x}^{(0)}_{n+1+\alpha}$ using Equation (4.64b) together with the original Newmark predictor given by Equation (4.52c) as

$$\mathbf{x}^{(0)}_{n+1+\alpha} = (1+\alpha)\mathbf{x}^{(0)}_{n+1} - \alpha\mathbf{x}_n. \tag{4.66}$$

‖ $\mathbf{v}_{n+1+\alpha}$ is only required if real viscous damping (as opposed to numerical dissipation/artificial viscosity) is included in the dynamics equilibrium equations as $\mathbf{C}\,\mathbf{v}_{n+1+\alpha}$.

The evaluation of the residual force within the Newton–Raphson procedure for iteration k is

$$\mathbf{R}\big(\mathbf{x}_{n+1+\alpha}^{(k)}, \mathbf{a}_{n+1}^{(k)}\big) = \mathbf{M}\mathbf{a}_{n+1}^{(k)} + \mathbf{T}\big(\mathbf{x}_{n+1+\alpha}^{(k)}\big) - \mathbf{F}(t_{n+1+\alpha}). \tag{4.67}$$

The update of the positions at iteration $k+1$ is evaluated using Equation (4.64b) as

$$\begin{aligned}\mathbf{x}_{n+1+\alpha}^{(k+1)} &= (1+\alpha)\mathbf{x}_{n+1}^{(k+1)} - \alpha\mathbf{x}_n \\ &= (1+\alpha)(\mathbf{x}_{n+1}^{(k)} + \Delta\mathbf{x}_{n+1}) - \alpha\mathbf{x}_n \\ &= (1+\alpha)\mathbf{x}_{n+1}^{(k)} - \alpha\mathbf{x}_n + (1+\alpha)\Delta\mathbf{x}_{n+1} \\ &= \mathbf{x}_{n+1+\alpha}^{(k)} + (1+\alpha)\Delta\mathbf{x}_{n+1},\end{aligned} \tag{4.68}$$

where $\Delta\mathbf{x}_{n+1}$ is given in Equation (4.56c). Using Equations (4.68) and (4.56c), the linearization of the residual force at iteration $k+1$ emerges as

$$\begin{aligned}&\mathbf{R}\big(\mathbf{x}_{n+1}^{(k+1)}, \mathbf{a}_{n+1+\alpha}^{(k+1)}\big) \\ &\quad\approx \mathbf{R}\big(\mathbf{x}_{n+1}^{(k)}, \mathbf{a}_{n+1+\alpha}^{(k)}\big) \\ &\qquad + \mathbf{M}\Delta\mathbf{a}_{n+1} + \mathbf{K}\big(\mathbf{x}_{n+1+\alpha}^{(k)}\big)(1+\alpha)\beta\Delta t^2\Delta\mathbf{a}_{n+1} = \mathbf{0}.\end{aligned} \tag{4.69}$$

Clearly, the difference between the above equation and Equations (4.57) and (4.58) used in the Newmark scheme is the appearance of the parameter α, through the term $(1+\alpha)$. Example 4.6 shows that the α-method introduces an artificial viscosity, which is a function of the tangent matrix.

EXAMPLE 4.6: Demonstration that the α-method produces an artificial viscosity term

This is achieved by examining the change in internal force caused by the introduction of the α-method; in other words, by observing how $\mathbf{T}(\mathbf{x}_{n+1})$ changes due to the time change from t_{n+1} to $t_{n+1+\alpha}$. From Equation (4.64a) this time change is $\alpha\Delta t$, giving an approximate position change as

$$\Delta\mathbf{x}_{n+1+\alpha}^k \approx \left.\frac{d\mathbf{x}}{dt}\right|_{t_{n+1}} \alpha\Delta t = \mathbf{v}_{n+1}\alpha\Delta t.$$

The internal force $\mathbf{T}(\mathbf{x}_{n+1+\alpha})$ in Equation (4.67) can be found as

$$\mathbf{T}(\mathbf{x}_{n+1+\alpha}) \approx \mathbf{T}(\mathbf{x}_{n+1}) + D\mathbf{T}(\mathbf{x}_{n+1})[\Delta\mathbf{x}_{n+1+\alpha}^k],$$

(continued)

Example 4.6: *(cont.)*

hence

$$\begin{aligned}\mathbf{T}(\mathbf{x}_{n+1+\alpha}) - \mathbf{T}(\mathbf{x}_{n+1}) &\approx D\mathbf{T}(\mathbf{x}_{n+1})[\alpha\Delta t\mathbf{v}_{n+1}] \\ &= \alpha\Delta t\mathbf{K}(\mathbf{x}_{n+1})[\mathbf{v}_{n+1}] \\ &= \mathbf{C}(\mathbf{x}_{n+1})\,\mathbf{v}_{n+1},\end{aligned}$$

where $\mathbf{C}(\mathbf{x}_{n+1}) = \alpha\Delta t\mathbf{K}(\mathbf{x}_{n+1})$ is an artificial viscosity matrix, which is a function of the tangent matrix $\mathbf{K}(\mathbf{x}_{n+1})$.

A compact implementation of the α-method algorithm can be formulated in the following steps. Using Equations (4.52a), (4.53), (4.64b), and (4.64c) gives the predictor as

$$\mathbf{a}_{n+1}^{(0)} = \mathbf{a}_n, \tag{4.70a}$$

$$\mathbf{v}_{n+1+\alpha}^{(0)} = \mathbf{v}_n + (1+\alpha)\Delta t\mathbf{a}_n, \tag{4.70b}$$

$$\mathbf{x}_{n+1+\alpha}^{(0)} = \mathbf{x}_n + (1+\alpha)\left[\Delta t\mathbf{v}_n + \frac{\Delta t^2}{2}\mathbf{a}_n\right]. \tag{4.70c}$$

Again, combination of Equations (4.56) with Equations (4.64b) and (4.64c) gives the corrector phase as

$$\mathbf{a}_{n+1}^{(k+1)} = \mathbf{a}_{n+1}^{(k)} + \Delta\mathbf{a}_{n+1}, \tag{4.71a}$$

$$\mathbf{v}_{n+1+\alpha}^{(k+1)} = \mathbf{v}_{n+1+\alpha}^{(k)} + (1+\alpha)\Delta\mathbf{v}_{n+1}\;; \quad \Delta\mathbf{v}_{n+1} = \gamma\Delta t\Delta\mathbf{a}_{n+1}, \tag{4.71b}$$

$$\mathbf{x}_{n+1+\alpha}^{(k+1)} = \mathbf{x}_{n+1+\alpha}^{(k)} + (1+\alpha)\Delta\mathbf{x}_{n+1}\;; \quad \Delta\mathbf{x}_{n+1} = \beta\Delta t^2\Delta\mathbf{a}_{n+1}, \tag{4.71c}$$

with $\Delta\mathbf{a}_{n+1}$ evaluated from Equation (4.61) as

$$[\mathbf{K}_\alpha]\Delta\mathbf{a}_{n+1} = -\mathbf{R}\left(\mathbf{x}_{n+1}^{(k)}, \mathbf{a}_{n+1+\alpha}^{(k)}\right) \tag{4.72a}$$

$$\mathbf{K}_\alpha = \left[\mathbf{M} + (1+\alpha)\,\beta\Delta t^2\mathbf{K}\left(\mathbf{x}_{n+1+\alpha}^{(k)}\right)\right]. \tag{4.72b}$$

Once convergence is achieved for the corrector phase, it is necessary to recover the positions and velocities at time t_{n+1} using Equations (4.64b) and (4.64c) as

$$\mathbf{x}_{n+1} = \frac{1}{(1+\alpha)}(\mathbf{x}_{n+1+\alpha} - \alpha\,\mathbf{x}_n), \tag{4.73a}$$

$$\mathbf{v}_{n+1} = \frac{1}{(1+\alpha)}(\mathbf{v}_{n+1+\alpha} - \alpha\,\mathbf{v}_n). \tag{4.73b}$$

The resulting α-method/Newton–Raphson algorithm is shown in Box 4.2.

BOX 4.2: Alpha-method algorithm

- INPUT geometry, material properties, and solution parameters
- INITIALIZE $\mathbf{F} = 0$, $\mathbf{x} = \mathbf{X}$, $\mathbf{v} = \mathbf{v}_0$
- INITIALIZE $\mathbf{a} = \mathbf{a}_0$ (4.62)
- WHILE $t < T$ (time steps)
  - SET $t_{n+1+\alpha}$ (4.64a)
  - SET $\mathbf{F} = \mathbf{F}(t_{n+1+\alpha})$
  - PREDICT $\mathbf{a}_{n+1}, \mathbf{v}_{n+1+\alpha}, \mathbf{x}_{n+1+\alpha}$ (4.70)
  - FIND $\mathbf{R}$ (4.67)
  - FIND $\mathbf{K}_\alpha$ (4.72)
  - WHILE ($\|\mathbf{R}\|/\|\mathbf{F}\| >$ tolerance)
    - SOLVE $\mathbf{K}_\alpha \Delta \mathbf{a}_{n+1} = -\mathbf{R}$ (4.72)
    - CORRECT $\mathbf{a}_{n+1}, \mathbf{v}_{n+1+\alpha}, \mathbf{x}_{n+1+\alpha}$ (4.71)
    - FIND $\mathbf{R}$ (4.67)
    - FIND $\mathbf{K}_\alpha$ (4.72)
  - ENDLOOP
  - RECOVER $\mathbf{x}_{n+1}, \mathbf{v}_{n+1}$ (4.73)
- ENDLOOP

4.7 VARIATIONAL INTEGRATORS

Chapter 2, in the context of trusses, presented the concept of variational time integrators and the conservation advantages that can be derived from their use. This section will show how the same ideas can be extended to a general finite element discretization of a solid. In Section 4.5.2, the Lagrangian of a moving solid discretized by finite elements was given in Equation (4.44) as the functional $\mathcal{L}$ of the vector of nodal positions $\mathbf{x}(t)$ and velocities $\dot{\mathbf{x}}(t)$. The action integral $\mathcal{A}\big(\mathbf{x}(t)\big)$ can be formulated as the sum of the Lagrangian integrated over each time step in turn:

$$\mathcal{A}\big(\mathbf{x}(t)\big) = \sum_{n=0}^{N_t-1} \int_{t_n}^{t_{n+1}} \mathcal{L}\big(\mathbf{x}(t), \dot{\mathbf{x}}(t)\big)\, dt. \tag{4.74}$$

The integral over each time step in this equation can be approximated in terms of the values of $\mathbf{x}$ at n and $n+1$, thus defining the approximate incremental Lagrangian $L_n(\mathbf{x}_n, \mathbf{x}_{n+1})$ as

$$L_n(\mathbf{x}_n, \mathbf{x}_{n+1}) \approx \int_{t_n}^{t_{n+1}} \mathcal{L}\big(\mathbf{x}(t), \dot{\mathbf{x}}(t)\big)\, dt. \tag{4.75}$$

The action integral may now be recovered as the sum of discrete time integrated Lagrangians as

$$\mathcal{A}\big(\mathbf{x}(t)\big) \approx \mathcal{S}(\mathbf{x}_0, \mathbf{x}_1, \ldots, \mathbf{x}_{N_t}) = \sum_{n=0}^{N_t-1} L_n(\mathbf{x}_n, \mathbf{x}_{n+1}). \tag{4.76}$$

A time-discrete version of Hamilton's principle can now be postulated, namely that the stationary conditions of the time-discrete action integral $\mathcal{S}$ with respect to a virtual perturbation of geometry $\delta\mathbf{x}_n$ at each time step t_n leads to a time-discrete version of the dynamic equilibrium equations pertaining to time step t_n.

Before proceeding, it is necessary to recall the definition of the directional derivative of a function; see NL-Statics, Section 2.3, for a full exposition. In general, if a function $\boldsymbol{\mathcal{F}}$ depends upon arguments $(\mathbf{a}_0, \mathbf{a}_1, \ldots, \mathbf{a}_i, \ldots, \mathbf{a}_N)$ then the directional derivative of $\boldsymbol{\mathcal{F}}$ in the direction of $\mathbf{a}_i$ is defined as

$$D_i\boldsymbol{\mathcal{F}}(\mathbf{a}_0, \mathbf{a}_1, \ldots, \mathbf{a}_i, \ldots, \mathbf{a}_N)[\delta\mathbf{a}_i] = \frac{d}{d\epsilon}\bigg|_{\epsilon=0} \boldsymbol{\mathcal{F}}(\mathbf{a}_0, \mathbf{a}_1, \ldots, \mathbf{a}_i + \epsilon\delta\mathbf{a}_i, \ldots, \mathbf{a}_N). \tag{4.77}$$

The directional derivative can now be used to determine the virtual variation of the time-discrete action integral $\mathcal{S}$ with respect to a virtual perturbation of geometry $\delta\mathbf{x}_n$ as

$$\delta\mathcal{S} = \sum_{n=0}^{N_t-1} D_n\mathcal{S}(\mathbf{x}_0, \mathbf{x}_1, \ldots, \mathbf{x}_{N_t})[\delta\mathbf{x}_n]. \tag{4.78}$$

Before determining the nth directional derivative $D_n\mathcal{S}$, note that in Equation (4.76) it is only the time-discrete integrated Lagrangians $L(\mathbf{x}_{n-1}, \mathbf{x}_n)$ and $L(\mathbf{x}_n, \mathbf{x}_{n+1})$ that contain $\mathbf{x}_n$ and therefore will contribute to $D_n\mathcal{S}$. Consequently,

$$\begin{aligned} D_n\mathcal{S}(\mathbf{x}_0, \mathbf{x}_1, \ldots, \mathbf{x}_{N_t})[\delta\mathbf{x}_n] &= D_2L_{n-1}(\mathbf{x}_{n-1}, \mathbf{x}_n)[\delta\mathbf{x}_n] \\ &\quad + D_1L_n(\mathbf{x}_n, \mathbf{x}_{n-1})[\delta\mathbf{x}_n]. \end{aligned} \tag{4.79}$$

The stationary conditions of the approximate time-discrete action integral given by Equation (4.76) can be established by setting Equation (4.78) to zero for any virtual variation $\delta\mathbf{x}_n$, compatible with the boundary, initial, and end conditions; that is,

$$D_n\mathcal{S}(\mathbf{x}_0, \mathbf{x}_1, \ldots, \mathbf{x}_{N_t})[\delta\mathbf{x}_n] = 0; \quad n = 1, \ldots, N_t - 1, \tag{4.80}$$

or, alternatively, noting the simplification given by Equation (4.79),

$$D_2L_{n-1}(\mathbf{x}_{n-1}, \mathbf{x}_n)[\delta\mathbf{x}_n] + D_1L_n(\mathbf{x}_n, \mathbf{x}_{n-1})[\delta\mathbf{x}_n] = 0; \quad n = 1, \ldots, N_t - 1. \tag{4.81}$$

Note that, in order to comply with the initial and end conditions at time steps $n = 0$ and $n = N_t$, the corresponding variation $\delta \mathbf{x}_n$ vanishes, that is, $\delta \mathbf{x}_0 = \delta \mathbf{x}_{N_t} = \mathbf{0}$. Among all the possible expressions that the time-discrete integrated Lagrangians can adopt, two popular options are those leading to the well-known leap-frog and mid-point time integrators.

4.7.1 Leap-Frog Time Integrator

Recall the definition of the Lagrangian given by Equation (4.44) and consider the following approximation to the time integrated Lagrangian $L_n(\mathbf{x}_n, \mathbf{x}_{n+1})$:

$$L_n^{LF}(\mathbf{x}_n, \mathbf{x}_{n+1}) = \Delta t K(\mathbf{v}_{n+1/2}) - \frac{\Delta t}{2}\left[\Pi_{\text{int}}(\mathbf{x}_n) + \Pi_{\text{ext}}(\mathbf{x}_n)\right] - \frac{\Delta t}{2}\left[\Pi_{\text{int}}(\mathbf{x}_{n+1}) + \Pi_{\text{ext}}(\mathbf{x}_{n+1})\right], \tag{4.82}$$

where the intermediate velocity, $\mathbf{v}_{n+1/2}$, is explicitly defined as

$$\mathbf{v}_{n+1/2} = \frac{\mathbf{x}_{n+1} - \mathbf{x}_n}{\Delta t}. \tag{4.83}$$

This approximation implies a mid-point rectangular integration for the kinetic energy term and a trapezoidal integration for the internal and external energy

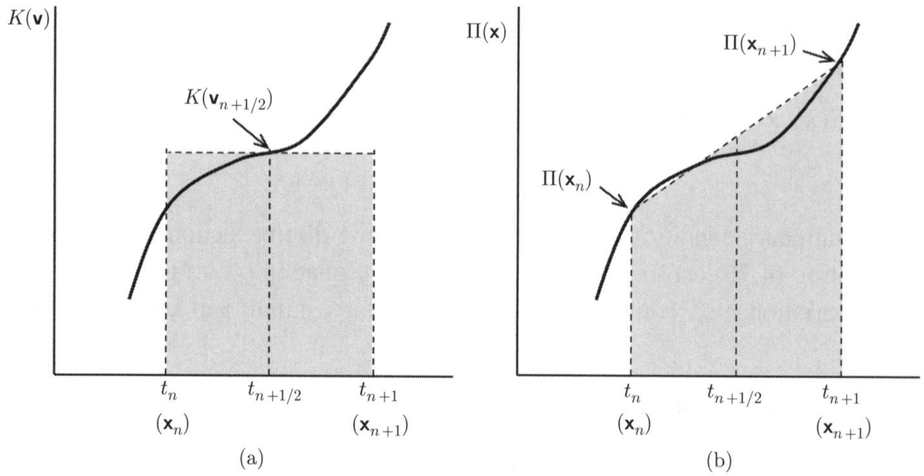

FIGURE 4.2 (a) Mid-point integration of kinetic energy; (b) Trapezoidal integration of internal and external energy.

terms;** see Figure 4.2. A similar expression to Equation (4.82) can be derived for the time integrated Lagrangian $L_n^{LF}(\mathbf{x}_{n-1}, \mathbf{x}_n)$.

It is now possible to obtain the two directional derivatives necessary to form the nth time-discrete expression of Hamilton's principle given by Equation (4.81). Using the kinetic energy Equation (4.37) and the external energy term from Equation (4.43), together with the internal energy derivation given in Example 4.7, the directional derivatives D_2 and D_1 are

$$D_2 L_{n-1}^{LF}(\mathbf{x}_{n-1}, \mathbf{x}_n)[\delta\mathbf{x}_n] = \delta\mathbf{x}_n^T \left[\mathbf{M}\left(\frac{\mathbf{x}_n - \mathbf{x}_{n-1}}{\Delta t}\right) - \frac{\Delta t}{2}\mathbf{T}(\mathbf{x}_n) + \frac{\Delta t}{2}\mathbf{F}(t_n)\right] \tag{4.84a}$$

and

$$D_1 L_n^{LF}(\mathbf{x}_n, \mathbf{x}_{n+1})[\delta\mathbf{x}_n] = \delta\mathbf{x}_n^T \left[-\mathbf{M}\left(\frac{\mathbf{x}_{n+1} - \mathbf{x}_n}{\Delta t}\right) - \frac{\Delta t}{2}\mathbf{T}(\mathbf{x}_n) + \frac{\Delta t}{2}\mathbf{F}(t_n)\right]. \tag{4.84b}$$

Adding the above two equations and noting that the virtual change in position $\delta\mathbf{x}_n$ is arbitrary yields

EXAMPLE 4.7: Demonstrate that $D\Pi_{\text{int}}(\mathbf{x})[\delta\mathbf{x}] = \delta\mathbf{x}^T\mathbf{T}(\mathbf{x})$

The internal energy $\Pi_{\text{int}}(\mathbf{x})$ can be formulated as the sum of the internal energy of all elements (m) in the mesh, which in turn is the sum of strain energies $\Psi(\boldsymbol{F})$:

$$\Pi_{\text{int}}(\mathbf{x}) = \sum_{e=1}^{M} \Pi_{\text{int}}^{(e)}(\mathbf{x}) = \sum_{e=1}^{M} \int_{V^{(e)}} \Psi(\boldsymbol{F})\, dV,$$

where $\boldsymbol{F}$ can be found in terms of the finite element discretization; see Equation (4.5). Observe that the directional derivative of the internal energy can be expressed as the sum over all nodes (N) in a mesh:

$$D\Pi_{\text{int}}(\mathbf{x})[\delta\mathbf{x}] = \sum_{a=1}^{N} D\Pi_{\text{int}}(\mathbf{x})[\delta\boldsymbol{x}_a]. \tag{a}$$

(continued)

** The next section will show that if mid-point integration is used for all terms the eventual outcome is the mid-point scheme.

Example 4.7: *(cont.)*

The contribution to the directional derivative from all elements connecting to node (a) can be written as

$$D\Pi_{\text{int}}(\mathbf{x})[\delta \boldsymbol{x}_a] = \sum_{\substack{e=1 \\ e \ni a}}^{M_a} \int_{V^{(e)}} D\Psi(\boldsymbol{F})[\delta \boldsymbol{x}_a]\, dV. \tag{b}$$

The integrand above can be expressed in terms of the first Piola–Kirchhoff stress tensor $\boldsymbol{P}$ as

$$\begin{aligned} D\Psi(\boldsymbol{F})[\delta \boldsymbol{x}_a] &= \frac{\partial \Psi}{\partial \boldsymbol{F}} : D\boldsymbol{F}[\delta \boldsymbol{x}_a] \\ &= \boldsymbol{P} : D\boldsymbol{F}[\delta \boldsymbol{x}_a], \end{aligned}$$

where, using Equation (4.5), it can be shown that

$$D\boldsymbol{F}[\delta \boldsymbol{x}_a] = \delta \boldsymbol{x}_a \otimes \boldsymbol{\nabla}_0 N_a,$$

which can be substituted back to give

$$D\Psi(\boldsymbol{F})[\delta \boldsymbol{x}_a] = \delta \boldsymbol{x}_a \cdot (\boldsymbol{P} \boldsymbol{\nabla}_0 N_a). \tag{c}$$

Substituting Equation (c) into (b) yields

$$\begin{aligned} D\Pi_{\text{int}}(\mathbf{x})[\delta \boldsymbol{x}_a] &= \sum_{\substack{e=1 \\ e \ni a}}^{M_a} \int_{V^{(e)}} \delta \boldsymbol{x}_a \cdot (\boldsymbol{P} \boldsymbol{\nabla}_0 N_a)\, dV \\ &= \delta \boldsymbol{x}_a \cdot \sum_{\substack{e=1 \\ e \ni a}}^{M_a} \boldsymbol{T}_a^{(e)} = \delta \boldsymbol{x}_a \cdot \boldsymbol{T}_a, \end{aligned}$$

where

$$\boldsymbol{T}_a^{(e)} = \int_{V^{(e)}} \boldsymbol{P} \boldsymbol{\nabla}_0 N_a\, dV \tag{d}$$

and $\boldsymbol{T}_a^{(e)}$ is the element contribution to the equivalent nodal force vector associated with node (a); it represents the total Lagrangian counterpart to the updated Lagrangian expression given by Equation (4.12a,b)$_a$. Finally, Equation (a) can now be written as

$$D\Pi_{\text{int}}(\mathbf{x})[\delta \mathbf{x}] = \sum_{a=1}^{N} \delta \boldsymbol{x}_a \cdot \boldsymbol{T}_a = \delta \mathbf{x}^T \mathbf{T}.$$

$$\mathbf{M}\left(\frac{\mathbf{x}_{n+1} - 2\mathbf{x}_n + \mathbf{x}_{n-1}}{\Delta t^2}\right) + \mathbf{T}(\mathbf{x}_n) = \mathbf{F}(t_n). \tag{4.85}$$

It would be possible to solve Equation (4.85) directly for $\mathbf{x}_{n+1}$, or alternatively rewrite this equation as

$$\mathbf{M}\mathbf{a}_n + \mathbf{T}(\mathbf{x}_n) = \mathbf{F}(t_n); \quad \mathbf{a}_n = \left(\frac{\mathbf{x}_{n+1} - 2\mathbf{x}_n + \mathbf{x}_{n-1}}{\Delta t^2}\right), \tag{4.86a,b}$$

where $\mathbf{a}_n$ is the central difference approximation of the acceleration at time t_n. Equation $(4.86\text{a,b})_a$ enables $\mathbf{a}_n$ to be determined directly and $\mathbf{x}_{n+1}$ to be found subsequently from Equation $(4.86\text{a,b})_b$. Observe that this matches the Newmark scheme (see Section 4.6) for the case when $\beta = 0$ and $\gamma = 1/2$.

To move toward the leap-frog implementation of this time integration scheme, the dynamic equilibrium equation is rewritten in terms of the half-time-step velocities as

$$\mathbf{M}\mathbf{a}_n + \mathbf{T}(\mathbf{x}_n) = \mathbf{F}(t_n); \quad \mathbf{a}_n = \frac{\mathbf{v}_{n+1/2} - \mathbf{v}_{n-1/2}}{\Delta t}, \tag{4.87a,b}$$

where Equations (4.83), for $\mathbf{v}_{n+1/2}$ and by implication $\mathbf{v}_{n-1/2}$, and $(4.86\text{a,b})_b$ have been employed. The leap-frog scheme now clearly emerges as the sequence of three consecutive steps: first, solve Equations $(4.86\text{a,b})_a$ for $\mathbf{a}_n$ followed by the half-time-step velocity and position updates as

$$\mathbf{v}_{n+1/2} = \mathbf{v}_{n-1/2} + \Delta t \mathbf{a}_n, \tag{4.88a}$$

$$\mathbf{x}_{n+1} \quad = \mathbf{x}_n + \Delta t \mathbf{v}_{n+1/2}. \tag{4.88b}$$

Observe in Equation $(4.87\text{a,b})_a$ that, because $\mathbf{T}(\mathbf{x}_n)$ is known prior to the computation of $\mathbf{a}_n$, the leap-frog scheme is explicit. Invariably, for computational efficiency, the mass matrix $\mathbf{M}$ in this equation is substituted by the lumped mass matrix $\mathbf{M}^L$ so that $\mathbf{a}_n$ can be obtained without recourse to solving a set of linear equations.

The leap-frog implementation presented above involves the evaluation of the velocities at the half time step. However, it is useful to be able to find the velocities at the full time step when outputting information such as energy. To enable full-time-step velocities to be found, the update of the velocity can be split into two sequential stages, namely

$$\mathbf{v}_{n+1/2} = \mathbf{v}_n + \frac{\Delta t}{2}\mathbf{a}_n\,; \quad \mathbf{v}_{n+1} = \mathbf{v}_{n+1/2} + \frac{\Delta t}{2}\mathbf{a}_{n+1}. \tag{4.89a,b}$$

A summary of the leap-frog algorithm is shown in Box 4.3.

BOX 4.3: Leap-frog algorithm

- INPUT geometry, material properties, and solution parameters
- INITIALIZE $\mathbf{F} = 0$, $\mathbf{x} = \mathbf{X}$, $\mathbf{v} = \mathbf{v}_0$
- INITIALIZE $\mathbf{a} = \mathbf{a}_0$ (4.62)
- WHILE $t < T$ (time steps)
  - SET $\mathbf{F} = \mathbf{F}(t)$
  - UPDATE $\mathbf{v}_{n+1/2}$ (4.88a)
  - UPDATE $\mathbf{x}_{n+1}$ (4.88b)
  - SOLVE $\mathbf{a}_{n+1}$ (4.87a,b)$_a$
  - UPDATE $\mathbf{v}_{n+1}$ (4.89a,b)$_b$
- ENDLOOP

4.7.2 Mid-Point Time Integrator

The mid-point time integrator can be derived using the following discrete time integrated Lagrangian:

$$L_n(\mathbf{x}_n, \mathbf{x}_{n+1}) = \Delta t K(\mathbf{v}_{n+1/2}) - \Delta t\big[\Pi_{\text{int}}(\mathbf{x}_{n+1/2}) + \Pi_{\text{ext}}(\mathbf{x}_{n+1/2})\big], \quad (4.90a)$$

where, by definition, the position $\mathbf{x}_{n+1/2}$ and velocity $\mathbf{v}_{n+1/2}$ (see Equation (4.83)) are

$$\mathbf{x}_{n+1/2} = \frac{1}{2}(\mathbf{x}_n + \mathbf{x}_{n+1})\ ; \quad \mathbf{v}_{n+1/2} = \frac{\mathbf{x}_{n+1} - \mathbf{x}_n}{\Delta t}. \quad (4.90b,c)$$

The only difference between the integrated Lagrangian given by Equation (4.90a) and its counterpart for the leap-frog algorithm, Equation (4.83), is in the integration of the internal and external energy contributions. The trapezoidal rule was used in the former whereas the mid-point rule is used in the latter. As mentioned in Chapter 2, differences between these schemes become apparent when dealing with nonquadratic energetic contributions.

The directional derivatives comprising the nth time-discrete version of Hamilton's equations can be obtained as

$$D_2 L_{n-1}^{MP}(\mathbf{x}_{n-1}, \mathbf{x}_n)[\delta\mathbf{x}_n] = \delta\mathbf{x}_n^T\left[\mathbf{M}\left(\frac{\mathbf{x}_n - \mathbf{x}_{n-1}}{\Delta t}\right)\right]$$

$$- \delta\mathbf{x}_n^T\left[\frac{\Delta t}{2}\mathbf{T}(\mathbf{x}_{n-1/2}) - \frac{\Delta t}{2}\mathbf{F}(t_{n-1/2})\right] \quad (4.91a)$$

and

$$D_1 L_n^{MP}(\mathbf{x}_n, \mathbf{x}_{n+1})[\delta\mathbf{x}_n] = \delta\mathbf{x}_n^T\left[-\mathbf{M}\left(\frac{\mathbf{x}_{n+1}-\mathbf{x}_n}{\Delta t}\right)\right]$$

$$-\delta\mathbf{x}_n^T\left[\frac{\Delta t}{2}\mathbf{T}(\mathbf{x}_{n+1/2}) - \frac{\Delta t}{2}\mathbf{F}(t_{n+1/2})\right]. \quad (4.91b)$$

Adding the above two expressions, noting the arbitrariness of $\delta\mathbf{x}_n$, and rearranging yields

$$\mathbf{M}\left(\frac{\mathbf{x}_{n+1} - 2\mathbf{x}_n + \mathbf{x}_{n-1}}{\Delta t^2}\right) + \frac{1}{2}\left[\mathbf{T}(\mathbf{x}_{n+1/2}) - \mathbf{F}(t_{n+1/2})\right]$$

$$+\frac{1}{2}\left[\mathbf{T}(\mathbf{x}_{n-1/2}) - \mathbf{F}(t_{n-1/2})\right] = \mathbf{0}. \quad (4.92)$$

This equation is identical to Equation (2.58) in Chapter 2. Following a similar procedure to that given in Chapter 2 (see Equations (2.59) to (2.64)), the mid-point time integrator can be summarized by the following dynamic equilibrium equation:

$$\mathbf{M}\mathbf{a}_{n+1/2} + \mathbf{T}(\mathbf{x}_{n+1/2}) - \mathbf{F}(t_{n+1/2}) = \mathbf{0}, \quad (4.93)$$

complemented by the update sequence for veleocities and positions as

$$\mathbf{v}_{n+1} = \mathbf{v}_n + \Delta t\mathbf{a}_{n+1/2}, \quad (4.94a)$$

$$\mathbf{x}_{n+1} = \mathbf{x}_n + \frac{\Delta t}{2}(\mathbf{v}_n + \mathbf{v}_{n+1}), \quad (4.94b)$$

$$\mathbf{x}_{n+1/2} = \frac{1}{2}(\mathbf{x}_n + \mathbf{x}_{n+1}). \quad (4.94c)$$

In contrast to the explicit leap-frog scheme, the mid-point scheme is implicit. This can be observed by the fact that the $\mathbf{x}_{n+1/2}$ term in Equation (4.93) is a function of the unknown term $\mathbf{x}_{n+1}$ given by Equation (4.90b,c)$_b$. As previously discussed, an implicit time integration scheme requires solution through the use of a Newton–Raphson iterative predictor–corrector procedure. Such a procedure has already been described in Chapter 2; see Equations (2.67) to (2.73). Here, a more compact algorithm is presented by simply combining some of the equations from Chapter 2. Consequently, the predictor step is defined by

$$\mathbf{a}_{n+1/2}^{(0)} = \mathbf{a}_{n-1/2}, \quad (4.95a)$$

$$\mathbf{v}_{n+1}^{(0)} = \mathbf{v}_n + \Delta t\, \mathbf{a}_{n+1/2}^{(0)}, \quad (4.95b)$$

$$\mathbf{x}^{(0)}_{n+1/2} = \mathbf{x}_n + \frac{\Delta t}{2}\mathbf{v}_n + \frac{\Delta t^2}{4}\mathbf{a}^{(0)}_{n+1/2}, \tag{4.95c}$$

where $\mathbf{a}_{n-1/2}$ is the converged acceleration obtained from the previous time step $t_{n-1/2}$. To initiate the corrector iterations, a residual force is evaluated using the above predicted values (or later iterative values as required) as

$$\mathbf{R}(\mathbf{x}^{(k)}_{n+1/2}, \mathbf{a}^{(k)}_{n+1/2}) \approx \mathbf{M}\mathbf{a}^{(k)}_{n+1/2} + \mathbf{T}(\mathbf{x}^{(k)}_{n+1/2}) - \mathbf{F}(t_{n+1/2}), \tag{4.96}$$

where k is the iteration number. In order to construct the Newton–Raphson iterative scheme, a suitable linearization of the above residual for iteration $k+1$ is formulated as

$$\mathbf{R}(\mathbf{x}^{(k+1)}_{n+1/2}, \mathbf{a}^{(k+1)}_{n+1/2}) \approx \mathbf{R}(\mathbf{x}^{(k)}_{n+1/2}, \mathbf{a}^{(k)}_{n+1/2}) + \mathbf{M}\Delta\mathbf{a}_{n+1/2} + D\mathbf{T}(\mathbf{x}^{(k)}_{n+1/2})[\Delta\mathbf{x}_{n+1/2}] = \mathbf{0}, \tag{4.97}$$

together with the following sequence of corrector steps:

$$\mathbf{a}^{(k+1)}_{n+1/2} = \mathbf{a}^{(k)}_{n+1/2} + \Delta\mathbf{a}_{n+1/2}, \tag{4.98a}$$

$$\mathbf{v}^{(k+1)}_{n+1} = \mathbf{v}^{(k)}_{n+1} + \Delta t\Delta\mathbf{a}_{n+1/2}, \tag{4.98b}$$

$$\mathbf{x}^{(k+1)}_{n+1/2} = \mathbf{x}^{(k)}_{n+1/2} + \frac{\Delta t^2}{4}\Delta\mathbf{a}_{n+1/2}. \tag{4.98c}$$

From Equation $(4.98)_c$ it is evident that the iteration change $\Delta\mathbf{x}_{n+1/2}$ in Equation (4.97) is equal to $(\Delta t^2\Delta\mathbf{a}_{n+1/2})/4$. Consequently, Equation (4.97) can be rewritten in terms of the single unknown $\Delta\mathbf{a}_{n+1/2}$ as

$$[\mathbf{K}_m]\Delta\mathbf{a}_{n+1/2} = -\mathbf{R}\left(\mathbf{x}^{(k)}_{n+1/2}, \mathbf{a}^{(k)}_{n+1/2}\right); \tag{4.99a}$$

$$\mathbf{K}_m = [\mathbf{M} + \frac{\Delta t^2}{4}\mathbf{K}\left(\mathbf{x}^{(k)}_{n+1/2}\right)]. \tag{4.99b}$$

Upon convergence, the position $\mathbf{x}_{n+1}$ at the full time step t_{n+1} can be recovered, see Equation $(4.90\text{b,c})_b$, as

$$\mathbf{x}_{n+1} = 2\mathbf{x}_{n+1/2} - \mathbf{x}_n. \tag{4.100}$$

It is interesting to note that no special procedure is required for the initialization of the mid-point algorithm. Once $\mathbf{x}_0$ and $\mathbf{v}_0$ are known, the predictor–corrector strategy can be used to obtain $\mathbf{a}_{1/2}$, $\mathbf{v}_1$, and $\mathbf{x}_{1/2}$ (and $\mathbf{x}_1$ via (4.100) if required). A summary of the mid-point algorithm is shown in Box 4.4.

BOX 4.4: Mid-point algorithm

- INPUT geometry, material properties, and solution parameters
- INITIALIZE $\mathbf{F} = 0$, $\mathbf{x} = \mathbf{X}$, $\mathbf{v} = \mathbf{v}_0$
- WHILE $t < T$ (time steps)
  - SET $\mathbf{F} = \mathbf{F}(t_{n+1/2})$
  - PREDICT $\mathbf{a}_{n+1/2}, \mathbf{v}_{n+1}, \mathbf{x}_{n+1/2}$ (4.95)
  - FIND $\mathbf{R}$ (4.96)
  - FIND $\mathbf{K}_m$ (4.99a,b)
  - DO WHILE ($\|\mathbf{R}\|/\|\mathbf{F}\| >$ tolerance)
    - SOLVE $\mathbf{K}_m \Delta \mathbf{a}_{n+1/2} = -\mathbf{R}$ (4.99a,b)
    - CORRECT $\mathbf{a}_{n+1/2}, \mathbf{v}_{n+1}, \mathbf{x}_{n+1/2}$ (4.98)
    - FIND $\mathbf{R}$ (4.96)
    - FIND $\mathbf{K}_m$ (4.99a,b)
  - ENDLOOP
  - RECOVER $\mathbf{x}_{n+1}$ (4.100)
- ENDLOOP

4.8 GLOBAL CONSERVATION LAWS

The conservation of global quantities such as momentum and energy in solid dynamics has been discussed in Chapter 2 in the context of trusses and in Chapter 3 in the context of a continuum solid. This section will revisit these issues from the point of view of a solid discretized spatially by finite elements with the motion followed in time both continuously and discretely via a time-stepping scheme. It will be shown that the finite element discretization in space retains the global conservation properties of the continuum, namely conservation of linear and angular momentum and energy. However, the introduction of time discretization schemes can lead to some or all of these variables not being conserved exactly. In particular, it will be shown that the leap-frog scheme and the mid-point rule conserve both momentum quantities but not the energy. Failure to conserve energy can have negative consequences in relation to the long term stability of the time integration of highly stiff, that is, very rigid, systems. Several methodologies have been developed by various authors to ensure conservation of both momentum and energy. The resulting techniques are known as *energy-momentum time integrators*, and an example of one such technique will be presented below in the context of hyperelasticity.

4.8.1 Conservation of Linear Momentum

The linear momentum of a solid discretized by finite elements can be computed by summing up the linear momentum from each element as

$$\boldsymbol{L} = \sum_{e=1}^{M} \boldsymbol{L}^{(e)} = \sum_{e=1}^{M} \int_{V^{(e)}} \rho_0 \boldsymbol{v}\, dV$$
$$= \sum_{e=1}^{M} \int_{V^{(e)}} \rho_0 \left(\sum_{a=1}^{n} N_a \boldsymbol{v}_a \right) dV, \tag{4.101}$$

where M is the total number of elements in a mesh and n is the number of nodes per element. By swapping the summation terms and placing the nodal velocity $\boldsymbol{v}_a$ outside the volume integral, the total linear momentum can be reformulated in terms of the lumped nodal masses m^L_{aa} or consistent mass coefficients m_{ab} as

$$\boldsymbol{L} = \sum_{a=1}^{N} \left(\sum_{\substack{e=1 \\ e \ni a}}^{M_a} \int_{V^{(e)}} \rho_0 N_a \, dV \right) \boldsymbol{v}_a$$
$$= \sum_{a=1}^{N} m^L_{aa} \boldsymbol{v}_a = \sum_{a,b=1}^{N} m_{ab}\, \boldsymbol{v}_b$$
$$= \sum_{a=1}^{N} \boldsymbol{M}^L_a \boldsymbol{v}_a = \sum_{a,b=1}^{N} \boldsymbol{M}_{ab}\, \boldsymbol{v}_b, \tag{4.102}$$

where M_a is the number of elements contaning node a and N is the total number of nodes in the mesh. In the above equation, use of the consistent (m_{ab}) and lumped (m^L_{aa}) mass coefficients and their relationship given in Equation (4.24a,b)$_b$ has been employed together with the symmetry property $m_{ab} = m_{ba}$.

The rate of change of the global linear momentum with respect to time can be obtained as

$$\frac{d\boldsymbol{L}}{dt} = \sum_{a,b=1}^{N} \boldsymbol{M}_{ab} \frac{\boldsymbol{v}_b}{dt} = \sum_{a,b=1}^{N} \boldsymbol{M}_{ab}\, \boldsymbol{a}_b = \sum_{a=1}^{N} \left(\sum_{b=1}^{N} \boldsymbol{M}_{ab}\, \boldsymbol{a}_b \right). \tag{4.103}$$

By making use of the dynamic equilibrium Equation (4.27) it is now possible to replace the term in brackets in the above equation to give

$$\frac{d\boldsymbol{L}}{dt} = \sum_{a=1}^{N} (\boldsymbol{F}_a - \boldsymbol{T}_a) = \sum_{a=1}^{N} \boldsymbol{F}_a - \sum_{a=1}^{N} \boldsymbol{T}_a. \tag{4.104}$$

Note that this expression is also valid in the case of a lumped mass matrix being used by virtue of the last line in Equation (4.102). The last term in the above equation involving the sum of the internal forces $\boldsymbol{T}_a$ can be shown to be equal to zero.

In order to show this, recall the definition of $\boldsymbol{T}_a$ from Equation (4.12a,b)$_a$ as

$$
\begin{aligned}
\sum_{a=1}^{N} \boldsymbol{T}_a = \sum_{a=1}^{N} \sum_{\substack{e=1 \\ e \ni a}}^{M_a} \boldsymbol{T}_a^{(e)} &= \sum_{a=1}^{N} \sum_{\substack{e=1 \\ e \ni a}}^{M_a} \int_{v^{(e)}} \boldsymbol{\sigma} \boldsymbol{\nabla} N_a \, dv \\
&= \sum_{e=1}^{m} \int_{v^{(e)}} \boldsymbol{\sigma} \left(\sum_{a=1}^{n} \boldsymbol{\nabla} N_a \right) dv \\
&= \sum_{e=1}^{m} \int_{v^{(e)}} \boldsymbol{\sigma} \boldsymbol{\nabla} \left(\sum_{a=1}^{n} N_a \right) dv \\
&= \sum_{e=1}^{m} \int_{v^{(e)}} \boldsymbol{\sigma} \boldsymbol{\nabla} (1) \, dv \\
&= \boldsymbol{0},
\end{aligned}
\tag{4.105}
$$

where the partition of unity property, that is, $\sum_a N_a = 1$, has been used. Finally, from Equation (4.104), the rate of change of linear momentum is given by the sum of external forces; in other words, global linear momentum is conserved in the absence of an external resultant force.

The above derivation relates to a finite element discretization but with time assumed to be continuous. However, a question remains: is the same conclusion valid when time itself is discretized, as in the case of the various time-stepping schemes introduced previously? The answer is affirmative for some time-stepping schemes, and this will be demonstrated for the case of the leap-frog scheme as follows. The change in linear momentum occurring over a time step $\Delta t = t_{n+1/2} - t_{n-1/2}$ is given by

$$
\begin{aligned}
\boldsymbol{L}_{n+1/2} - \boldsymbol{L}_{n-1/2} &= \sum_{a,b=1}^{N} \boldsymbol{M}_{ab} \, \boldsymbol{v}_{b,n+1/2} - \sum_{ab} \boldsymbol{M}_{ab} \, \boldsymbol{v}_{b,n-1/2} \\
&= \sum_{a,b=1}^{N} \Delta t \boldsymbol{M}_{ab} \, \boldsymbol{a}_{b,n} \\
&= \Delta t \left[\sum_{a=1}^{N} \left(\boldsymbol{F}_a(t_n) - \boldsymbol{T}_a(\mathbf{x}_n) \right) \right] \\
&= \Delta t \sum_{a=1}^{N} \boldsymbol{F}_a(t_n),
\end{aligned}
\tag{4.106}
$$

where again it is noted (see Equation (4.105)) that the sum of the internal forces $\boldsymbol{T}_a$ is equal to zero. Clearly, when the resultant force vanishes, $\sum_a^N \boldsymbol{F}_a(t_n) = \boldsymbol{0}$ (or there are no external forces), the global linear momentum is preserved over a typical time step Δt.

EXAMPLE 4.8: Demonstration that the leap-frog scheme preserves global linear momentum over a time step $\Delta t = t_{n+1} - t_n$

$$\boldsymbol{L}_{n+1} - \boldsymbol{L}_n = \sum_{a,b=1}^{N} M_{ab}\boldsymbol{v}_{b,n+1} - \sum_{a,b=1}^{N} M_{ab}\boldsymbol{v}_{b,n}$$

$$= \sum_{a,b=1}^{N} M_{ab}(\boldsymbol{v}_{b,n+1} - \boldsymbol{v}_{b,n}). \qquad \text{(a)}$$

Note that

$$\boldsymbol{v}_{b,n+1} = \boldsymbol{v}_{b,n+1/2} + \frac{\Delta t}{2}\boldsymbol{a}_{b,n+1},$$

$$\boldsymbol{v}_{b,n} \quad = \boldsymbol{v}_{b,n-1/2} + \frac{\Delta t}{2}\boldsymbol{a}_{b,n}.$$

Subtracting the above equations yields

$$\boldsymbol{v}_{b,n+1} - \boldsymbol{v}_{b,n} = (\boldsymbol{v}_{b,n+1/2} - \boldsymbol{v}_{b,n-1/2}) + \frac{\Delta t}{2}(\boldsymbol{a}_{b,n+1} - \boldsymbol{a}_{b,n})$$

$$= \Delta t\boldsymbol{a}_{b,n} + \frac{\Delta t}{2}(\boldsymbol{a}_{b,n+1} - \boldsymbol{a}_{b,n})$$

$$= \frac{\Delta t}{2}(\boldsymbol{a}_{b,n+1} + \boldsymbol{a}_{b,n}),$$

which upon substitution into Equation (a), and noting Equation (4.105), gives the change in linear momentum over the time step n to $n + 1$ as

$$\boldsymbol{L}_{n+1} - \boldsymbol{L}_n = \sum_{a,b=1}^{N} M_{ab}\frac{\Delta t}{2}(\boldsymbol{a}_{b,n+1} + \boldsymbol{a}_{b,n})$$

$$= \frac{\Delta t}{2}\left[\sum_{a=1}^{N} (\boldsymbol{F}_a(t_{n+1}) + \boldsymbol{F}_a(t_n))\right],$$

which obviously vanishes in the absence of external forces.

4.8.2 Conservation of Angular Momentum

The angular momentum can be obtained by integrating the angular momenta $d\boldsymbol{A} = \boldsymbol{x} \times \rho_R \boldsymbol{v}\, dV$ over the reference volume of the solid. Introducing the disretization of $\boldsymbol{x}$ and $\boldsymbol{v}$ and summing over all elements m gives the total angular momentum as

$$
\begin{aligned}
\boldsymbol{A} = \sum_{e=1}^{m} \boldsymbol{A}^{(e)} &= \sum_{e=1}^{m} \int_{V^{(e)}} \left(\sum_{a=1}^{n} N_a \boldsymbol{x}_a \right) \times \rho_R \left(\sum_{b=1}^{n} N_a \boldsymbol{v}_b \right) dV \\
&= \sum_{a,b=1}^{N} \left(\sum_{\substack{e=1 \\ e \ni a,b}}^{M_{a,b}} \int_{V^{(e)}} \rho_R N_a N_b \, dV \right) \boldsymbol{x}_a \times \boldsymbol{v}_b \\
&= \sum_{a,b=1}^{N} m_{ab} \, \boldsymbol{x}_a \times \boldsymbol{v}_b \\
&= \sum_{a,b=1}^{N} \boldsymbol{M}_{ab} (\boldsymbol{x}_a \times \boldsymbol{v}_b),
\end{aligned} \tag{4.107}
$$

where $\boldsymbol{M}_{a,b} = m_{a,b} \boldsymbol{I}$ is the consistent mass matrix connecting nodes a and b; see Equation (4.19). The rate of change of angular momentum can be formulated as

$$
\frac{d\boldsymbol{A}}{dt} = \sum_{a,b=1}^{N} \boldsymbol{M}_{ab} (\boldsymbol{v}_a \times \boldsymbol{v}_b) + \sum_{a,b=1}^{N} \boldsymbol{M}_{ab} (\boldsymbol{x}_a \times \boldsymbol{a}_b). \tag{4.108}
$$

The first term on the right-hand side of this equation vanishes given that the consistent mass matrix coefficients are symmetric ($\boldsymbol{M}_{ab} = \boldsymbol{M}_{ba}$), the cross product term in the first parenthesis is skew symmetric ($\boldsymbol{v}_a \times \boldsymbol{v}_b = -\boldsymbol{v}_b \times \boldsymbol{v}_a$), and given the collinearity property ($\boldsymbol{v}_a \times \boldsymbol{v}_a = \boldsymbol{0}$). Consequently, introducing the dynamic equilibrium Equation (4.27) enables Equation (4.108) to be reformulated as

$$
\begin{aligned}
\frac{d\boldsymbol{A}}{dt} &= \sum_{a=1}^{N} \boldsymbol{x}_a \times \left(\sum_{b=1}^{N} m_{ab} \, \boldsymbol{a}_b \right) \\
&= \sum_{a=1}^{N} \boldsymbol{x}_a \times (\boldsymbol{F}_a - \boldsymbol{T}_a) \\
&= \sum_{a=1}^{N} \boldsymbol{x}_a \times \boldsymbol{F}_a - \sum_{a=1}^{N} \boldsymbol{x}_a \times \boldsymbol{T}_a.
\end{aligned} \tag{4.109}
$$

The first term in Equation (4.109) represents the external moment generated by the external forces $\boldsymbol{F}_a$, whereas the second term is the moment generated by the internal forces $\boldsymbol{T}_a$. It will now be demonstrated that this latter internal moment is zero. Recalling the definition of the equivalent nodal forces $\boldsymbol{T}_a$ given by Equation (4.12a,b)$_a$, it is possible to show that

$$\sum_{a=1}^{N} \boldsymbol{x}_a \times \boldsymbol{T}_a = \sum_{a=1}^{N} \boldsymbol{x}_a \times \left(\sum_{\substack{e=1 \\ e \ni a}}^{M_a} \boldsymbol{T}_a^{(e)} \right)$$
$$= \sum_{e=1}^{m} \int_{v^{(e)}} \sum_{a=1}^{n} \boldsymbol{x}_a \times (\boldsymbol{\sigma} \boldsymbol{\nabla} N_a) \, dv$$
$$= -\sum_{e=1}^{m} \int_{v^{(e)}} \boldsymbol{\mathcal{E}} : \left[\sum_{a=1}^{n} \boldsymbol{x}_a \otimes \boldsymbol{\sigma} \boldsymbol{\nabla} N_a \right] dv$$
$$= -\sum_{e=1}^{m} \int_{v^{(e)}} \boldsymbol{\mathcal{E}} : \left[\Big(\sum_{a=1}^{n} \boldsymbol{x}_a \otimes \boldsymbol{\nabla} N_a \Big) \boldsymbol{\sigma} \right] dv, \tag{4.110}$$

where $\boldsymbol{\mathcal{E}}$ is the third-order alternating tensor, that is, $\mathcal{E}_{ijk} = \boldsymbol{e}_i \cdot (\boldsymbol{e}_j \times \boldsymbol{e}_k)$, which allows a convenient way of re-expressing the cross product operation. The summation term above involving the tensor product is simply the spatial gradient of the discretized geometry, that is,

$$\sum_{a=1}^{n} \boldsymbol{x}_a \otimes \boldsymbol{\nabla} N_a = \boldsymbol{\nabla} \left(\sum_{a=1}^{n} N_a \boldsymbol{x}_a \right)$$
$$= \boldsymbol{\nabla} \boldsymbol{x} = \boldsymbol{I}. \tag{4.111}$$

This identity, combined with the symmetry of the stress tensor, leads to the desired outcome of a vanishing internal moment as

$$\sum_{a=1}^{N} \boldsymbol{x}_a \times \boldsymbol{T}_a = -\sum_{e=1}^{m} \int_{v^{(e)}} \boldsymbol{\mathcal{E}} : \left[\left(\sum_{a=1}^{n} \boldsymbol{x}_a \otimes \boldsymbol{\nabla} N_a \right) \boldsymbol{\sigma} \right] dv$$
$$= -\sum_{e=1}^{m} \int_{v^{(e)}} \boldsymbol{\mathcal{E}} : \boldsymbol{\sigma} \, dv$$
$$= \boldsymbol{0}. \tag{4.112}$$

Consequently, the rate of change of the angular momentum in Equation (4.109) is

$$\frac{d\boldsymbol{A}}{dt} = \sum_{a=1}^{N} \boldsymbol{x}_a \times \boldsymbol{F}_a, \tag{4.113}$$

from which it is now evident that if either $\boldsymbol{F}_a$ is zero or the external forces do not generate any moments, the angular momentum is globally conserved.

Again it is necessary to consider whether the same conclusion is valid when time itself is discretized. It is possible to demonstrate that any variational time integration scheme conserves global angular momentum. Unfortunately, the same cannot be guaranteed for other time integrators. In Chapter 2 the conservation of angular momentum was shown for the mid-point scheme; here it will be demonstrated for the leap-frog scheme by first considering the change in angular momentum from time step t_n to t_{n+1} as

$$\begin{aligned}\boldsymbol{A}_{n+1} - \boldsymbol{A}_n &= \sum_{a,b=1}^{N} \boldsymbol{M}_{ab}(\boldsymbol{x}_{a,n+1} \times \boldsymbol{v}_{b,n+1}) - \sum_{a,b=1}^{N} \boldsymbol{M}_{ab}(\boldsymbol{x}_{a,n} \times \boldsymbol{v}_{b,n}) \\ &= \sum_{a,b=1}^{N} \boldsymbol{M}_{ab}(\boldsymbol{x}_{a,n+1} \times \boldsymbol{v}_{b,n+1} - \boldsymbol{x}_{a,n} \times \boldsymbol{v}_{b,n}). \end{aligned} \tag{4.114}$$

The leap-frog update expressions given by Equation (4.89a,b)$_b$ from a half to a full time step are written for each node b as

$$\boldsymbol{v}_{b,n+1} = \boldsymbol{v}_{b,n+1/2} + \frac{\Delta t}{2} \boldsymbol{a}_{b,n+1}, \tag{4.115a}$$

$$\boldsymbol{v}_{b,n} = \boldsymbol{v}_{b,n+1/2} - \frac{\Delta t}{2} \boldsymbol{a}_{b,n}. \tag{4.115b}$$

Substituting these update equations into Equation (4.114) yields, after some re-arrangement,

$$\begin{aligned}\boldsymbol{A}_{n+1} - \boldsymbol{A}_n &= \sum_{a,b=1}^{N} \boldsymbol{M}_{ab}\Big[(\boldsymbol{x}_{a,n+1} - \boldsymbol{x}_{a,n}) \times \boldsymbol{v}_{b,n+1/2} \\ &\qquad + \frac{\Delta t}{2}(\boldsymbol{x}_{a,n+1} \times \boldsymbol{a}_{b,n+1} + \boldsymbol{x}_{a,n} \times \boldsymbol{a}_{b,n})\Big] \\ &= \sum_{a,b=1}^{N} \boldsymbol{M}_{ab} \Delta t (\boldsymbol{v}_{a,n+1/2} \times \boldsymbol{v}_{b,n+1/2}) \\ &\qquad + \sum_{a,b=1}^{N} \boldsymbol{M}_{ab} \frac{\Delta t}{2}(\boldsymbol{x}_{a,n+1} \times \boldsymbol{a}_{b,n+1} + \boldsymbol{x}_{a,n} \times \boldsymbol{a}_{b,n}). \end{aligned} \tag{4.116}$$

Using a similar development to that described after Equation (4.108), the first term in the last equality above vanishes, leaving

$$\begin{aligned}\boldsymbol{A}_{n+1} - \boldsymbol{A}_n = \frac{\Delta t}{2}\Big[&\sum_{a=1}^{N} \boldsymbol{x}_{a,n+1} \times \Big(\sum_{b=1}^{N} m_{ab}\boldsymbol{a}_{b,n+1}\Big) \\ &+ \sum_{a=1}^{N} \boldsymbol{x}_{a,n} \times \Big(\sum_{b=1}^{N} m_{ab}\boldsymbol{a}_{b,n}\Big)\Big]\end{aligned}$$

$$= \frac{\Delta t}{2}\left[\sum_{a=1}^{N} \boldsymbol{x}_{a,n+1} \times \big(\boldsymbol{F}_a(t_{n+1}) - \boldsymbol{T}_a(\boldsymbol{x}_{n+1})\big) + \sum_{a=1}^{N} \boldsymbol{x}_{a,n} \times \big(\boldsymbol{F}_a(t_n) - \boldsymbol{T}_a(\boldsymbol{x}_n)\big)\right]. \tag{4.117}$$

Following similar considerations to those in Equation (4.110), which determined that both $\sum_{a=1}^{N} \boldsymbol{x}_{a,n+1} \times \boldsymbol{T}_a(\boldsymbol{x}_{n+1}) = \boldsymbol{0}$ and $\sum_{a=1}^{N} \boldsymbol{x}_{a,n} \times \boldsymbol{T}_a(\boldsymbol{x}_n) = \boldsymbol{0}$, Equation (4.117) simplifies to

$$\boldsymbol{A}_{n+1} - \boldsymbol{A}_n = \frac{\Delta t}{2}\left[\sum_{a=1}^{N} \boldsymbol{x}_{a,n+1} \times \boldsymbol{F}_a(t_{n+1}) + \sum_{a=1}^{N} \boldsymbol{x}_{a,n} \times \boldsymbol{F}_a(t_n)\right]. \tag{4.118}$$

In the absence of external force moment contributions, the right-hand side of the above equation vanishes, resulting in the time-discrete conservation of angular momentum for the leap-frog scheme.

EXAMPLE 4.9: Demonstration that the mid-point scheme conserves global angular momentum over a time step $\Delta t = t_{n+1} - t_n$

Using Equation (4.114) and the distributive property of the cross product, the change in angular momentum between time steps n and $n+1$ is written as

$$\boldsymbol{A}_{n+1} - \boldsymbol{A}_n = \sum_{a,b=1}^{N} \frac{1}{2}\boldsymbol{M}_{ab}\Big[(\boldsymbol{x}_{a,n+1} - \boldsymbol{x}_{a,n}) \times (\boldsymbol{v}_{b,n+1} + \boldsymbol{v}_{b,n}) + (\boldsymbol{x}_{a,n+1} + \boldsymbol{x}_{a,n}) \times (\boldsymbol{v}_{b,n+1} - \boldsymbol{v}_{b,n})\Big]. \tag{a}$$

From Equations (4.94) the relevant kinematic relations for the mid-point scheme are

$$\begin{aligned}
\boldsymbol{v}_{b,n+1} - \boldsymbol{v}_{b,n} &= \Delta t \boldsymbol{a}_{b,n+1/2},\\
\boldsymbol{x}_{a,n+1} - \boldsymbol{x}_{a,n} &= \frac{\Delta t}{2}(\boldsymbol{v}_{a,n+1} + \boldsymbol{v}_{a,n}),\\
\boldsymbol{x}_{a,n+1} + \boldsymbol{x}_{a,n} &= 2\boldsymbol{x}_{a,n+1/2}.
\end{aligned}$$

(continued)

Example 4.9: *(cont.)*

Substituting these update expressions into Equation (a) gives, after some rearrangement,

$$\begin{aligned}\boldsymbol{A}_{n+1}-\boldsymbol{A}_n &= \Delta t\sum_{a=1}^{N}\boldsymbol{x}_{a,n+1/2}\times\left(\sum_{b=1}^{N} m_{ab}\boldsymbol{a}_{b,n+1/2}\right)\\ &= \Delta t\sum_{a=1}^{N}\boldsymbol{x}_{a,n+1/2}\times\left(\boldsymbol{F}_a(t_{n+1/2})-\boldsymbol{T}_a(t_{n+1/2})\right)\\ &= \Delta t\sum_{a=1}^{N}\boldsymbol{x}_{a,n+1/2}\times\boldsymbol{F}_a(t_{n+1/2}),\end{aligned}$$

which vanishes in the absence of external forces.

4.8.3 Conservation of Total Energy

The total energy associated with the deforming body has been introduced in terms of the finite element discretization and comprises the sum of the kinetic, internal, and external energies (see Equation (4.38)), repeated here for convenience as

$$\Pi(\mathbf{x},\mathbf{v}) = K(\mathbf{v}) + \Pi_{\text{int}}(\mathbf{x}) + \Pi_{\text{ext}}(\mathbf{x}), \tag{4.119}$$

where the kinetic energy K, internal energy Π_{int}, and external energy Π_{ext} are given by Equations (4.37), (4.39), and (4.43) respectively. To determine the rate of change of the total energy, Π, consideration is given to each individual component, beginning with the rate of kinetic energy as

$$\frac{dK}{dt} = \frac{d}{dt}\left(\frac{1}{2}\mathbf{v}^T\mathbf{M}\mathbf{v}\right) = \mathbf{v}^T\mathbf{M}\mathbf{a}. \tag{4.120}$$

The rate of change of the internal energy can be expressed as

$$\begin{aligned}\frac{d\Pi_{\text{int}}}{dt} &= \frac{d}{dt}\sum_{e=1}^{M}\int_{V^{(e)}}\Psi(\boldsymbol{F})\,dV\\ &= \sum_{e=1}^{M}\int_{V^{(e)}}\frac{\partial\Psi}{\partial\boldsymbol{F}}:\dot{\boldsymbol{F}}\,dV\\ &= \sum_{e=1}^{M}\int_{V^{(e)}}\boldsymbol{P}:\boldsymbol{\nabla}_0\boldsymbol{v}\,dV\end{aligned}$$

$$
\begin{aligned}
&= \sum_{e=1}^{M} \int_{V^{(e)}} \boldsymbol{P} : \left(\sum_{a=1}^{n} \boldsymbol{v}_a \otimes \boldsymbol{\nabla}_0 N_a \right) dV \\
&= \sum_{a=1}^{N} \boldsymbol{v}_a \cdot \sum_{\substack{e=1 \\ e \ni a}}^{M_a} \int_{V^{(e)}} \boldsymbol{P} \boldsymbol{\nabla}_0 N_a \, dV \\
&= \sum_{a=1}^{N} \boldsymbol{v}_a \cdot \boldsymbol{T}_a \\
&= \mathbf{v}^T \mathbf{T},
\end{aligned}
\tag{4.121}
$$

where Equation (d) in Example 4.7 is used to give the equivalent nodal force $\boldsymbol{T}_a$ in terms of the first Piola–Kirchhoff stress tensor $\boldsymbol{P}$.

Finally, the rate of change of the external energy is obtained easily from Equation (4.43) as

$$
\frac{d\Pi_{\text{ext}}}{dt} = \frac{d}{dt}\left(-\mathbf{x}^T \mathbf{F}\right) = -\mathbf{v}^T \mathbf{F}. \tag{4.122}
$$

Adding together the three preceding equations gives the rate of change of total energy as

$$
\frac{d\Pi}{dt} = \mathbf{v}^T (\mathbf{Ma} + \mathbf{T} - \mathbf{F}). \tag{4.123}
$$

The term in parenthesis is clearly the dynamic equilibrium equation, and consequently the rate of change of the total energy is zero, that is, $d\Pi/dt = 0$.

Having discovered that the above continuum rate of change of the total energy is zero, it is pertinent to enquire again whether or not the same conclusion can be reached for the time-discrete counterpart. As already mentioned in Chapter 2, variational integrators do not in general guarantee the discrete conservation of energy. Fortunately, it is sometimes possible to introduce appropriate algorithmic modifications in order to enforce energy conservation. In this section this will be studied in relation to the mid-point time integrator, following similar reasoning to that given in Chapter 2; see Equations (2.99) and (2.103). It was shown in Chapter 2 that energy is conserved over a time step t_n to t_{n+1} provided the following condition was met:

$$
K(\mathbf{v}_{n+1}) + \Pi_{\text{int}}(\mathbf{x}_{n+1}) + \Pi_{\text{ext}}(\mathbf{x}_{n+1}) = K(\mathbf{v}_n) + \Pi_{\text{int}}(\mathbf{x}_n) + \Pi_{\text{ext}}(\mathbf{x}_n). \tag{4.124}
$$

Following the same procedure outlined in Chapter 2, Equations (2.98) and (2.100), it can be shown that for the above energy conservation expression to be satisfied, between time steps n and $n+1$, when a mid-point integration scheme is used, it is necessary for the internal force vector at the mid-point to satisfy the following scalar constraint:

$$(\mathbf{x}_{n+1} - \mathbf{x}_n)^T \mathbf{T}(\mathbf{x}_{n+1/2}) = \Pi_{\text{int}}(\mathbf{x}_{n+1}) - \Pi_{\text{int}}(\mathbf{x}_n). \tag{4.125}$$

This condition is not necessarily true for a general hyperelastic material, but it can be enforced by introducing a suitable modification to the stress tensor used to evaluate the internal force $\mathbf{T}$. This can be accomplished by manipulating the left-hand side of Equation (4.125) to extract the first Piola–Kirchhoff stress tensor $\boldsymbol{P}$. Using the discretized deformation gradient given by Equation (4.5), and again noting Equation (d) in Example 4.7, the manipulation proceeds as follows:††

$$\begin{aligned}
(\mathbf{x}_{n+1} - \mathbf{x}_n)^T \mathbf{T}(\mathbf{x}_{n+1/2}) &= \sum_{a=1}^{N} (\boldsymbol{x}_{a,n+1} - \boldsymbol{x}_{a,n}) \cdot \sum_{\substack{e=1 \\ e \ni a}}^{M_a} \boldsymbol{T}_a^{(e)}(\mathbf{x}_{n+1/2}) \\
&= \sum_{a=1}^{N} \sum_{\substack{e=1 \\ e \ni a}}^{M_a} (\boldsymbol{x}_{a,n+1} - \boldsymbol{x}_{a,n}) \cdot \int_{V^{(e)}} \boldsymbol{P}(\boldsymbol{F}_{n+1/2}) \boldsymbol{\nabla}_0 N_a \, dV \\
&= \sum_{e=1}^{m} \int_{V^{(e)}} \boldsymbol{P}(\boldsymbol{F}_{n+1/2}) : \sum_{a=1}^{n} (\boldsymbol{x}_{a,n+1} - \boldsymbol{x}_{a,n}) \otimes \boldsymbol{\nabla}_0 N_a \, dV \\
&= \sum_{e=1}^{m} \int_{V^{(e)}} \boldsymbol{P}(\boldsymbol{F}_{n+1/2}) : (\boldsymbol{F}_{n+1} - \boldsymbol{F}_n) \, dV,
\end{aligned} \tag{4.126}$$

where $\boldsymbol{F}_{n+1/2}$ denotes the deformation gradient at position $\mathbf{x}_{n+1/2}$, that is, $\boldsymbol{F}_{n+1/2} = \boldsymbol{\nabla}_0 \boldsymbol{x}_{n+1/2}$. Combining Equations (4.126) and (4.125) gives the requirement

$$\Pi_{\text{int}}(\mathbf{x}_{n+1}) - \Pi_{\text{int}}(\mathbf{x}_n) = \sum_{e=1}^{m} \int_{V^{(e)}} \boldsymbol{P}(\boldsymbol{F}_{n+1/2}) : (\boldsymbol{F}_{n+1} - \boldsymbol{F}_n) \, dV. \tag{4.127}$$

A sufficient condition for satisfaction of this equation can be established at the element level as

$$\Pi_{\text{int}}^{(e)}(\mathbf{x}_{n+1}) - \Pi_{\text{int}}^{(e)}(\mathbf{x}_n) = \int_{V^{(e)}} \boldsymbol{P}(\boldsymbol{F}_{n+1/2}) : (\boldsymbol{F}_{n+1} - \boldsymbol{F}_n) \, dV. \tag{4.128}$$

Equation (4.39) for the internal energy of an element suggests an even stronger local condition,

$$\Psi(\boldsymbol{F}_{n+1}) - \Psi(\boldsymbol{F}_n) = \boldsymbol{P}(\boldsymbol{F}_{n+1/2}) : (\boldsymbol{F}_{n+1} - \boldsymbol{F}_n), \tag{4.129}$$

†† As it can get confusing, recall: $\boldsymbol{F}$ is the deformation gradient, $\boldsymbol{F}_a$ is the equivalent nodal force at node a, and $\mathbf{F}$ is the list of all equivalent nodal forces.

where the equality is enforced at a continuum level or at the quadrature point level in a finite element spatial discretization. Equation (4.129) describes the so-called *directionality property*, referring to energy-momentum time integrators which establish a constraint on the evaluation of the first Piola–Kirchhoff stress $\boldsymbol{P}_{n+1/2} = \partial\Psi/\partial\boldsymbol{F}|_{n+1/2}$. Note that Equation (4.129) is simply an incremental version of the rate of energy expression $\dot{\Psi} = \boldsymbol{P} : \dot{\boldsymbol{F}}$.

For most general nonlinear materials, the correct evaluation of the second Piola–Kirchhoff stress as the derivative of the strain energy function will not lead to an expression that satisfies the directionality constraint given by Equation (4.129). To resolve this issue, an algorithmic approximate stress tensor $\boldsymbol{P}^*_{n+1/2}$ is derived such that, by construction, the directionality constraint is satisfied. In so doing, it is important to ensure that the symmetry of the Cauchy stress is not lost. Recall the relationship between the various stress measures, namely the Cauchy stress $\boldsymbol{\sigma}$, the first Piola–Kirchhoff stress $\boldsymbol{P}$, and the second Piola–Kirchhoff stress $\boldsymbol{S}$ as given in NL-Statics, Section 5.5, as

$$J\boldsymbol{\sigma} = \boldsymbol{P}\boldsymbol{F}^T = \boldsymbol{F}\boldsymbol{S}\boldsymbol{F}^T. \tag{4.130}$$

The symmetry of $\boldsymbol{\sigma}$ is ensured if the algorithmic stress $\boldsymbol{P}^*$ is derived from an algorithmic $\boldsymbol{S}^*$ as

$$\boldsymbol{P}^*_{n+1/2} = \boldsymbol{F}_{n+1/2}\boldsymbol{S}^*_{n+1/2}. \tag{4.131}$$

To find a suitable expression for $\boldsymbol{S}^*_{n+1/2}$, it is first necessary to rewrite the directionality constraint in terms of $\boldsymbol{S}^*$ instead of $\boldsymbol{P}^*$. To this end, substitute Equation (4.131) for $\boldsymbol{P}^*$ into Equation (4.129) to give

$$\begin{aligned}\Psi_{n+1} - \Psi_n &= \left(\boldsymbol{F}_{n+1/2}\boldsymbol{S}^*_{n+1/2}\right) : (\boldsymbol{F}_{n+1} - \boldsymbol{F}_n)\\ &= \boldsymbol{S}^*_{n+1/2} : \left[\boldsymbol{F}^T_{n+1/2}(\boldsymbol{F}_{n+1} - \boldsymbol{F}_n)\right]\\ &= \frac{1}{2}\boldsymbol{S}^*_{n+1/2} : \left[(\boldsymbol{F}_{n+1} + \boldsymbol{F}_n)^T(\boldsymbol{F}_{n+1} - \boldsymbol{F}_n)\right],\end{aligned} \tag{4.132}$$

where $\left(\boldsymbol{F}_{n+1/2}\boldsymbol{S}^*_{n+1/2}\right)$ replaces $\boldsymbol{P}$ in Equation (4.129) and $\Psi_{n+1} = \Psi(\boldsymbol{F}_{n+1})$ and $\Psi_n = \Psi(\boldsymbol{F}_n)$. Expansion of the last term in the above equation yields

$$(\boldsymbol{F}_{n+1}+\boldsymbol{F}_n)^T(\boldsymbol{F}_{n+1}-\boldsymbol{F}_n)=(\boldsymbol{C}_{n+1}-\boldsymbol{C}_n)+(\boldsymbol{F}^T_n\boldsymbol{F}_{n+1}-\boldsymbol{F}^T_{n+1}\boldsymbol{F}_n). \tag{4.133}$$

The second Piola–Kirchhoff stress $\boldsymbol{S}^*_{n+1/2}$ is symmetric, and the second term in Equation (4.133) is skew symmetric; consequently, substitution of Equation (4.133) into (4.132) gives the energy constraint equation in terms of $\boldsymbol{S}^*_{n+1/2}$ as

$$\Psi_{n+1} - \Psi_n = \frac{1}{2}\boldsymbol{S}^*_{n+1/2} : \Delta\boldsymbol{C}\,; \quad \Delta\boldsymbol{C} = \boldsymbol{C}_{n+1} - \boldsymbol{C}_n. \tag{4.134a,b}$$

This equation now defines a new expression for the directionality property in terms of the second Piola–Kirchhoff stress. Hence, the use of an expression for the second Piola–Kirchhoff stress which complies with Equation (4.134a,b) will ensure overall discrete conservation of energy. A suitable expression which, by construction, satisfies Equation (4.134a,b) is

$$\boldsymbol{S}^*_{n+1/2} = \boldsymbol{S}_{n+1/2} + 2\left(\frac{\Psi_{n+1} - \Psi_n - \frac{1}{2}\boldsymbol{S}_{n+1/2} : \Delta\boldsymbol{C}}{\Delta\boldsymbol{C} : \Delta\boldsymbol{C}}\right)\Delta\boldsymbol{C}, \tag{4.135a}$$

where

$$\boldsymbol{S}_{n+1/2} = \left.\frac{\partial\Psi}{\partial\boldsymbol{C}}\right|_{\boldsymbol{C}=\boldsymbol{F}^T_{n+1/2}\boldsymbol{F}_{n+1/2}}. \tag{4.135b}$$

Indeed, it it straightforward to demonstrate that the previous equation fulfils the directionality property given by Equation (4.134a,b) as

$$\begin{aligned}
\frac{1}{2}\boldsymbol{S}^*_{n+1/2} : \Delta\boldsymbol{C} &= \frac{1}{2}\boldsymbol{S}_{n+1/2} : \Delta\boldsymbol{C} \\
&\quad + \frac{1}{2}2\left(\frac{\Psi_{n+1} - \Psi_n - \frac{1}{2}\boldsymbol{S}_{n+1/2} : \Delta\boldsymbol{C}}{\Delta\boldsymbol{C} : \Delta\boldsymbol{C}}\right)\Delta\boldsymbol{C} : \Delta\boldsymbol{C} \\
&= \Psi_{n+1} - \Psi_n.
\end{aligned} \tag{4.136}$$

Use of $\boldsymbol{P}^*_{n+1/2} = \boldsymbol{F}_{n+1/2}\boldsymbol{S}^*_{n+1/2}$ instead of $\boldsymbol{P}_{n+1/2}$ provides discrete conservation of energy as well as of linear and angular momentum. Unfortunately, an immediate implication is the introduction of further complexity in the evaluation of the tangent stiffness matrix. This is a consequence of the presence of the second term in Equation (4.135a), which leads to a far more elaborate expression for the tangent operator.

Exercises

1. Compute the consistent and lumped mass matrices for a generic two-dimensional three-node (triangular) finite element. (Hint: see Examples 4.4 and 4.5 and Equation (4.16a,b).)
2. For the Newmark time integration scheme:

   (a) Show that the time update equations can be rewritten as

$$\mathbf{a}_{n+1} = \left(1 - \frac{1}{2\beta}\right)\mathbf{a}_n - \frac{1}{\beta\Delta t}\mathbf{v}_n + \frac{1}{\beta\Delta t^2}(\mathbf{x}_{n+1} - \mathbf{x}_n),$$

$$\mathbf{v}_{n+1} = \left(1 - \frac{\gamma}{\beta}\right)\mathbf{v}_n + \left(1 - \frac{\gamma}{2\beta}\right)\Delta t\mathbf{a}_n + \frac{\gamma}{\beta\Delta t}(\mathbf{x}_{n+1} - \mathbf{x}_n).$$

(b) Derive the so-called Newmark u-method scheme in terms of displacement updates $\Delta\mathbf{x}_{n+1}$, with the predictor phase defined as

$$\mathbf{x}_{n+1}^{(0)} = \mathbf{x}_n,$$

$$\mathbf{v}_{n+1}^{(0)} = \left(1 - \frac{\gamma}{\beta}\right)\mathbf{v}_n + \left(1 - \frac{\gamma}{2\beta}\right)\Delta t\mathbf{a}_n + \frac{\gamma}{\beta\Delta t}(\mathbf{x}_{n+1}^{(0)} - \mathbf{x}_n),$$

$$\mathbf{a}_{n+1}^{(0)} = \left(1 - \frac{1}{2\beta}\right)\mathbf{a}_n - \frac{1}{\beta\Delta t}\mathbf{v}_n + \frac{1}{\beta\Delta t^2}(\mathbf{x}_{n+1}^{(0)} - \mathbf{x}_n).$$

(c) Show that the Newton–Raphson corrector equation now takes the following form:

$$\left[\mathbf{K}(\mathbf{x}_{n+1}^{(k)}) + \frac{1}{\beta\Delta t^2}\mathbf{M}\right]\Delta\mathbf{x}_{n+1} = -\mathbf{R}(\mathbf{x}_{n+1}^{(k)}, \mathbf{a}_{n+1}^{(k)}).$$

3. Demonstrate that the leap-frog time integrator can also be obtained if instead of using a trapezoidal integration for the internal and external energy terms, a left (or right) rectangular integration is employed, namely

$$L_n^{LF*}(\mathbf{x}_n, \mathbf{x}_{n+1}) = \Delta t K(\mathbf{v}_{n+1/2}) - \Delta t\left[\Pi_{\text{int}}(\mathbf{x}_n) + \Pi_{\text{ext}}(\mathbf{x}_n)\right]$$

or

$$L_n^{LF**}(\mathbf{x}_n, \mathbf{x}_{n+1}) = \Delta t K(\mathbf{v}_{n+1/2}) - \Delta t\left[\Pi_{\text{int}}(\mathbf{x}_{n+1}) + \Pi_{\text{ext}}(\mathbf{x}_{n+1})\right].$$

4. An elegant way to obtain Equation (4.135a) is via the stationary point of the following functional:

$$\begin{aligned}\mathcal{G}(\boldsymbol{S}_{n+1/2}^*, \lambda) &= \frac{1}{2}(\boldsymbol{S}_{n+1/2}^* - \boldsymbol{S}_{n+1/2}) : (\boldsymbol{S}_{n+1/2}^* - \boldsymbol{S}_{n+1/2}) \\ &\quad - \lambda(\Psi_{n+1} - \Psi_n - \frac{1}{2}\boldsymbol{S}_{n+1/2}^* : \Delta\boldsymbol{C}),\end{aligned}$$

where λ is a scalar Lagrange multiplier used to enforce the directionality property. In order to do this, clearly the following conditions apply:

$$D\mathcal{G}(\boldsymbol{S}_{n+1/2}^*, \lambda)[\Delta\boldsymbol{S}^*] = 0; \qquad D\mathcal{G}(\boldsymbol{S}_{n+1/2}^*, \lambda)[\Delta\lambda] = 0.$$

CHAPTER FIVE

CONSERVATION LAWS IN SOLID DYNAMICS

5.1 INTRODUCTION

Chapter 3 presented the equations describing the motion of a general solid in terms of integral equations using the principle of virtual work or through local equilibrium equations governed by partial differential equations. The latter, involving second time derivatives of the position vector, namely the acceleration, and spatial gradients of the stress tensors. In particular, the principle of virtual work has been the basis of the standard finite element discretization of dynamic equilibrium presented in Chapter 4. The present chapter aims to reformulate the dynamic equilibrium of a solid in the form of a system of first-order conservation laws for a certain set of physical and geometric variables describing the motion and deformation of the solid. These conservation laws will be presented both in integral form expressing the rate of change of the global quantities and locally through first-order partial differential equations relating time rates to spatial gradients of the set of physical quantities.

The reason for introducing first-order conservation laws is that they are used extensively in computational fluid dynamics and other fields of computational modeling, such as electromagnetics. Such laws provide an intuitive representation of the physical processes being considered. The discretization of systems of conservation laws via the finite element method, or other techniques such as finite volume methods, is well understood and routinely applied to simulate the motion of fluids and electromagnetics. Casting the governing equations describing the motion of solids in the form of a set of first-order conservation laws therefore has advantages. First, it allows the use of a wide variety of robust discretization techniques developed in other areas of computational physics. Second, it makes it possible to introduce additional variables and physical features such as thermal effects within a unified formulation. A computational framework based on these conservation laws for the simulation of solid dynamics will be presented in detail in Chapter 7 together with an associated program described in Chapter 9.

The following sections will begin with a description of the general form of a conservation law before applying this to both the Lagrangian and Eulerian descriptions of motion.

5.2 THE GENERAL FORM OF A CONSERVATION LAW

Before discussing in detail the conservation laws governing solid dynamics, it is useful to consider the general form of such laws in a three-dimensional space described by Cartesian coordinates x_1, x_2, and x_3. No distinction is made at this juncture between reference or spatial coordinates, so that x_1, x_2, and x_3 will be simply replaced by X_1, X_2, and X_3 when formulating the conservation laws in a Lagrangian setting.

Consider now a variable $\mathcal{U}(\boldsymbol{x}, t)$, representing the amount per unit volume of a physical quantity $Q(t)$ at a spatial point $\boldsymbol{x}$ and time t. In essence, $\mathcal{U}$ represents the concentration or spatial density of Q, which can therefore be calculated over an arbitrary volume v as

$$Q(t) = \int_v \mathcal{U}(\boldsymbol{x}, t)\, dv. \tag{5.1}$$

For instance, if Q denotes the mass m, then $\mathcal{U}(\boldsymbol{x}, t)$ represents the density ρ. The rate of change of Q can be decomposed into a volume component measuring the amount of Q that is inserted or removed at each spatial point and a surface component measuring the amount of Q that flows in or out through the surface boundary of v. In this way a global conservation law for $\mathcal{U}(\boldsymbol{x}, t)$ is defined by an expression giving the rate of change of Q as

$$\frac{d}{dt}\int_v \mathcal{U}(\boldsymbol{x}, t)\, dv = \int_v \mathcal{S}(\boldsymbol{x}, t)\, dv - \int_{\partial v} \mathcal{F}_n(\mathcal{U}, \boldsymbol{x}, t, \boldsymbol{n})\, da. \tag{5.2}$$

In this expression the first term on the right-hand side represents the global or resultant source term measuring the amount of Q that is introduced or removed from the volume v per unit time due to known external sources $\mathcal{S}(\boldsymbol{x}, t)$. The last term measures the amount of Q that enters or leaves the volume through its surface and is known as the resultant surface flux. It is obtained by integrating a function $\mathcal{F}_n(\mathcal{U}, \boldsymbol{x}, t, \boldsymbol{n})$ which measures the amount of Q per unit area and time that leaves the surface with normal $\boldsymbol{n}$. Insofar as the surface flux $\mathcal{F}_n$ in the above equation is generic, its functional dependence upon $\mathcal{U}, \boldsymbol{x}, t$, and $\boldsymbol{n}$ allows for a wide variety of candidates for particular choices of surface flux and for their associated functional dependencies.

Note that the sign of the last term in Equation (5.2) implies that the sign of $\mathcal{F}_n$ has been arbitrarily chosen to be positive as an outflow, that is, following the outward direction of $\boldsymbol{n}$. In this way Equation (5.2) can be interpreted as simply stating that the rate of change of Q in volume v is given by the resultant source term, measured by the integral of $\mathcal{S}$, minus the net outflow, measured by the surface integral of $\mathcal{F}_n$.

It is simple to show (see Example 5.1) that the conservation law itself implies the existence of a flux vector $\boldsymbol{\mathcal{F}}(\mathcal{U})$ such that

$$\mathcal{F}_n = \boldsymbol{\mathcal{F}} \cdot \boldsymbol{n}, \tag{5.3}$$

so that Equation (5.2) can be rewritten as

$$\frac{d}{dt}\int_v \mathcal{U}\, dv = \int_v \mathcal{S}\, dv - \int_{\partial v} \boldsymbol{\mathcal{F}} \cdot \boldsymbol{n}\, da, \tag{5.4}$$

where the explicit functional dependencies of $\mathcal{U}$ on $\{\boldsymbol{x}, t\}$, $\boldsymbol{\mathcal{F}}$ on $\mathcal{U}$, and $\mathcal{S}$ on $\{\boldsymbol{x}, t\}$ have been omitted in order to simplify the notation. In general, $\mathcal{S}$ will be a known source term, the functional dependency of $\boldsymbol{\mathcal{F}}$ on $\mathcal{U}$ will be established via physical relationships, such as the constitutive equations, and the conservation law given by Equation (5.4) will enable the evaluation of changes in Q or its concentration $\mathcal{U}$.

A local version of the conservation equation for $\mathcal{U}$ can be derived by employing the divergence theorem on the surface term in Equation (5.4) to give

$$\frac{d}{dt}\int_v \mathcal{U}\, dv + \int_v \operatorname{div} \boldsymbol{\mathcal{F}}\, dv = \int_v \mathcal{S}\, dv. \tag{5.5}$$

Assuming that the volume v does not change in time, that is, it is fixed in the Cartesian space under consideration, then the first term in the above equation can be rewritten as

$$\frac{d}{dt}\int_v \mathcal{U}\, dv = \int_v \frac{\partial \mathcal{U}}{\partial t}\, dv. \tag{5.6}$$

Combining Equations (5.5) and (5.6), and assuming that the volume v is arbitrary, that is, the same expression can be written for any volume v, gives the local version of the conservation law as

$$\frac{\partial \mathcal{U}}{\partial t} + \operatorname{div} \boldsymbol{\mathcal{F}} = \mathcal{S}. \tag{5.7}$$

The relationship between $\boldsymbol{\mathcal{F}}$ and $\mathcal{U}$ determines the physical behavior of the process being described by the generic conservation law given by Equation (5.7); see Example 5.2.

The following sections will particularize the generic equation for different physical quantities such as mass, momentum, and energy, both in Lagrangian and Eulerian settings where some of these quantities are vectors. For example, in the case of $\mathcal{U}$ being a vector $\boldsymbol{\mathcal{U}}$ it is worth noting that $\boldsymbol{\mathcal{F}}_n$ is also a vector and therefore $\boldsymbol{\mathcal{F}}$ will be a tensor, denoted $\boldsymbol{\mathscr{F}}$, giving $\boldsymbol{\mathcal{F}}_n = \boldsymbol{\mathscr{F}}\boldsymbol{n}$.

TABLE 5.1 Physical variables associated with generic conservation variables and fluxes

Total Lagrangian						
Variable	Q	$\mathcal{U}$	$\boldsymbol{\mathcal{U}}$	$\boldsymbol{\mathcal{F}}$	$\boldsymbol{\mathscr{F}}$	$\mathcal{S}$
Mass	m	ρ_R	–	0	–	0
Linear momentum	$\boldsymbol{L}$	–	$\boldsymbol{p}_R$	–	$-\boldsymbol{P}$	$\boldsymbol{f}_R$
Energy	Π	E	–	$\boldsymbol{Q} - \boldsymbol{P}^T\boldsymbol{v}$	–	$\boldsymbol{f}_R \cdot \boldsymbol{v} + r_R$
Deformation gradient	–	–	$\boldsymbol{F}$	–	$-\boldsymbol{v} \otimes \boldsymbol{I}$	$\boldsymbol{0}$
Updated Lagrangian						
Mass	m	ρ	–	0	–	0
Linear momentum	$\boldsymbol{L}$	–	$\boldsymbol{p}$	–	$-\boldsymbol{\sigma}$	$\boldsymbol{f}$
Energy[a]	Π	e	–	$\boldsymbol{q} - \boldsymbol{\sigma}\boldsymbol{v}$	–	$\boldsymbol{f} \cdot \boldsymbol{v} + \boldsymbol{r}$
Deformation gradient[a]	–	–	$\boldsymbol{F}$	–	$-\boldsymbol{v} \otimes J^{-1}\boldsymbol{F}^T$	$\boldsymbol{0}$
Eulerian						
Mass	m	ρ	–	$\rho\boldsymbol{v}$	–	0
Linear momentum	$\boldsymbol{L}$	–	$\boldsymbol{p}$	–	$-\boldsymbol{\sigma} + \boldsymbol{p} \otimes \boldsymbol{v}$	$\boldsymbol{f} \cdot \boldsymbol{v} + r$
Energy[a]	Π	e	–	$\boldsymbol{q} - \boldsymbol{\sigma}\boldsymbol{v} + e\boldsymbol{v}$	–	$\boldsymbol{f} \cdot \boldsymbol{v} + r$
Deformation gradient[a]	–	–	$\boldsymbol{F}^{-1}$	–	$\boldsymbol{F}^{-1}\boldsymbol{v} \otimes \boldsymbol{I}$	$\boldsymbol{0}$

[a] See Exercise 5.

Table 5.1 lists for Lagrangian and Eulerian settings, particular cases of the various generic quantities, for instance mass, linear momentum or energy. The respective flux terms will be derived throughout this chapter.

EXAMPLE 5.1: Flux vector

The procedure to prove that the normal fluxes $\mathcal{F}_n$ must be obtained from a flux vector $\boldsymbol{\mathcal{F}}$ is fundamentally the same as the process used in Section 5.2.1

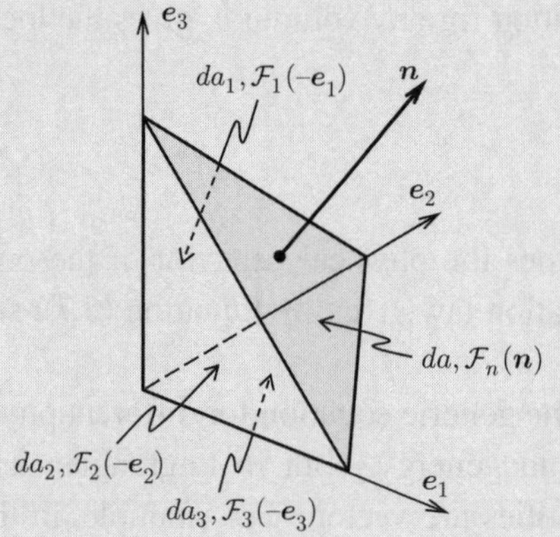

(continued)

Example 5.1: *(cont.)*

of NL-Statics to prove the existence of the Cauchy stress tensor $\boldsymbol{\sigma}$ such that the traction vector $\boldsymbol{t} = \boldsymbol{\sigma n}$. The proof relies on the use of the conservation law on an elemental tetrahedron, as shown below.

The conservation law applied to the elemental tetrahedron in the diagram implies

$$\frac{d}{dt}\mathcal{U}\,dv = \mathcal{S}\,dv - \mathcal{F}_n(\boldsymbol{n})\,da - \sum_{i=1}^{3}\mathcal{F}_i(-\boldsymbol{e}_i)\,da_i,$$

where $\mathcal{F}_i(\boldsymbol{e}_i)$ is the normal flux on the three sides, with normals along the three main axes but in a negative direction. Dividing the above equation by da and noting that $dv/da \to 0$ gives

$$\mathcal{F}_n(\boldsymbol{n}) = \sum_{i=1}^{3} -\mathcal{F}_i(-\boldsymbol{e}_i)\frac{da_i}{da}.$$

The area ratio da_i/da can be expressed by the product of the corresponding unit vectors as $da_i/da = \boldsymbol{n}\cdot\boldsymbol{e}_i$. In addition, the inflow on a surface with normal $\boldsymbol{e}_i$ is identical to the outflow of the other side of the same surface with normal $-\boldsymbol{e}_i$, that is, $\mathcal{F}_i(-\boldsymbol{e}_i) = -\mathcal{F}_i(\boldsymbol{e}_i)$; consequently,

$$\begin{aligned}\mathcal{F}_n(\boldsymbol{n}) &= \sum_{i=1}^{3} -\mathcal{F}_i(-\boldsymbol{e}_i)\boldsymbol{n}\cdot\boldsymbol{e}_i \\ &= \left(\sum_{i=1}^{3}\mathcal{F}_i(\boldsymbol{e}_i)\boldsymbol{e}_i\right)\cdot\boldsymbol{n} \\ &= \boldsymbol{\mathcal{F}}\cdot\boldsymbol{n},\end{aligned}$$

where

$$\boldsymbol{\mathcal{F}} = \sum_{i=1}^{3}\mathcal{F}_i(\boldsymbol{e}_i)\boldsymbol{e}_i.$$

EXAMPLE 5.2: Convective and conductive fluxes

The relationship between $\boldsymbol{\mathcal{F}}$ and $\mathcal{U}$ determines the physical behavior of the process being described by the conservation law given by Equation (5.7). In this context two types of simple fluxes are usually identified, namely convective and conductive fluxes.

(continued)

Example 5.2: *(cont.)*

Convective fluxes are defined by relationships between $\mathcal{F}$ and $\mathcal{U}$ given by

$$\mathcal{F} = \boldsymbol{\nu}\mathcal{U},$$

where $\boldsymbol{\nu}$, in its simplest form, is a constant vector. Physically this represents the transport of $\mathcal{U}$ in the direction of $\boldsymbol{\nu}$ so that the amount crossing a given surface per unit area and time is

$$\mathcal{F}_n = (\boldsymbol{\nu} \cdot \boldsymbol{n})\mathcal{U}.$$

The local (convective flux) conservation law now becomes

$$\frac{\partial \mathcal{U}}{\partial t} + \boldsymbol{\nu} \cdot \boldsymbol{\nabla}\mathcal{U} = \mathcal{S}.$$

Conductive fluxes are defined by the function $\mathcal{F}$ being proportional to the gradient of $\mathcal{U}$ as

$$\mathcal{F} = -\nu\boldsymbol{\nabla}\mathcal{U}.$$

The negative sign is typically a physical imperative that ensures that the material constant ν is positive. The resulting (conductive flux) conservation law now becomes

$$\frac{\partial \mathcal{U}}{\partial t} = \nu\nabla^2\mathcal{U} + \mathcal{S},$$

where $\nabla^2\mathcal{U} = \text{div}\,(\boldsymbol{\nabla}\mathcal{U})$ is the Laplacian of $\mathcal{U}$. This produces a very different behavior compared to the local convective flux conservation law. Now, instead of $\mathcal{U}$ being transported along, it diffuses through the domain. (See Exercises 5.1 and 5.2)

5.3 TOTAL LAGRANGIAN CONSERVATION LAWS IN SOLID DYNAMICS

This section will present the Lagrangian conservation laws that govern solid dynamics as particular instances of the general conservation laws expressed in Section 5.2. Consider for this purpose the motion of a solid defined by a mapping $\boldsymbol{\phi}$, as shown in Figure 5.1. The total Lagrangian nature of a set of conservation laws is established by the fact that the control volume under consideration represents an arbitrary component of the solid defining a volume V^* in the initial configuration, as shown in Figure 5.1. A general quantity Q can be measured in volume V^* via a density function $\mathcal{U}_R$ as

$$Q = \int_{V^*} \mathcal{U}_R \, dV, \tag{5.8}$$

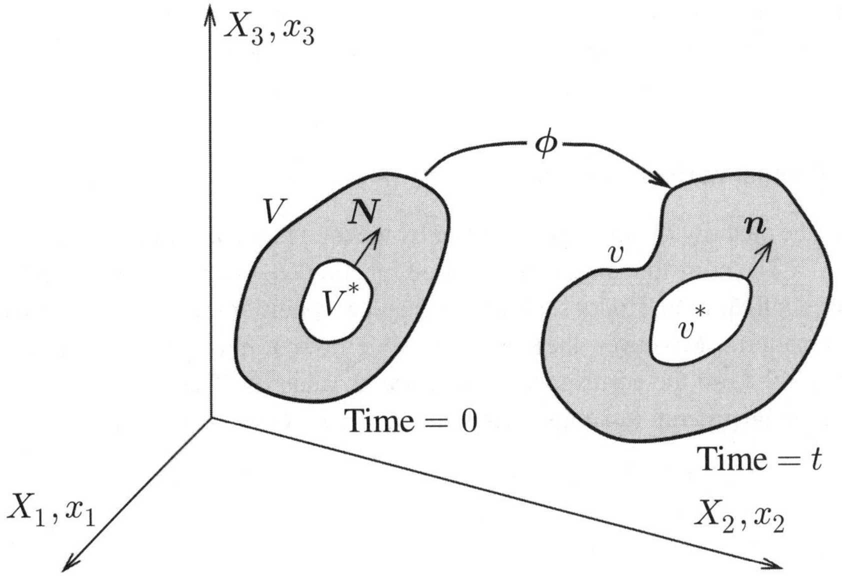

FIGURE 5.1 Lagrangian control volume V^* with surface normal $\boldsymbol{N}$.

where the subscript R is used to denote that $\mathcal{U}_R$ measures the amount of Q per unit reference volume. The corresponding conservation law can be derived by applying Equation (5.4) in the reference configuration to give

$$\frac{d}{dt}\int_{V^*} \mathcal{U}_R \, dV = \int_{V^*} \mathcal{S}_R \, dV - \int_{\partial V^*} \boldsymbol{\mathcal{F}}_R \cdot \boldsymbol{N} \, dA, \tag{5.9}$$

where dA is the element of the boundary surface on V^* with normal $\boldsymbol{N}$, and $\boldsymbol{\mathcal{F}}_R$ denotes the total Lagrangian version of the flux vector. Similarly, $\mathcal{S}_R$ denotes the external source of Q per unit reference volume. The local version of this global conservation law becomes

$$\frac{\partial \mathcal{U}_R}{\partial t} + \text{DIV}\boldsymbol{\mathcal{F}}_R = \mathcal{S}_R, \tag{5.10}$$

where now the divergence operator DIV is taken at the reference configuration and the partial time derivative of $\mathcal{U}_R$ is understood in a material sense, that is, the rate of change of $\mathcal{U}_R$ at a fixed particle with initial coordinates $\boldsymbol{X}$. A similar expression for the case of a vector variable $\boldsymbol{\mathcal{U}}_R$ can be established by replacing $\boldsymbol{\mathcal{F}}_R$ by $\mathscr{F}_R$ and $\mathcal{S}_R$ by $\boldsymbol{\mathcal{S}}_R$ to give the global and local total Lagrangian generic conservation law for a vector quantity $\boldsymbol{\mathcal{U}}$ as

$$\frac{d}{dt}\int_{V^*} \boldsymbol{\mathcal{U}}_R \, dV = \int_{V^*} \boldsymbol{\mathcal{S}}_R \, dV - \int_{\partial V^*} \mathscr{F}_R \boldsymbol{N} \, dA, \tag{5.11a}$$

$$\frac{\partial \boldsymbol{\mathcal{U}}_R}{\partial t} + \text{DIV}\mathscr{F}_R = \boldsymbol{\mathcal{S}}_R. \tag{5.11b}$$

In the next three subsections these general conservation laws will be applied to physical quantities such as mass, linear momentum, and energy.

5.3.1 Conservation of Mass

Consider the quantity Q, now being defined by the total mass of material inside the volume V^*. The corresponding concentration function $\mathcal{U}_R$ is the density ρ_R in the reference configuration. Under standard mechanical conditions there is no external mass source term. Moreover, there is no possible flow of mass across the physical surface ∂V^*, so the corresponding flux vector vanishes. The resulting global conservation law therefore simply states that the mass of V^* is constant in time, that is,

$$\frac{d}{dt}\int_{V^*} \rho_R \, dV = 0. \tag{5.12}$$

This in turn implies that the density at the reference configuration cannot be a function of time, hence

$$\frac{\partial \rho_R}{\partial t} = 0. \tag{5.13}$$

As a consequence, ρ_R is given by the initial conditions of the solid, it remains constant throughout the motion, and does not need to be considered as an unknown to be determined in time. In the present Lagrangian setting both Equations (5.12) and (5.13) were obvious a priori and do not lead to any useful insights or equations. In the Eulerian setting, explored later in this chapter, this will not be the case and a useful conservation law will be derived.

5.3.2 Conservation of Linear Momentum

The conservation of linear momentum of a solid in motion was presented in Chapter 3 but will be revisited here in the context of a conservation law. The quantity being conserved is a total linear momentum vector defined by

$$\boldsymbol{L} = \int_{V^*} \boldsymbol{v}(\boldsymbol{X},t)\, dm = \int_{V^*} \boldsymbol{p}_R \, dV; \quad \boldsymbol{p}_R = \rho_R \boldsymbol{v}. \tag{5.14a,b}$$

It needs to be emphasized that $\boldsymbol{v}$ is a function of the material coordinates $\boldsymbol{X}$ as $\boldsymbol{v}(\boldsymbol{X},t) = \partial\boldsymbol{\phi}(\boldsymbol{X},t)/\partial t$; see NL-Statics, Section 4.11. In the above equation the generic term associations are as follows: the total quantity being conserved is the vector $\boldsymbol{L}$ and its density in the reference domain is $\boldsymbol{p}_R$.

The conservation law for $\boldsymbol{L}$ has already been presented in Chapter 3, Equation (3.12), as the balance between inertial, volume, and traction forces as

$$\frac{d}{dt}\int_{V^*} \boldsymbol{p}_R \, dV = \int_{V^*} \boldsymbol{f}_R \, dV + \int_{\partial V^*} \boldsymbol{t}_R \, dA. \tag{5.15}$$

This equation follows the same format as the generic total Lagrangian global conservation law given by Equation (5.11a), where the external source term $\boldsymbol{\mathcal{S}}_R$ is given by $\boldsymbol{f}_R$ and the normal flux term $\mathscr{F}_R\boldsymbol{N}$ can now be identified as minus the traction vector $\boldsymbol{t}_R$. The traction vector can, in turn, be related to the first Piola–Kirchhoff stress tensor to give $\mathscr{F}_R\boldsymbol{N} = -\boldsymbol{t}_R = -\boldsymbol{P}\boldsymbol{N}$. This implies that the corresponding flux tensor $\mathscr{F}_R$ is minus the first Piola–Kirchhoff tensor $\boldsymbol{P}$ and leads to the global conservation of linear momentum equation as

$$\frac{d}{dt}\int_{V^*} \boldsymbol{p}_R \, dV = \int_{V^*} \boldsymbol{f}_R \, dV + \int_{\partial V^*} \boldsymbol{P}\boldsymbol{N} \, dA; \tag{5.16}$$

the corresponding local conservation law, see derivatives in Equation (5.10), is

$$\frac{\partial \boldsymbol{p}_R}{\partial t} - \mathrm{DIV}\boldsymbol{P} = \boldsymbol{f}_R. \tag{5.17}$$

Note that this is simply a re-expression of Equation (3.22b). Note also that the body force $\boldsymbol{f}_R$ plays the role of a momentum source per unit reference volume and that $-\boldsymbol{P}$ plays the role of the flux tensor $\mathscr{F}_R$, rather than the flux vector, given that the quantity under consideration is a vector. The negative sign may appear counterintuitive at first but it is due to the fact that $\mathscr{F}_R$ has been arbitrarily defined positive when it represents an outflow, that is, a loss of the quantity under consideration. A positive $\boldsymbol{P}$ implies a traction in tension and hence a gain in momentum of the body.

5.3.3 Conservation of Energy

A further important physical quantity for which it is possible to establish a conservation law is the total energy of a solid. The resulting principle is nothing more than the first law of thermodynamics. For this purpose let E denote the total energy per unit reference volume, which may include kinetic, elastic, and thermal components, so that the total energy of the solid defined by volume V^* is

$$\Pi = \int_{V^*} E \, dV. \tag{5.18}$$

The external source terms that introduce or remove energy from the solid can be of mechanical or thermal nature. Similarly the surface or flux terms can also be of mechanical or thermal nature. The total mechanical component is given by the combined work rate of the external body forces and surface traction forces given as

$$\mathcal{W} = \int_{V^*} \boldsymbol{f}_R \cdot \boldsymbol{v} \, dV + \int_{\partial V^*} \boldsymbol{v} \cdot \boldsymbol{P}\boldsymbol{N} \, dA. \tag{5.19}$$

The thermal source can be defined by an external heat energy source rate r_R per unit time and reference volume. Finally, the thermal surface term is given by a heat flux vector $\boldsymbol{Q}$ representing the flow of thermal energy per unit time and reference area due to conduction effects in the material. The total rate of change of energy due to thermal effects is then given by

$$\mathcal{Q} = \int_{V^*} r_R \, dV - \int_{\partial V^*} \boldsymbol{Q} \cdot \boldsymbol{N} \, dA. \tag{5.20}$$

The minus sign in the above equation implies that $\boldsymbol{Q}$ has been defined following the positive outflow convention used for the generic fluxes $\boldsymbol{\mathcal{F}}$. The resulting energy conservation law is obtained by combining Equations (5.18) to (5.20) to give the so-called first law of thermodynamics as

$$\frac{d}{dt}\Pi = \mathcal{W} + \mathcal{Q}, \tag{5.21}$$

or, more specifically,

$$\frac{d}{dt}\int_{V^*} E \, dV = \int_{V^*} (r_R + \boldsymbol{f}_R \cdot \boldsymbol{v}) \, dV - \int_{\partial V^*} (\boldsymbol{Q} - \boldsymbol{P}^T\boldsymbol{v}) \cdot \boldsymbol{N} \, dA. \tag{5.22}$$

Note the fact that $\boldsymbol{v} \cdot \boldsymbol{P}\boldsymbol{N} = (\boldsymbol{P}^T\boldsymbol{v}) \cdot \boldsymbol{N}$ has been used in the last term in order to collect the mechanical and thermal flux vectors into a single resultant energy flux vector $\boldsymbol{\mathcal{F}} = (\boldsymbol{Q} - \boldsymbol{P}^T\boldsymbol{v})$. Similarly, the resultant external energy rate source term $\mathcal{S} = (r_R + \boldsymbol{f}_R \cdot \boldsymbol{v})$ comprises thermal and mechanical components.

The local energy conservation law corresponding to the global Equation (5.22) is obtained by recalling the general local conservation law (Equation (5.10)) to give

$$\frac{\partial E}{\partial t} + \text{DIV}(\boldsymbol{Q} - \boldsymbol{P}^T\boldsymbol{v}) = r_R + \boldsymbol{f}_R \cdot \boldsymbol{v}. \tag{5.23}$$

This is the local version of the first law of thermodynamics, or energy conservation, expressed in the reference configuration. This equation will be the point of departure in Chapter 6 for the development of a number of concepts in thermodynamics, such as internal energy, temperature, and entropy, and their relationships with stress and deformation. Finally, the definition of the heat energy flux vector $\boldsymbol{Q}$ above is somewhat vague and is simply based upon the amount of heat energy flowing out of the surface with normal $\boldsymbol{N}$ per unit reference area and time. Its relationship with the thermal gradient via a heat conduction coefficient will be explored in the next chapter.

5.3.4 Geometric Conservation Law

The conservation laws developed so far relate to physical quantities such as mass, momentum, and energy. The rate equations obtained rely on the evaluation of the first Piola–Kirchhoff stress tensor $\boldsymbol{P}$. In elasticity, $\boldsymbol{P}$ is primarily a function of the deformation gradient $\boldsymbol{F}$ via a constitutive equation. Consequently, in order to have a complete system of equations it is necessary to derive an evolution law for $\boldsymbol{F}$, that is, how $\boldsymbol{F}$ changes with time. In fact, it is actually possible to do so in the form of a conservation law. To achieve this, note that $\boldsymbol{F}$ is defined as follows (see Section 4.4, NL-Statics):

$$\boldsymbol{F} = \boldsymbol{\nabla}_0 \boldsymbol{\phi}. \tag{5.24}$$

Integrating this equation over an arbitrary reference volume V^* and using the gradient theorem (see Equation (2.137) in NL-Statics) gives

$$\begin{aligned} \int_{V^*} \boldsymbol{F}\, dV &= \int_{V^*} \boldsymbol{\nabla}_0 \boldsymbol{\phi}\, dV \\ &= \int_{\partial V^*} \boldsymbol{\phi} \otimes \boldsymbol{N}\, dA. \end{aligned} \tag{5.25}$$

Taking the time derivative of this expression, and noting that $\boldsymbol{v} = \partial \boldsymbol{\phi}/\partial t$, gives a conservation law for $\boldsymbol{F}$ as

$$\frac{d}{dt} \int_{V^*} \boldsymbol{F}\, dV = \int_{\partial V^*} \boldsymbol{v} \otimes \boldsymbol{N}\, dA. \tag{5.26}$$

This conservation law is characterized by the absence of an external source term and by a very simple surface flux term $-(\boldsymbol{v} \otimes \boldsymbol{I})\boldsymbol{N}$ which is linearly dependent upon the velocity. Employing the divergence theorem, the corresponding local conservation law is simply

$$\frac{\partial \boldsymbol{F}}{\partial t} - \text{DIV}(\boldsymbol{v} \otimes \boldsymbol{I}) = 0, \tag{5.27}$$

which is nothing more than another way of writing the evolution law for $\boldsymbol{F}$ as

$$\frac{\partial \boldsymbol{F}}{\partial t} = \boldsymbol{\nabla}_0 \boldsymbol{v} \tag{5.28}$$

(see Equation (4.89) in NL-Statics) in a form that matches the generic conservation law.

5.3.5 Summary of the Total Lagrangian Conservation Laws

The set of conservation laws derived in Sections 5.3.2 to 5.3.4 are collected as a single set of equations as follows:

$$\frac{\partial \boldsymbol{p}_R}{\partial t} - \text{DIV}\boldsymbol{P} = \boldsymbol{f}_R, \tag{5.29a}$$

$$\frac{\partial \boldsymbol{F}}{\partial t} - \text{DIV}(\boldsymbol{v} \otimes \boldsymbol{I}) = 0, \tag{5.29b}$$

$$\frac{\partial E}{\partial t} + \text{DIV}(\boldsymbol{Q} - \boldsymbol{P}^T\boldsymbol{v}) = r_R + \boldsymbol{f}_R \cdot \boldsymbol{v}, \tag{5.29c}$$

where the conservation of mass $\partial \rho_R/\partial t = 0$ has been omitted as it does not add any useful information.

Equations (5.29) will be combined with the thermoelastic equations defined in Chapter 6, and then in Chapter 7 they will be discretized using a Petrov–Galerkin technique originally developed for fluid mechanics problems but now employed to solve solid mechanics problems.

5.4 UPDATED LAGRANGIAN AND EULERIAN CONSERVATION LAWS

The set of total Lagrangian conservation laws derived in the preceding sections will form the basis of the finite element discretization presented in Chapter 7. It is possible to derive and discretize the conservation laws in a spatial setting rather than with respect to the reference configuration. This can be achieved by using updated Lagrangian or Eulerian formulations. The difference between these two formulations requires a subtle refinement of what is meant by the "current" configuration.

In an updated Lagrangian formulation the various equations are expressed in terms of current coordinates $\boldsymbol{x} = \boldsymbol{\phi}(\boldsymbol{X}, t)$ of the particles comprising the body. By contrast, in an Eulerian formulation the various equations are expressed in terms of the spatial coordinates $\boldsymbol{x}$ (not a function of time) through which the body is flowing. In the updated Lagrangian scenario it is as though the particles $\boldsymbol{X}$ *acquire* the temporary coordinate labels $\boldsymbol{x} = \boldsymbol{\phi}(\boldsymbol{X}, t)$ as they arrive at positions $\boldsymbol{x}$ at time t; whereas in the Eulerian formulation the focus is on the region in space coincidentally occupied by the body at time t. It is as though the body were flowing through this region in space but not remaining in it, which is the concept regularly employed in fluid dynamics equations.

As the reader can no doubt appreciate, these two alternative formulations can be a source of confusion, and the following sections aim to present these alternatives in a clear and rigorous manner. For simplicity, only the conservation of mass and momentum will be considered, but energy and geometric conservation laws can also be derived, as indicated in the Exercises section.

Following on from Section 5.2, both the updated Lagrangian and Eulerian conservation laws will be initially established in a generic form before considering the specific examples of conservation of mass and momentum.

5.4.1 Updated Lagrangian Conservation Laws

Consider again the solid in motion depicted in Figure 5.1. A generic conservation law for a quantity Q with Lagrangian density $\mathcal{U}_R$, source term $\mathcal{S}_R$, and total Lagrangian flux $\boldsymbol{\mathcal{F}}_R$ is expressed in terms of reference volume integrals, see Equation (5.9), as

$$\frac{d}{dt}\int_{V^*} \mathcal{U}_R \, dV = \int_{V^*} \mathcal{S}_R \, dV - \int_{\partial V^*} \boldsymbol{\mathcal{F}}_R \cdot \boldsymbol{N} \, dA. \tag{5.30}$$

The original volume V^* currently occupies a spatial volume v^* which , unlike V^*, will change in time as the solid moves. The integrals above can be evaluated with respect to the current volume by noting that the volume and area elements are transformed (see Equations (4.57) and (4.68) in NL-Statics) as

$$dv = JdV; \quad \boldsymbol{n}\, da = J\boldsymbol{F}^{-T}\boldsymbol{N}\, dA. \tag{5.31a,b}$$

Consequently, the generic conservation law given by Equation (5.30) can be expressed in the spatial setting as

$$\begin{aligned}\frac{d}{dt}\int_{v^*(t)} J^{-1}\mathcal{U}_R \, dv &= \int_{v^*(t)} J^{-1}\mathcal{S}_R \, dv - \int_{\partial v^*(t)} \boldsymbol{\mathcal{F}}_R \cdot (J^{-1}\boldsymbol{F}^T\boldsymbol{n})\, da \\ &= \int_{v^*(t)} \mathcal{S}_R J^{-1} \, dv - \int_{\partial v^*(t)} (J^{-1}\boldsymbol{F}\boldsymbol{\mathcal{F}}_R)\cdot \boldsymbol{n}\, da, \end{aligned} \tag{5.32}$$

where the notation $\int_{v^*(t)}$ has been introduced in order to emphasize that the domain of integration v^* is changing with time as the body moves. That is, v^* represents the current volume occupied by the particles that originally were contained in $\boldsymbol{V}^*$. To be more precise, the particles originally contained in volume V^* now occupy a region in space represented by the volume v^*. In Equation (5.32) the term $J^{-1}\mathcal{U}_R$ gives the concentration of Q per unit current volume and will be denoted $\mathcal{U}$. Similarly, $J^{-1}\mathcal{S}_R$ is the external input of Q per unit spatial volume and time and will be denoted as $\mathcal{S}$. Finally the term $J^{-1}\boldsymbol{F}\boldsymbol{\mathcal{F}}_R$ represents the so-called push-forward of $\boldsymbol{\mathcal{F}}_R$ to the spatial setting and will be denoted as the updated Lagrangian flux $\boldsymbol{\mathcal{F}}_U$. The updated Lagrangian generic conservation law now emerges as

$$\frac{d}{dt}\int_{v^*(t)} \mathcal{U} \, dv = \int_{v^*(t)} \mathcal{S} \, dv - \int_{\partial v^*(t)} \boldsymbol{\mathcal{F}}_U \cdot \boldsymbol{n}\, da, \tag{5.33}$$

with

$$\mathcal{U} = J^{-1}\mathcal{U}_R; \quad \mathcal{S} = J^{-1}\mathcal{S}_R; \quad \boldsymbol{\mathcal{F}}_U = J^{-1}\boldsymbol{F}\boldsymbol{\mathcal{F}}_R. \tag{5.34a,b,c}$$

Remark 5.1: As previously explained for vector quantities $\boldsymbol{\mathcal{U}}$, the corresponding flux is the tensor $\mathscr{F}$. The push-forward transformation from the total Lagrangian flux to the updated Lagrangian flux is given by

$$\int_{\partial V^*} \mathscr{F}_R \boldsymbol{N}\, dA = \int_{\partial v^*(t)} J^{-1}\mathscr{F}_R \boldsymbol{F}^T \boldsymbol{n}\, da, \tag{5.35}$$

and therefore the updated Lagrangian flux $\mathscr{F}_U$ is obtained as

$$\mathscr{F}_U = J^{-1}\mathscr{F}_R \boldsymbol{F}^T. \tag{5.36}$$

To obtain an updated Lagrangian local version of the global conservation law given by Equation (5.33), note first that, using the divergence theorem on the flux term, Equation (5.33) can be transformed as

$$\frac{d}{dt}\int_{v^*(t)} \mathcal{U}\, dv = \int_{v^*(t)} \mathcal{S}\, dv - \int_{v^*(t)} \operatorname{div} \boldsymbol{\mathcal{F}}_U\, dv. \tag{5.37}$$

The fact that the spatial volume v^* and therefore the volume element dv are functions of time makes the evaluation of the time derivative in the left-hand side of the above equation complicated. A simple procedure employed to overcome this problem is to revert back to the original fixed volume V^* before differentiating and then to transform the result to the current volume as

$$\begin{aligned}
\frac{d}{dt}\int_{v^*(t)} \mathcal{U}\, dv &= \frac{d}{dt}\int_{V^*} (J\mathcal{U})\, dV \\
&= \int_{V^*} \left.\frac{\partial (J\mathcal{U})}{\partial t}\right|_{\boldsymbol{X}} dV \\
&= \int_{v^*(t)} J^{-1} \left.\frac{\partial (J\mathcal{U})}{\partial t}\right|_{\boldsymbol{X}} dv,
\end{aligned} \tag{5.38}$$

where the notation $(\partial/\partial t)|_{\boldsymbol{X}}$ has been used to denote that this partial derivative is a material derivative, that is, the rate of change in $J\mathcal{U}$ experienced by a fixed particle originally at $\boldsymbol{X}$ but currently at the spatial position $\boldsymbol{x} = \boldsymbol{\phi}(\boldsymbol{X}, t)$. Combining Equations (5.38) and (5.37) gives the generic local updated Lagrangian conservation law as

$$J^{-1}\left.\frac{\partial (J\mathcal{U})}{\partial t}\right|_{\boldsymbol{X}} + \operatorname{div} \boldsymbol{\mathcal{F}}_U = \mathcal{S}. \tag{5.39}$$

This is in essence the same Lagrangian conservation law for $\mathcal{U}_R = J\mathcal{U}$ but expressed in the current setting through the spatial divergence of the updated

Lagrangian fluxes $\mathcal{F}_U$. As an example of this generic local conservation equation, consider the application to the conservation of linear momentum $\boldsymbol{L}$. The corresponding density of momentum per unit reference volume is $\boldsymbol{p}_R = \rho_R \boldsymbol{v}$, whereas the density per unit deformed volume is $\boldsymbol{p}$ given as

$$\boldsymbol{p} = J^{-1}\boldsymbol{p}_R = J^{-1}\rho_R \boldsymbol{v} = \rho \boldsymbol{v}. \tag{5.40}$$

The local Lagrangian conservation law given by Equation (5.17) with source term $\boldsymbol{f}_R$ and flux term $-\boldsymbol{P}$ becomes

$$J^{-1}\left.\frac{\partial(J\boldsymbol{p})}{\partial t}\right|_{\boldsymbol{X}} = \operatorname{div}(J^{-1}\boldsymbol{P}\boldsymbol{F}^T) + \boldsymbol{f}, \tag{5.41}$$

where $\boldsymbol{f} = J^{-1}\boldsymbol{f}_R$ is the external force per unit deformed volume and the term $J^{-1}\boldsymbol{P}\boldsymbol{F}^T$ arising from the push-forward of the total Lagrangian flux $\boldsymbol{P}$, via Equation (5.36), is simply the Cauchy stress tensor $\boldsymbol{\sigma}$. In addition, the left-hand side of the above equation can be simplified using simple algebra as

$$\begin{aligned}
J^{-1}\left.\frac{\partial(J\boldsymbol{p})}{\partial t}\right|_{\boldsymbol{X}} &= J^{-1}\left.\frac{\partial(J\rho\boldsymbol{v})}{\partial t}\right|_{\boldsymbol{X}} \\
&= J^{-1}\left.\frac{\partial(\rho_R\boldsymbol{v})}{\partial t}\right|_{\boldsymbol{X}} \\
&= J^{-1}\,\rho_R\left.\frac{\partial\boldsymbol{v}}{\partial t}\right|_{\boldsymbol{X}} \\
&= J^{-1}\rho_R\boldsymbol{a} \\
&= \rho\boldsymbol{a},
\end{aligned} \tag{5.42}$$

where $\boldsymbol{a}$ is the particle acceleration. The final local form of the updated Lagrangian conservation law of linear momentum emerges as the well-known Equation (3.22a), previously introduced in Chapter 3, as

$$\rho\boldsymbol{a} = \operatorname{div}\boldsymbol{\sigma} + \boldsymbol{f}. \tag{5.43}$$

The next section will now emphasize the difference, alluded to in the introduction to this section, between this updated Lagrangian form of the equilibrium equation and a true Eulerian form of the same physical principle.

5.4.2 Eulerian Conservation Laws

Eulerian conservation laws, unlike total or updated Lagrangian formulations, are rarely used in solid mechanics but commonly employed in fluid dynamics. The exceptions are cases of fluid structure interaction problems where the solid is fully immersed in a fluid. Nevertheless, it is instructive to develop the Eulerian counterpoint to Lagrangian conservation laws both for completeness and to be able to draw comparisons with the conservation laws commonly used in fluid mechanics.

The starting point required to develop an Eulerian consevation law is the generic updated Lagrangian Equation (5.37). As explained at the beginning of Section 5.4, the key difference is that in an Eulerian context the volume v^* is fixed in space rather than moving with the body and being associated with a constant initial material volume V^*. In essence, an Eulerian conservation law will provide an equation for the rate of change of the total amount of Q in a fixed control volume as

$$Q = \int_{v_x^*} \mathcal{U} \, dv, \tag{5.44}$$

where the notation v_x^* has been used to denote that this integration volume is fixed in the coordinate system $\boldsymbol{x}$ and therefore is not a function of time; consequently, it is intrinsically different from $v^*(t)$, although crucially at a given time instant t they coincide in space. Equation (5.37) can now be written as

$$\frac{d}{dt} \int_{v^*(t)} \mathcal{U} \, dv = \int_{v_x^*} \mathcal{S} \, dv - \int_{\partial v_x^*} \boldsymbol{\mathcal{F}}_U \cdot \boldsymbol{n} \, da. \tag{5.45}$$

Note that the change from $v^*(t)$ to v_x^* on the right-hand side is possible in the absence of a time derivative on the basis that, at a given fixed time t, $v_x^* = v^*(t)$. The term on the left-hand side is more complex as $v^*(t)$ will change in time, thus affecting the time derivative. It is possible to transform this integral into an integral over a fixed volume v_x^* by employing the following sequence of transformations leading to the so-called Reynolds transport theorem:

$$\begin{aligned} \frac{d}{dt} \int_{v^*(t)} \mathcal{U} \, dv &= \int_{V^*} \left(\left. \frac{\partial \mathcal{U}}{\partial t} \right|_{\boldsymbol{X}} J \, dV + \mathcal{U} \left. \frac{\partial J}{\partial t} \right|_{\boldsymbol{X}} dV \right) \\ &= \int_{v^*(t)} \left(\left. \frac{\partial \mathcal{U}}{\partial t} \right|_{\boldsymbol{X}} dv + \mathcal{U} \left. \frac{\partial J}{\partial t} \right|_{\boldsymbol{X}} J^{-1} \, dv \right). \end{aligned} \tag{5.46}$$

In order to proceed, observe that the time derivative of the Jacobian J is given as

$$\left. \frac{\partial J}{\partial t} \right|_{\boldsymbol{X}} = J \operatorname{div} \boldsymbol{v} \tag{5.47}$$

(see Equation (4.128) in NL-Statics). In addition, the material time derivative of $\mathcal{U}$ can be related to the spatial time derivative as

$$\begin{aligned} \frac{\partial \mathcal{U}(\boldsymbol{X}, t)}{\partial t} &= \frac{\partial \mathcal{U}(\boldsymbol{x}, t)}{\partial t} + \frac{\partial \mathcal{U}(\boldsymbol{x}, t)}{\partial \boldsymbol{x}} \cdot \frac{\partial \boldsymbol{x}}{\partial t} \\ &= \left. \frac{\partial \mathcal{U}}{\partial t} \right|_{\boldsymbol{x}} + (\boldsymbol{\nabla} \mathcal{U}) \cdot \boldsymbol{v}. \end{aligned} \tag{5.48}$$

Substituting Equations (5.47) and (5.48) into Equation (5.46) gives the Reynolds transport equation as

$$
\begin{aligned}
\frac{d}{dt}\int_{v^*(t)} \mathcal{U}\, dv &= \int_{v_x^*} \left(\left.\frac{\partial \mathcal{U}}{\partial t}\right|_{\boldsymbol{x}} + \boldsymbol{\nabla}\mathcal{U} \cdot \boldsymbol{v} + \mathcal{U}\,\mathrm{div}\,\boldsymbol{v} \right) dv \\
&= \int_{v_x^*} \left(\left.\frac{\partial \mathcal{U}}{\partial t}\right|_{\boldsymbol{x}} + \mathrm{div}\,(\mathcal{U}\boldsymbol{v}) \right) dv \\
&= \frac{d}{dt}\int_{v_x^*} \mathcal{U}\, dv + \int_{\partial v_x^*} \mathcal{U}\boldsymbol{v} \cdot \boldsymbol{n}\, da.
\end{aligned}
\tag{5.49}
$$

A graphical interpretation of the above Reynolds transportation theorem is shown in Figure 5.2. The difference between the rate of change of quantity Q measured on the fixed volume v_x^* as opposed to the moving volume $v^*(t)$ is given by the amount of Q that flows in and out of v_x^* through its surface as a result of the motion of the material. This amount is quantified by the surface integral of the convective flux term $\mathcal{U}\boldsymbol{v} \cdot \boldsymbol{n}$, which measures the amount of Q entering or leaving the volume v_x^* per unit area and time simply due to the material particles flowing into or out of this volume.

Using the Reynolds transport theorem encapsulated by Equation (5.49), the updated Lagrangian conservation law given by Equation (5.33) can be rewritten as

$$
\frac{d}{dt}\int_{v_x^*} \mathcal{U}\, dv = \int_{v_x^*} \mathcal{S}\, dv - \int_{\partial v_x^*} (\boldsymbol{\mathcal{F}}_U + \mathcal{U}\boldsymbol{v}) \cdot \boldsymbol{n}\, da. \tag{5.50}
$$

The term $\boldsymbol{\mathcal{F}}_U + \mathcal{U}\boldsymbol{v}$ is known as the Eulerian flux $\boldsymbol{\mathcal{F}}_E$ and is related to the total Lagrangian flux $\boldsymbol{\mathcal{F}}_R$ as

$$
\boldsymbol{\mathcal{F}}_E = \boldsymbol{\mathcal{F}}_U + \mathcal{U}\boldsymbol{v} = J^{-1}\boldsymbol{F}\boldsymbol{\mathcal{F}}_R + \mathcal{U}\boldsymbol{v}. \tag{5.51}
$$

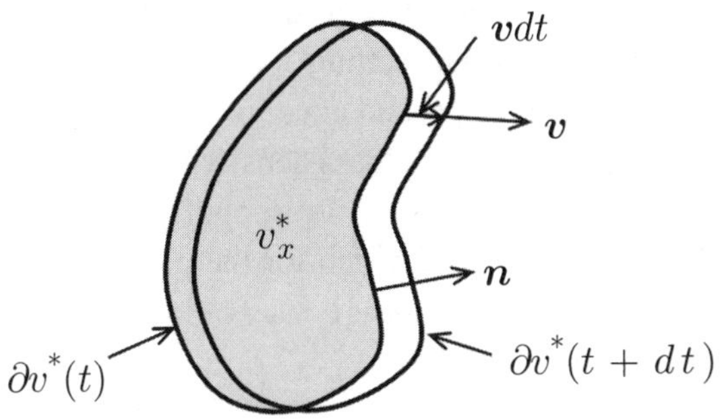

FIGURE 5.2 Fixed Eulerian volume v_x^* and time-dependent updated Lagrangian volume $v^*(t)$.

The term $\mathcal{U}\boldsymbol{v}$ in above expression represents a convection flux, that is, a flux proportional to the main variable $\mathcal{U}$, which is typical of Eulerian formulations. The global generic conservation law for Q can now be established in Eulerian form as

$$\frac{d}{dt}\int_{v_x^*} \mathcal{U}\, dv = \int_{v_x^*} \mathcal{S}\, dv - \int_{\partial v_x^*} \boldsymbol{\mathcal{F}}_E \cdot \boldsymbol{n}\, da, \tag{5.52}$$

which has a local equivalent as

$$\left.\frac{\partial \mathcal{U}}{\partial t}\right|_{\boldsymbol{x}} + \operatorname{div} \boldsymbol{\mathcal{F}}_E = \mathcal{S}. \tag{5.53}$$

Note that, unlike the total and updated Lagrangian conservation laws, the time derivative above is a spatial derivative, that is, taken at a constant $\boldsymbol{x}$ rather than $\boldsymbol{X}$. It assumes that $\mathcal{U}$ is given as a function of $\boldsymbol{x}$ and t as $\mathcal{U}(\boldsymbol{x}, t)$ rather than as $\mathcal{U}(\boldsymbol{X}, t)$, which is the Lagrangian case.

Finally, observe also that for vector quantities $\boldsymbol{\mathcal{U}}$ the equivalent transformation from Lagrangian to Eulerian fluxes is obtained with the help of Equation (5.36) as

$$\mathscr{F}_E = \mathscr{F}_U + \boldsymbol{\mathcal{U}} \otimes \boldsymbol{v} = J^{-1}\mathscr{F}_R \boldsymbol{F}^T + \boldsymbol{\mathcal{U}} \otimes \boldsymbol{v}, \tag{5.54}$$

where $\mathscr{F}_R$ represents the total Lagrangiam flux, with $\mathscr{F}_U$ being its updated Lagrangian counterpart and $\mathscr{F}_E$ being the Eulerian flux tensor.

5.4.3 Eulerian Conservation of Mass and Momentum

Scalar and vector examples of Eulerian conservation laws will now be considered. The conservation of energy and an Eulerian geometric conservation law will be developed as exercises at the end of this chapter.

It was shown in Section 5.3.1 that the Lagrangian conservation of mass simply implies the obvious fact that the reference mass density ρ_R is constant in time. Additionally, the corresponding Lagrangian source term and flux both vanish. However, the same conservation law in the Eulerian setting leads to the well–known continuity equation. In order to derive this important equation, observe that, following the generic Equation (5.34a,b,c)$_a$, the Eulerian mass density is the spatial density $\rho(\boldsymbol{x}, t) = J^{-1}\rho_R$. Also, from Equation (5.51), the corresponding Eulerian flux, with $\boldsymbol{\mathcal{F}}_R = 0$, is simply the convective term $\mathcal{U}\boldsymbol{v}$, which for the present case where $\mathcal{U} = \rho$ becomes $\rho\boldsymbol{v}$. The resulting local Eulerian mass conservation equation (the continuity equation) therefore emerges as

$$\frac{\partial \rho}{\partial t} + \operatorname{div} \rho\boldsymbol{v} = 0. \tag{5.55}$$

In the above expression, ρ is a function of $\boldsymbol{x}$ and time t, and consequently the partial time derivative is taken at constant $\boldsymbol{x}$, yielding a spatial time rate. The global integral conservation law associated with Equation (5.55) is given as

$$\frac{d}{dt}\int_{v_x^*} \rho\, dv + \int_{\partial v_x^*} \rho \boldsymbol{v} \cdot \boldsymbol{n}\, da = 0. \tag{5.56}$$

This simply states that the change in mass occupying a fixed volume v_x^* is given by the amount of material flowing in and out through the surface, as depicted in Figure 5.2.

The Eulerian conservation of linear momentum can be derived following the generic transformation rules for a vector quantity given in Section 5.4.2. In this case, the Lagrangian variable is the momentum density $\boldsymbol{p}_R = \rho_R \boldsymbol{v}$; consequently, the Eulerian variable is $\boldsymbol{p} = J^{-1}\boldsymbol{p}_R = \rho\boldsymbol{v}$. The corresponding updated Lagrangian flux is minus the Cauchy stress tensor, as shown in Section 5.4.1, thus the Eulerian flux becomes

$$\mathscr{F}_E = -\boldsymbol{\sigma} + \boldsymbol{p} \otimes \boldsymbol{v}, \tag{5.57}$$

which leads from the application of Equation (5.53) to a local Eulerian conservation of linear momentum equation (the equilibrium equation) as

$$\frac{\partial \boldsymbol{p}}{\partial t} + \text{div}\,(\boldsymbol{p} \otimes \boldsymbol{v} - \boldsymbol{\sigma}) = \boldsymbol{f}. \tag{5.58}$$

It is useful for comparison to gather together the three versions of the local conservation of linear momentum equations, that is, the local equilibrium equations, namely the total Lagrangian, updated Lagrangian, and Eulerian as follows:

$$\text{total Lagrangian:} \quad \frac{\partial \boldsymbol{p}_R}{\partial t} - \text{DIV}\boldsymbol{P} = \boldsymbol{f}_R, \tag{5.59a}$$

$$\text{updated Lagrangian:} \quad \rho\boldsymbol{a} = \text{div}\,\boldsymbol{\sigma} + \boldsymbol{f}, \tag{5.59b}$$

$$\text{Eulerian:} \quad \frac{\partial \boldsymbol{p}}{\partial t} + \text{div}\,(\boldsymbol{p} \otimes \boldsymbol{v} - \boldsymbol{\sigma}) = \boldsymbol{f}, \tag{5.59c}$$

where the time derivative in Equation (5.59a) is a material rate, whereas the time derivative in Equation (5.59c) denotes a spatial rate. Observe that, in the absence of velocity, that is, the static case, the updated Lagrangian and Eulerian local equilibrium equations are identical.

A summary of various conservation laws and their fluxes and source terms in the total Lagrangian, updated Lagrangian, and Eulerian formulations is shown in Table 5.1.

EXAMPLE 5.3: Conservation of the volume ratio J

In the same way that the conservation of a constant Lagrangian quantity, namely the reference density ρ_R, leads to a meaningful equation in the Eulerian setting, see Equation (5.55), the reverse is also true. The conservation of a constant in the Eulerian setting leads to a useful equation for the rate of the Jacobian J in the Lagrangian setting. This can become useful in cases where the material is nearly incompressible and a very accurate determination of the Jacobian J is necessary. To show this, consider a constant Eulerian function, for example unity across the spatial domain. The trivial Eulerian conservation can be established locally or globally as

$$\frac{\partial 1}{\partial t} = 0; \qquad \frac{d}{dt}\int_{v_x^*} 1\, dv = 0.$$

Note that both the Eulerian source and flux in the generic conservation law vanish. The relationship between Eulerian and Lagrangian conservation variables $\mathcal{U} = J^{-1}\mathcal{U}_R$ implies that for $\mathcal{U} = 1, \mathcal{U}_R = J$. The updated Lagrangian and total Lagrangian fluxes are (see Equation (5.34a,b,c)$_c$)

$$\boldsymbol{\mathcal{F}}_U = \boldsymbol{\mathcal{F}}_E - \mathcal{U}\boldsymbol{v} = -\boldsymbol{v}; \qquad \boldsymbol{\mathcal{F}}_R = J\boldsymbol{F}^{-1}\boldsymbol{\mathcal{F}}_U = -J\boldsymbol{F}^{-1}\boldsymbol{v}.$$

The local updated Lagrangian and total Lagrangian conservation laws for J emerge from Equations (5.10) and (5.39) as

$$J^{-1}\left.\frac{\partial J}{\partial t}\right|_{\boldsymbol{X}} = \operatorname{div}\boldsymbol{v} \quad \text{and} \quad \left.\frac{\partial J}{\partial t}\right|_{\boldsymbol{X}} = \text{DIV}(J\boldsymbol{F}^{-1}\boldsymbol{v}),$$

or in global form as

$$\frac{d}{dt}\int_{v^*(t)} dv = \int_{\partial v^*} \boldsymbol{v}\cdot\boldsymbol{n}\, da \quad \text{and} \quad \frac{d}{dt}\int_{V^*} J\, dV = \int_{\partial V^*} (J\boldsymbol{F}^{-1}\boldsymbol{v})\cdot\boldsymbol{N}\, dA.$$

Note that the total generic quantity Q being conserved by this equation is the spatial volume v_x^*.

Exercises

1. Show that in the absence of external source terms the analytical solution of the conservation law with constant convective fluxes $\boldsymbol{\mathcal{F}} = c\boldsymbol{\nu}\mathcal{U}$, where $\boldsymbol{\nu}$ is a unit direction vector (see Example 5.2), with intial conditions $\mathcal{U}(\boldsymbol{x},0) = \mathcal{U}_0(\boldsymbol{x})$ is

$$\mathcal{U}(\boldsymbol{x},t) = \mathcal{U}_0(\boldsymbol{x} - c\boldsymbol{\nu}t).$$

Explain graphically how this expression represents the transport of $\mathcal{U}_0(\boldsymbol{x})$ in the direction $\boldsymbol{\nu}$ with speed c.

2. Consider now the case of conductive fluxes $\mathcal{F} = -\nu\boldsymbol{\nabla}\mathcal{U}$ (see Example 5.2) also in the absence of source terms where the conservation law becomes

$$\frac{\partial \mathcal{U}}{\partial t} = \nu\nabla^2\mathcal{U}.$$

Assuming a two-dimensional case with boundary conditions in a rectangle $[0,a]\times[0,b]$ as

$$\mathcal{U}(0,y,t) = \mathcal{U}(x,0,t) = 0,$$
$$\mathcal{U}(a,y,t) = \mathcal{U}(x,b,t) = 0,$$

and initial conditions expressed as a harmonic series as

$$\mathcal{U}_0(x,y,0) = \sum_{k=1}^{\infty} A_k \sin\left(\frac{k\pi x}{a}\right)\sin\left(\frac{k\pi y}{b}\right).$$

Show that the analytical solution for $\mathcal{U}(x,y,t)$ is

$$\mathcal{U}(x,y,t) = \sum_{k=1}^{\infty} A_k e^{-\nu\lambda_k t}\sin\left(\frac{k\pi x}{a}\right)\sin\left(\frac{k\pi y}{b}\right),$$

where

$$\lambda_k = k^2\pi^2\left(\frac{1}{a^2}+\frac{1}{b^2}\right).$$

3. Show that the local energy conservation equation in an updated Lagrangian formulation becomes

$$J^{-1}\frac{\partial(Je)}{\partial t} + \mathrm{div}\,(\boldsymbol{q}-\boldsymbol{\sigma v}) = r + \boldsymbol{f}\cdot\boldsymbol{v},$$

where $e = J^{-1}E$ is the energy per unit deformed volume, r is the external heat energy source per unit current volume and $\boldsymbol{q}$ is the heat flux vector in the reference configuration related to $\boldsymbol{Q}$ via the transformation

$$\boldsymbol{q} = J^{-1}\boldsymbol{FQ}.$$

4. Derive the Eulerian local and global energy conservation laws as given in the final column of Table 5.1.

5. Using similar procedures to those employed in section 5.3.4 to derive a conservation law for the deformation gradient $\boldsymbol{F}$, derive an Eulerian conservation law for $\boldsymbol{F}^{-1}$ starting from the relationship

$$\boldsymbol{F}^{-1} = \frac{\partial \boldsymbol{X}}{\partial \boldsymbol{x}},$$

where $\boldsymbol{X}(\boldsymbol{x}, t)$ is the inverse of the motion map $\boldsymbol{x} = \boldsymbol{\phi}(\boldsymbol{X}, t)$, that is, $\boldsymbol{X}(\boldsymbol{\phi}(\boldsymbol{X}, t), t) = \boldsymbol{X}$. Prove that the resulting local conservation law becomes

$$\left.\frac{\partial \boldsymbol{F}^{-1}}{\partial t}\right|_{\boldsymbol{x}} + \text{div}\left(\boldsymbol{F}^{-1}\boldsymbol{v} \otimes \boldsymbol{I}\right) = \boldsymbol{0}.$$

CHAPTER SIX

THERMODYNAMICS

6.1 INTRODUCTION

The previous chapter established a set of conservation laws governing the motion and deformation of a solid. One of these laws establishes the conservation of energy for a given material or spatial volume. This law is known as the first law of thermodynamics and it is fundamental to the study of the inter-relationships between mechanical, thermal, and other possible physical phenomena. Thermodynamics is often approached with some hesitation by engineers, but a good working understanding is required in order to describe many important phenomena in solids, such as heat expansion, the dissipation of energy caused by inelastic effects, and the propagation of shocks.

The aim of this chapter is to present, in the simplest possible form, the basic concepts of thermodynamics, such as temperature, entropy, and internal energy, and the equations that relate them to each other and to stress and deformation. These relationships constitute the fundamental framework upon which constitutive equations can be established. We will avoid making the often repeated reference to vague physical descriptions such as "measure of disorder" when presenting the concept of entropy as they are not very illuminating in the context of a rational discussion of these concepts. Instead we will rely heavily on the concept of energy conjugacy, which has already been employed to good effect to describe conjugate pairs of stress and strain measures. We will also restrict the development to simple thermoelastic models, such as entropic elasticity and the Mie–Grüneisen equations of state. These provide simple coupling models between thermal and mechanical effects which are widely used and general enough to provide a comprehensive description of the physics involved.

6.2 THE FIRST LAW OF THERMODYNAMICS

6.2.1 Internal Energy

The first law of thermodynamics is nothing more than the statement of energy conservation described in the material setting in Section 5.3.3, Equation (5.23), namely

$$\frac{\partial E}{\partial t} + \mathrm{DIV}(\boldsymbol{Q} - \boldsymbol{P}^T \boldsymbol{v}) = \boldsymbol{f}_R \cdot \boldsymbol{v} + r_R. \tag{6.1}$$

This equation can be transformed into a more physically meaningful expression of energy balances at a material point through combination with the conservation of linear momentum and deformation gradient equations given by Equations (5.17) and (5.28). To achieve this, first multiply the conservation of linear momentum Equation (5.17) by $\boldsymbol{v}$ to give

$$\boldsymbol{v} \cdot \frac{\partial \boldsymbol{p}_R}{\partial t} - \boldsymbol{v} \cdot \mathrm{DIV}\boldsymbol{P} = \boldsymbol{f}_R \cdot \boldsymbol{v}. \tag{6.2}$$

Noting that $\boldsymbol{p}_R = \rho_R \boldsymbol{v}$ and using the product rule for the divergence, namely $\mathrm{DIV}(\boldsymbol{P}^T \boldsymbol{v}) = \boldsymbol{v} \cdot \mathrm{DIV}\boldsymbol{P} + \boldsymbol{P} : \nabla_0 \boldsymbol{v}$ gives,

$$\frac{\partial}{\partial t}\left(\frac{1}{2}\rho_R \boldsymbol{v} \cdot \boldsymbol{v}\right) - \mathrm{DIV}(\boldsymbol{P}^T \boldsymbol{v}) + \boldsymbol{P} : \nabla_0 \boldsymbol{v} = \boldsymbol{f}_R \cdot \boldsymbol{v}. \tag{6.3}$$

Now, multiplying the conservation law for the deformation gradient, Equation (5.28), by the first Piola–Kirchhoff stress tensor yields

$$\boldsymbol{P} : \frac{\partial \boldsymbol{F}}{\partial t} - \boldsymbol{P} : \nabla_0 \boldsymbol{v} = 0. \tag{6.4}$$

Substituting Equations (6.3) and (6.4) into Equation (6.1) gives

$$\frac{\partial}{\partial t}\left(E - \frac{1}{2}\rho_R \boldsymbol{v} \cdot \boldsymbol{v}\right) + \mathrm{DIV}\boldsymbol{Q} = r_R + \boldsymbol{P} : \frac{\partial \boldsymbol{F}}{\partial t}. \tag{6.5}$$

At this juncture, it is important to reflect on the meaning of the term in brackets in the above equation: E denotes the total amount of energy per unit undeformed volume (not accounting for potential energy due to external forces such as gravity) and $\frac{1}{2}\rho_R \boldsymbol{v} \cdot \boldsymbol{v}$ measures the amount of kinetic energy due to the motion of an undeformed unit volume. The difference between these two quantities represents the amount of energy stored in the material itself, either through mechanical phenomena, like elasticity, or in the form of thermal energy. Hence, the term in brackets is referred to as the "internal energy" $\mathcal{E}$. This definition allows Equation (6.5) to be expressed in the classical "first law of thermodynamics" form as

$$\frac{\partial \mathcal{E}}{\partial t} + \mathrm{DIV}\boldsymbol{Q} = r_R + \boldsymbol{P} : \frac{\partial \boldsymbol{F}}{\partial t}. \tag{6.6}$$

This expression simply states that the rate of change of internal energy $\dot{\mathcal{E}}$ at a material point is given by the external heat source rate r_R, plus the mechanical work $\boldsymbol{P}: \dot{\boldsymbol{F}}$, minus the net outflow of energy DIV$\boldsymbol{Q}$. It is instructive to consider simple cases of this equation. For instance, take the case of an elastic material in the absence of thermal effects for which $\boldsymbol{Q} = 0$ and $r_R = 0$. The internal energy is simply given by the mechanical energy, hence its rate is determined by the product of the first Piola–Kirchhoff stress and the rate of the deformation gradient:

$$\frac{\partial \mathcal{E}}{\partial t} = \boldsymbol{P} : \frac{\partial \boldsymbol{F}}{\partial t}. \tag{6.7}$$

However, for an elastic material in the absence of thermal effects, the mechanical energy $\mathcal{E}$ is a function of the deformation gradient $\boldsymbol{F}$, that is, $\mathcal{E} = \mathcal{E}(\boldsymbol{F})$, and therefore

$$\frac{\partial \mathcal{E}}{\partial t} = \frac{\partial \mathcal{E}}{\partial \boldsymbol{F}} : \frac{\partial \boldsymbol{F}}{\partial t}. \tag{6.8}$$

Comparing Equations (6.7) and (6.8) gives the first Piola–Kirchhoff stress as

$$\boldsymbol{P} = \frac{\partial \mathcal{E}}{\partial \boldsymbol{F}}. \tag{6.9}$$

In other words, in the absence of thermal and inelastic effects, the internal energy $\mathcal{E}$ is simply the elastic potential Ψ used in Chapter 6 of NL-Statics to define a hyperelastic material; that is, $\mathcal{E}(\boldsymbol{F}) = \Psi(\boldsymbol{F})$.

Now consider the converse case of a rigid material that stores and transmits heat. In this case the mechanical component in Equation (6.6) vanishes, as does the rate of change of the deformation gradient $\boldsymbol{F}$; hence,

$$\frac{\partial \mathcal{E}}{\partial t} + \text{DIV}\boldsymbol{Q} = r_R. \tag{6.10}$$

This is the classical heat transfer equation expressed in terms of internal energy $\mathcal{E}$, heat fluxes $\boldsymbol{Q}$, and heat source r_R.

In order to derive more useful forms of the energy balance Equation (6.6), it is necessary to identify an energy conjugate thermal pair similar to the stress-strain pair given by the first Piola–Kirchhoff stress $\boldsymbol{P}$ and deformation gradient $\boldsymbol{F}$ used to describe the mechanical energy. To achieve this we need to introduce the concepts of temperature and entropy.

6.2.2 Temperature and Entropy

It is physically intuitive to think that temperature θ is a variable associated with the thermal energy state of a body. Heating a body through the application of heat energy will lead to a rise in its temperature. The term “heat” will be used interchangeably with “thermal energy.” However, the term heat can often be confused with temperature. A large body of water will contain more heat (thermal energy) than a cup of water *at the same* temperature. In general, the temperature θ at a material particle $\boldsymbol{X}$ is a function of time and can be described in the Lagrangian or reference setting as $\theta(\boldsymbol{X}, t)$, or in a spatial setting by $\theta(\boldsymbol{x}, t)$, where

$$\theta(\boldsymbol{X}, t) = \theta\big(\boldsymbol{x}(\boldsymbol{X}, t), t\big). \tag{6.11}$$

In the same way that the magnitude of stress is related to the elastic energy stored at a point in a solid, the temperature is likewise related to the amount of energy stored in the solid during a thermal process. Like stress, temperature can be measured with relative ease in experimental settings, and, for convenience, we will assume that it is given in absolute terms using the Kelvin scale, with units denoted in kelvin (K). Hence θ can only take values greater than zero. Having accepted the existence of temperature as a stress-like variable associated with the thermal energy, the next step is to introduce its energy conjugate variable that will mirror the role played by the strain in the mechanical process. Consequently, in the same way that the energy rate is given by $\boldsymbol{P} : \dot{\boldsymbol{F}}$ in a mechanical process, in a thermal process it is possible to define an energy conjugate variable to θ, to be denoted entropy, η, so that, in a pure thermal process,

$$\left.\frac{\partial \mathcal{E}}{\partial t}\right|_{\boldsymbol{F}=\text{constant}} = \theta \frac{\partial \eta}{\partial t}. \tag{6.12}$$

Observe that, in essence, this relationship is a definition of entropy which avoids the concepts of "measure of disorder" sometimes employed in the literature.

In more general thermodynamic processes, the energy can now be expressed in terms of both deformation and entropy as $\mathcal{E}(\boldsymbol{F}, \eta)$ so that the energy rate contains both mechanical and thermal components:

$$\frac{\partial \mathcal{E}}{\partial t} = \frac{\partial \mathcal{E}(\boldsymbol{F}, \eta)}{\partial \boldsymbol{F}} : \dot{\boldsymbol{F}} + \frac{\partial \mathcal{E}(\boldsymbol{F}, \eta)}{\partial \eta} \dot{\eta}. \tag{6.13}$$

Comparing the above expression to Equations (6.9) and (6.12) gives the energy conjugacy relationships for the first Piola–Kirchhoff stress and temperature as

$$\boldsymbol{P} = \frac{\partial \mathcal{E}}{\partial \boldsymbol{F}}; \quad \theta = \frac{\partial \mathcal{E}}{\partial \eta}. \tag{6.14a,b}$$

It is now possible to rewrite the first law of thermodynamics in terms of the entropy η by combining Equation (6.6) with Equations (6.13) and (6.14a,b) to give a thermal expression in which mechanical terms have been conveniently eliminated to yield

$$\theta \frac{\partial \eta}{\partial t} + \mathrm{DIV}\boldsymbol{Q} = r_R. \tag{6.15}$$

Note that the only assumption made to arrive at this expression is that the internal energy of the material can be expressed in terms of deformation and entropy. Materials that satisfy this assumption are known as "thermoelastic." For more general materials, the internal energy may also be a function of a further set of variables known as "internal state variables," which can describe phenomena such as plastic deformation. For instance, in elasto-plastic materials the internal energy is a function of $\boldsymbol{F}$ via its elastic component $\boldsymbol{F}_e = \boldsymbol{F}\boldsymbol{F}_p^{-1}$, where $\boldsymbol{F}_p$ is the plastic deformation gradient (see Chapter 7 of NL-Statics), enabling $\mathcal{E}$ to be expressed more generally as $\mathcal{E}(\boldsymbol{F}, \eta, \boldsymbol{F}_p)$. The effect of these variables on the energy balance statement Equation (6.6) will be discussed in the next section.

6.3 THE SECOND LAW OF THERMODYNAMICS

The first law of thermodynamics, which describes the energy balance in a body, is invariably presented in conjunction with the second law, or entropy "production inequality," which broadly states that the total entropy of a closed system must increase. Any legitimate constitutive model that describes the relationships between thermomechanical variables in real materials must be such that this statement is never violated. This section explains this law with reference to simple principles, for example the fact that heat (thermal energy) flows from hotter to colder points of a continuum.

6.3.1 Entropy Production in Thermoelasticity

Consider thermoelastic Equation (6.15), but now divide the expression by the absolute temperature θ to give

$$\frac{\partial \eta}{\partial t} + \frac{1}{\theta}\mathrm{DIV}\boldsymbol{Q} = \frac{r_R}{\theta}. \tag{6.16}$$

Using the product rule it is possible to express the second term on the left-hand side as

$$\frac{1}{\theta}\mathrm{DIV}\boldsymbol{Q} = \mathrm{DIV}\left(\frac{\boldsymbol{Q}}{\theta}\right) + \frac{1}{\theta^2}\boldsymbol{Q}\cdot\nabla_0\theta. \tag{6.17}$$

Equation (6.16) can now be rewritten to give an alternative expression involving the rate of entropy:

$$\frac{\partial \eta}{\partial t} + \mathrm{DIV}\left(\frac{\boldsymbol{Q}}{\theta}\right) = \frac{r_R}{\theta} - \frac{1}{\theta^2}\boldsymbol{Q}\cdot\nabla_0\theta. \tag{6.18}$$

Except for the last term, this equation has the form of a conservation law for the total entropy of the solid, where $\boldsymbol{Q}/\theta$ represents the flux of entropy and the right-hand side represents the entropy source per unit undeformed volume. In fact, integration over an arbitrary solid volume V together with use of the divergence theorem gives

$$\frac{d}{dt}\int_V \eta \, dV + \int_{\partial V} \frac{Q_N}{\theta} \, dA = \int_V \frac{r_R}{\theta} \, dV - \int_V \frac{1}{\theta^2} \boldsymbol{Q} \cdot \nabla_0 \theta \, dV, \tag{6.19}$$

where $Q_N = \boldsymbol{Q} \cdot \boldsymbol{N}$ is the heat flow normal to the surface ∂V. For a closed system subject to no external heat source inputs $r_R = 0$ and no fluxes across its surface (that is, $Q_N = 0$ over ∂V), the total entropy in the body varies according to the last term in Equation (6.19). Of course, it is not possible to know the value of this term for any given problem, as it will depend on the actual distribution of temperature θ. However, its sign can be predicted given that heat must flow from hotter to colder regions of the solid, and consequently the direction of $\boldsymbol{Q}$ must oppose the direction of $\nabla_0 \theta$; that is,

$$\boldsymbol{Q} \cdot \nabla_0 \theta \leq 0. \tag{6.20}$$

This implies that Equation (6.19) can be written in the form of an inequality:

$$\frac{d}{dt}\int_V \eta \, dV + \int_{\partial V} \frac{Q_N}{\theta} \, dA - \int_V \frac{r_R}{\theta} \, dV \geq 0. \tag{6.21}$$

This inequality is known as the second law of thermodynamics. It is often presented as a fundamental principle from which Equation (6.20), expressing the directionality of the heat flow, can then be derived. The physical implication of this equation is that the total entropy of a closed system (one without Q_N and r_R) can only grow.

The equivalent differential form of Equation (6.21) is

$$\frac{\partial \eta}{\partial t} + \text{DIV}\left(\frac{\boldsymbol{Q}}{\theta}\right) \geq \frac{r_R}{\theta}. \tag{6.22}$$

Note that the physical consequence of this inequality, that is, the fact that $\boldsymbol{Q}$ must oppose the thermal gradient $\nabla_0 \theta$, implies that thermoelastic processes cannot in general be reversed. Typically, a process can only be reversed if the inequality becomes an equality when $\boldsymbol{Q}$ vanishes. Such processes are known as adiabatic. If, in addition, there are no heat sources, the entropy value will be conserved, in which case the process is known as isentropic.

6.3.2 Internal Dissipation and the Clausius-Duhem Inequality

The expression for the second law of thermodynamics presented in the previous section applies to processes more general than those exclusively associated with thermoelasticity. In fact, from the preceding two equations a fundamental principle can be postulated a priori that, for *all* thermomechanical processes, the following global and equivalent local inequalities are satisfied:

$$\frac{d}{dt}\int_v \eta\, dV + \int_{\partial V} \frac{Q_N}{\theta}\, dA - \int_V \frac{r_R}{\theta}\, dV \geq 0, \tag{6.23a}$$

$$\frac{\partial \eta}{\partial t} + \text{DIV}\left(\frac{\boldsymbol{Q}}{\theta}\right) \geq \frac{r_R}{\theta}. \tag{6.23b}$$

These are usually referred to as the Clausius–Duhem inequalities. It is possible to combine the above local form given by Equation (6.23b) with the general energy conservation Equation (6.6) to give

$$\begin{aligned} 0 &\leq \frac{\partial \eta}{\partial t} + \text{DIV}\frac{\boldsymbol{Q}}{\theta} - \frac{r_R}{\theta} \\ &\leq \theta\frac{\partial \eta}{\partial t} + \theta\,\text{DIV}\frac{\boldsymbol{Q}}{\theta} - r_R \\ &\leq \theta\frac{\partial \eta}{\partial t} + \text{DIV}\boldsymbol{Q} - \frac{1}{\theta}\boldsymbol{Q}\cdot\nabla_0\theta - r_R \\ &\leq \left(\theta\frac{\partial \eta}{\partial t} - \frac{\partial \mathcal{E}}{\partial t} + \boldsymbol{P} : \frac{\partial \boldsymbol{F}}{\partial t}\right) - \frac{1}{\theta}\boldsymbol{Q}\cdot\nabla_0\theta. \end{aligned} \tag{6.24}$$

The term in brackets represents the amount of mechanical work (that is, $\boldsymbol{P} : \dot{\boldsymbol{F}}$) that is not accumulated into internal energy ($\partial\mathcal{E}/\partial t$) or accounted for in the thermal process ($\theta\partial\eta/\partial t$). It is known as the *internal dissipation* and defined as

$$D_{\text{int}} = \boldsymbol{P} : \dot{\boldsymbol{F}} + \theta\dot{\eta} - \dot{\mathcal{E}}. \tag{6.25}$$

For thermoelastic materials, this term vanishes since $\mathcal{E} = \mathcal{E}(\boldsymbol{F}, \eta)$ and $\dot{\mathcal{E}} = \theta\dot{\eta} + \boldsymbol{P} : \dot{\boldsymbol{F}}$. However, for more general materials the internal dissipation measures the amount of energy being lost through internal friction or viscous-type behavior. In these cases, the general Clausius–Duhem inequality can be expressed through Equation (6.24) as

$$D_{\text{int}} - \frac{1}{\theta}\boldsymbol{Q}\cdot\nabla_0\theta \geq 0. \tag{6.26}$$

Furthermore, since the two physical phenomena being accumulated in the above equation, namely heat flow and internal friction, can occur independently of each other, it is invariably accepted that each term must be independently positive; hence,

$$D_{\text{int}} \geq 0; \quad \frac{1}{\theta}\boldsymbol{Q}\cdot\nabla_0\theta \leq 0. \tag{6.27a,b}$$

In order to provide a better understanding of the internal dissipation term, consider the internal energy generally expressed in terms of entropy, deformation, and a set of state variables such as plastic deformation or similar. Assume that these variables are collected in the form of a tensor $\boldsymbol{\alpha}$ so that

$$\mathcal{E} = \mathcal{E}(\eta, \boldsymbol{F}, \boldsymbol{\alpha}). \tag{6.28}$$

As an example of this tensor $\boldsymbol{\alpha}$ consider the case of elastoplasticity, where the elastic energy is expressed in terms of the elastic left Cauchy–Green tensor $\boldsymbol{b}_e = \boldsymbol{F}\boldsymbol{C}_p^{-1}\boldsymbol{F}^T$; hence, in this case, $\mathcal{E} = \mathcal{E}(\eta, \boldsymbol{F}, \boldsymbol{C}_p^{-1})$ so that $\boldsymbol{\alpha} = \boldsymbol{C}_p^{-1}$.

With the inclusion of the state variable tensor $\boldsymbol{\alpha}$ and noting Equation (6.14a,b), the internal dissipation given in Equation (6.25) now becomes

$$D_{\text{int}} = \boldsymbol{P} : \dot{\boldsymbol{F}} + \theta\dot{\eta} - \dot{\mathcal{E}} \tag{6.29a}$$

$$= \boldsymbol{P} : \dot{\boldsymbol{F}} + \theta\frac{\partial \eta}{\partial t} - \frac{\partial \mathcal{E}}{\partial \eta}\dot{\eta} - \frac{\partial \mathcal{E}}{\partial \boldsymbol{F}} : \dot{\boldsymbol{F}} - \frac{\partial \mathcal{E}}{\partial \boldsymbol{\alpha}} : \dot{\boldsymbol{\alpha}} \tag{6.29b}$$

$$= \left(\boldsymbol{P} - \frac{\partial \mathcal{E}}{\partial \boldsymbol{F}}\right) : \dot{\boldsymbol{F}} - \frac{\partial \mathcal{E}}{\partial \boldsymbol{\alpha}} : \dot{\boldsymbol{\alpha}}. \tag{6.29c}$$

The term $(\boldsymbol{P} - \partial\mathcal{E}/\partial\boldsymbol{F})$ vanishes in elasticity but can become nonzero in the presence of internal viscous effects. It is defined as the viscous component of the first Piola–Kirchhoff stress tensor,

$$\boldsymbol{P}_v = \boldsymbol{P} - \frac{\partial \mathcal{E}}{\partial \boldsymbol{F}}, \tag{6.30}$$

and is generally a function of the velocity gradient (or $\dot{\boldsymbol{F}}$). With this definition the internal dissipation becomes

$$D_{\text{int}} = \boldsymbol{P}_v : \dot{\boldsymbol{F}} - \frac{\partial \mathcal{E}}{\partial \boldsymbol{\alpha}} : \dot{\boldsymbol{\alpha}}. \tag{6.31}$$

EXAMPLE 6.1: Spatial form of the second law of thermodynamics

As with the conservation laws given in Chapter 5, the entropy inequality describing the second law of thermodynamics can be written in an updated Lagrangian form and Eulerian representation. In order to derive the updated form, note the usual volume and surface integral transformations

$$\frac{d}{dt}\int_V \eta\, dV = \frac{d}{dt}\int_{v(t)} \eta_x\, dv\ ; \quad \eta_x = J^{-1}\eta,$$

$$\int_{\partial V} \frac{1}{\theta}\boldsymbol{Q}\cdot d\boldsymbol{A} = \int_{\partial v(t)} \frac{1}{\theta}\boldsymbol{q}\cdot d\boldsymbol{a}\ ; \quad \boldsymbol{q} = J^{-1}\boldsymbol{F}\boldsymbol{Q},$$

$$\int_V \frac{r_R}{\theta}\, dV = \int_{v(t)} \frac{r}{\theta}\, dv\ ; \quad r = J^{-1}r_R,$$

(continued)

Example 6.1: *(cont.)*

where η_x is the entropy per unit deformed volume, $v(t)$ denotes the spatial volume occupied by the deformed original Lagrangian volume V, and $\boldsymbol{Q}$ represents the thermal energy flux per unit deformed surface area. With the help of these transformations, Equation (6.21) can be expressed in an updated Lagrangian manner as

$$\frac{d}{dt}\int_{v(t)} \eta_x \, dv + \int_{\partial v(t)} \frac{q_n}{\theta} \, dA - \int_{v(t)} \frac{r}{\theta} \, dv \geq 0\,; \quad q_n = \boldsymbol{q} \cdot \boldsymbol{n}.$$

The full Eulerian equivalent can be derived by making use of the Reynolds transportation theorem, to give

$$\frac{d}{dt}\int_{v(t)} \eta_x \, dv = \frac{d}{dt}\int_{v_x} \eta_x \, dv + \int_{\partial v_x} \eta_x \, \boldsymbol{v} \cdot d\boldsymbol{a},$$

where v_x is now a fixed Eulerian control volume. Consequently the Eulerian version of the second law of thermodynamics becomes

$$\frac{d}{dt}\int_{v_x} \eta_x \, dv + \int_{\partial v_x} \left(\frac{q_n}{\theta} + \eta_x \boldsymbol{v} \cdot \boldsymbol{n}\right) da - \int_{v_x} \frac{r}{\theta} \, dv \geq 0.$$

6.4 THERMOELASTICITY

6.4.1 The Reference Configuration

Sections 6.2.1 and 6.2.2 developed the basic concepts used in thermoelasticity, namely internal energy, temperature, and entropy. The aim of this section is to establish the constitutive equations that relate thermal and mechanical variables. To achieve this it is essential that a reference state for the body be defined both from a mechanical *and* thermal point of view, as different starting temperatures, prior to deformation, clearly imply different states of expansion for most materials. Hence it will be assumed that there is a reference state of the body occupying a volume V with density ρ_R at a uniform reference temperature θ_R. Typically this could be room temperature (approximately 293 K equivalent to 20 °C) at which most experimental testing is performed. In principle, the reference temperature could be taken as absolute zero (0 K), but practical considerations make this choice unhelpful. For instance, any process occurring at room temperature would require consideration of a substantial amount of expansion from the reference state. A general thermomechanical process is then represented in Figure 6.1 by a motion ϕ and

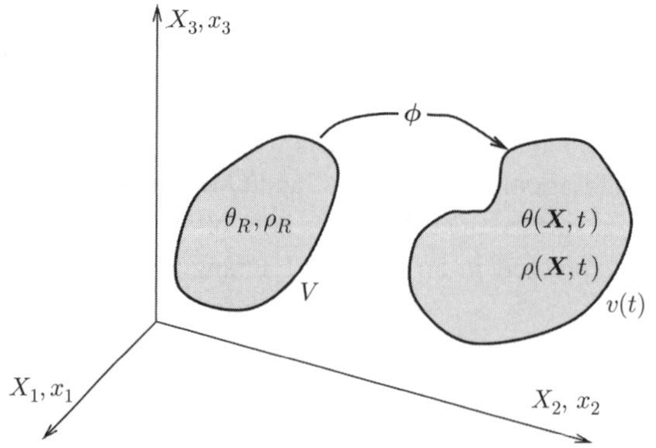

FIGURE 6.1 Reference system.

a thermal change from initial volume V to current volume v, where the changing temperature is measured in Lagrangian form as $\theta(\boldsymbol{X}, t)$ or in Eulerian form as $\theta(\boldsymbol{x}, t)$, where $\theta(\boldsymbol{x}, t) = \theta(\boldsymbol{x}(\boldsymbol{X}, t), t)$. The thermal changes in the body can be governed either by internal processes or by changes in the temperature of the outside medium, or both.

For simplicity it will be assumed that the reference state is stress free throughout the solid. Note that, in the presence of a surrounding fluid, e.g. air, this implies that pressure is measured in a relative sense, where the origin is taken as the pressure of the surrounding fluid at the reference temperature. Similarly, the internal energy will be measured with respect to the origin defined by the reference state. In other words, the internal energy will be zero at the reference configuration. From a strictly physical standpoint this is only true if $\theta_R = 0$ K, but it has no more consequence than a simple shift in the origin of the energy scale.

6.4.2 Energy-Temperature-Entropy Relationships

Experimentally, it is possible to measure the amount of energy required to increase the temperature of a unit mass of material by 1 K. When the volume and shape of the body is kept constant during this process, the resulting physical coefficient is known as the *specific heat at constant volume*, denoted C_v. This definition can be expressed as a relationship between the change in energy and temperature as follows:

$$\left.\frac{d\mathcal{E}}{d\theta}\right|_{\boldsymbol{F}=\text{constant}} = c_v; \qquad c_v = \rho_R C_v. \tag{6.32a,b}$$

Observe that the product by ρ_R is needed as $\mathcal{E}$ has been defined as the energy per unit reference volume, where C_v is typically given as energy per unit mass. In addition, since the deformation is held constant during the determination of C_v, the density does not change as the material is heated.

Equation (6.32a,b) forms the basis of the energy-temperature-entropy constitutive models. In thermoelasticity, the internal energy $\mathcal{E}$ can be expressed as a function of the entropy η and $\boldsymbol{F}$, as given in Equation (6.28); consequently, Equation (6.32a,b) can be recast using the chain rule to yield

$$\left.\frac{\partial \mathcal{E}}{\partial \eta}\frac{d\eta}{d\theta}\right|_{\boldsymbol{F}=\text{constant}} = c_v, \tag{6.33}$$

or, given that $\partial\mathcal{E}/\partial\eta = \theta$,

$$\left.\frac{d\eta}{d\theta}\right|_{\boldsymbol{F}=\text{constant}} = \frac{c_v}{\theta} \quad \text{or} \quad \left.\frac{d\theta}{d\eta}\right|_{\boldsymbol{F}=\text{constant}} = \frac{\theta}{c_v}. \tag{6.34a,b}$$

Equations (6.34a,b) establish a constitutive relationship between the entropy η and the temperature θ at constant deformation gradient. In general, the specific heat coefficient c_v will be a function of θ, but for most materials, at temperatures not too far away from θ_R (but not necessarily infinitesimally near either), this coefficient can be assumed constant for a reasonably large range of temperature variations. Restricting the derivations to this simple constant heat coefficient case enables Equations (6.34a,b) to be integrated analytically with respect to the entropy or temperature changes as

$$\int_{\eta(\boldsymbol{F},\theta_R)}^{\eta(\boldsymbol{F},\theta)} d\eta = \int_{\theta_R}^{\theta} \frac{c_v}{\theta}\, d\theta, \tag{6.35}$$

which leads to a simple relationship between entropy and temperature:

$$\eta(\boldsymbol{F},\theta) = \eta_R(\boldsymbol{F}) + c_v \ln\frac{\theta}{\theta_R}\,; \qquad \eta_R(\boldsymbol{F}) = \eta(\boldsymbol{F},\theta_R). \tag{6.36a,b}$$

In this expression, $\eta_R(\boldsymbol{F})$ represents the value of the entropy per unit reference volume at temperature θ_R, but subject to an arbitrary deformation $\boldsymbol{F}$. Establishing a specific function for $\eta_R(\boldsymbol{F})$ will be a key component of the definition of a thermoelastic constitutive model, and some examples will be given in Section 6.4.8 for a particular thermoelastic model. In general, at the reference configuration, where $\boldsymbol{F} = \boldsymbol{I}$, it is convenient to set $\eta_R(\boldsymbol{I}) = 0$ as the origin for the entropy value, as this simplifies the algebra considerably.

Using Equation (6.36a,b), the reverse relationship yielding the temperature as a function of entropy and deformation is given by

$$\begin{aligned}\theta(\boldsymbol{F},\eta) &= \theta_R e^{(\eta-\eta_R(\boldsymbol{F}))/c_v}\\ &= \theta_R e^{-\eta_R(\boldsymbol{F})/c_v} e^{\eta/c_v}\\ &= \theta_0(\boldsymbol{F}) e^{\eta/c_v}; \quad \theta_0(\boldsymbol{F}) = \theta(\boldsymbol{F},0) = \theta_R e^{-\eta_R(\boldsymbol{F})/c_v},\end{aligned} \tag{6.37a,b}$$

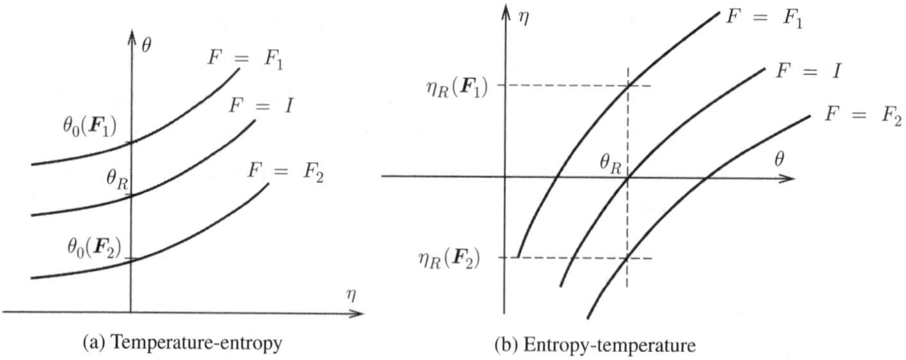

FIGURE 6.2 Temperature-entropy relationships for a material with constant specific heat coefficient.

where $\theta_0(\boldsymbol{F})$ is the temperature that the material takes if entropy is kept constant at zero while the deformation gradient $\boldsymbol{F}$ is changed. Note that if $\boldsymbol{F} = \boldsymbol{I}$ then $\theta_0(\boldsymbol{I}) = \theta_R$ as expected, due to the fact that $\eta_R(\boldsymbol{I}) = 0$. The relationships between entropy and temperature are shown graphically in Figure 6.2.

The relationship between the entropy at the reference temperature $\eta_R(\boldsymbol{F})$ and the temperature at constant entropy $\theta_0(\boldsymbol{F})$ can be reversed using Equation (6.37a,b)$_b$ to give

$$\eta_R(\boldsymbol{F}) = -c_v \ln \frac{\theta_0(\boldsymbol{F})}{\theta_R}. \tag{6.38}$$

The function $\theta_0(\boldsymbol{F})$ can be used in place of $\eta_R(\boldsymbol{F})$ to define the thermal component of the constitutive model. Experimentally, it is simpler to measure temperature than entropy so it is more practical to determine $\theta_0(\boldsymbol{F})$. The function $\theta_0(\boldsymbol{F})$ describes the way the temperature of a material changes when subject to deformation but when it is not permitted to change entropy. Recalling that entropy changes are determined by Equation (6.16) in terms of the heat flow $\boldsymbol{Q}$, maintaining a constant zero entropy implies the absence of heat flow. In practical experimental terms, this means taking measurements immediately after deformation takes place, before the heat flow $\boldsymbol{Q}$ has had time to alter the entropy values. The obvious example is a gas quickly expanding or contracting, leading to an instantaneous decrease or increase in temperature, respectively. These instantaneous changes in temperature are measured by the function $\theta_0(\boldsymbol{F})$.

Finally, for the simple case of a thermoelastic material with a constant specific heat coefficient, it is possible to write an explicit relationship for the internal energy by integrating Equation (6.37a,b)$_a$ expressed using Equation (6.14a,b)$_b$ as

$$\theta(\boldsymbol{F}, \eta) = \left.\frac{d\mathcal{E}}{d\eta}\right|_{\boldsymbol{F}=\text{constant}} = \theta_0(\boldsymbol{F}) e^{\eta/c_v}. \tag{6.39}$$

Integrating the above equation with respect to η between the limits $\eta = 0$ and a given value η yields

$$\int_{\mathcal{E}(\boldsymbol{F},0)}^{\mathcal{E}(\boldsymbol{F},\eta)} d\mathcal{E} = \theta_0(\boldsymbol{F}) \int_0^{\eta} e^{\eta/c_v}\, d\eta, \tag{6.40a}$$

$$\mathcal{E}(\boldsymbol{F},\eta) = \mathcal{E}_0(\boldsymbol{F}) + c_v\theta_0(\boldsymbol{F}) \left[e^{\eta/c_v} - 1\right]; \quad \mathcal{E}_0(\boldsymbol{F}) = \mathcal{E}(\boldsymbol{F},0). \tag{6.40b,c}$$

The term $\mathcal{E}_0(\boldsymbol{F})$ represents the amount of internal energy per unit reference volume accumulated when the body is deformed in an isentropic manner, that is, keeping the entropy constant at its reference value zero.

In order to provide further physical insights into Equation (6.40b,c)$_b$ it is helpful to rewrite it with the help of Equation (6.37a,b) as

$$\mathcal{E}(\boldsymbol{F},\eta) = \mathcal{E}_0(\boldsymbol{F}) + c_v \left[\theta(\boldsymbol{F},\eta) - \theta_0(\boldsymbol{F})\right], \tag{6.41}$$

which restates the starting assumption that $d\mathcal{E}/d\theta$ at a constant $\boldsymbol{F}$ is given by a constant material coefficient c_v.

6.4.3 Stress Evaluation in Thermoelasticity

Recalling Equation (6.9), the first Piola–Kirchhoff stress tensor is evaluated by differentiating the internal energy expression given by Equations (6.40b,c) with respect to the deformation gradient to give

$$\begin{aligned}\boldsymbol{P}(\boldsymbol{F},\eta) &= \frac{\partial \mathcal{E}}{\partial \boldsymbol{F}} = \frac{\partial \mathcal{E}_0}{\partial \boldsymbol{F}} + c_v \frac{\partial \theta_0}{\partial \boldsymbol{F}} \left[e^{\eta/c_v} - 1\right] \\ &= \boldsymbol{P}_0(\boldsymbol{F}) + c_v \frac{\partial \theta_0}{\partial \boldsymbol{F}} \left[e^{\eta/c_v} - 1\right]; \qquad \boldsymbol{P}_0 = \frac{\partial \mathcal{E}_0}{\partial \boldsymbol{F}}.\end{aligned} \tag{6.42a,b}$$

The term $\boldsymbol{P}_0(\boldsymbol{F})$ in the above equation represents the Piola–Kirchhoff stresses when the entropy remains constant at the zero reference value, and the second terms represents the thermal coupling due to changes in entropy. It is possible to combine the above equation with the relationship between temperature and entropy given by Equation (6.37a,b)$_a$ to give, after simple algebra,

$$\boldsymbol{P}(\boldsymbol{F},\eta) = \boldsymbol{P}_0(\boldsymbol{F}) + c_v \frac{\partial \theta_0}{\partial \boldsymbol{F}} \left[\frac{\theta(\boldsymbol{F},\eta) - \theta_0(\boldsymbol{F})}{\theta_0(\boldsymbol{F})}\right]. \tag{6.43}$$

This expression suggests the possibility of direct constitutive relationships between stress and temperature values, as well as the deformation gradient, without the need to evaluate entropy. These types of relationships are considered in detail in Section 6.4.4 via the definition of the Helmholtz free energy.

EXAMPLE 6.2: Rigid heat conductor

A particular case of the energy Equation (6.6) which is of wide use in engineering concerns heat conduction in a rigid material. In this case the energy is a function of temperature alone, and therefore

$$\frac{\partial \mathcal{E}}{\partial t} = c_v \frac{\partial \theta}{\partial t}.$$

In addition, the heat flow $\boldsymbol{Q}$ usually follows Fourier's law of thermal conduction, whereby

$$\boldsymbol{Q} = -k\nabla\theta,$$

where k is a property of the material, known as the thermal conductivity coefficient, that describes how well the material allows the transmission of thermal energy through conduction. Its units are watts per square meter per kelvin (W/m^2 K). For a uniform material the energy equation can now be expressed in the familiar heat conduction form as

$$\frac{\partial \theta}{\partial t} = h\nabla^2\theta + s,$$

where $s = r_R/c_v$ is the heat source term and $h = k/c_v$ is the diffusivity coefficient of the material. Note that h describes how quickly a material returns to a uniform temperature state from an initial disturbed distribution of temperature. High values of h require high thermal conductivity and low specific heat coefficients. The differential equation for the temperature given above is accompanied by boundary conditions for θ or $Q_N = \boldsymbol{Q} \cdot \boldsymbol{N}$ on the external body surface.

6.4.4 Isothermal Processes and the Helmholtz Free Energy

In the constitutive formulations derived above, the main variables defining the state of the system have been the deformation gradient and the entropy. For instance, the internal energy was expressed as a function $\mathcal{E}(\boldsymbol{F}, \eta)$. In order to observe directly the resulting constitutive equations through thermoelastic testing, it would be necessary to conduct experiments in a fast manner so that the effects of thermal diffusion via fluxes $\boldsymbol{Q}$ is minimal and so different states can be observed at constant entropy. In many practical situations, it is more feasible to proceed in a quasi-static manner, so that testing is performed at a given temperature and any increases of temperature due to deformation are allowed to dissipate before taking measurements. In this way it is possible to establish stress-strain relationships for a material at different values of the temperature. In order to facilitate this scenario, it is useful to rewrite the constitutive model as functions of $\boldsymbol{F}$ and θ rather than $\boldsymbol{F}$ and η. Consider first the internal energy. Using the relationship between entropy and temperature given by Equation (6.36a,b), it is possible to express the internal energy as a function of $\boldsymbol{F}$ and θ as

$$\tilde{\mathcal{E}}(\boldsymbol{F},\theta) = \mathcal{E}\big(\boldsymbol{F},\eta(\boldsymbol{F},\theta)\big). \tag{6.44}$$

For instance, at the reference temperature, this relationship can be written, with the help of Equation (6.41), as

$$\tilde{\mathcal{E}}(\boldsymbol{F},\theta_R) = \mathcal{E}_R(\boldsymbol{F}) = \mathcal{E}_0(\boldsymbol{F}) + c_v\big(\theta_R - \theta_0(\boldsymbol{F})\big). \tag{6.45}$$

The internal energy term $\mathcal{E}_R(\boldsymbol{F})$ represents the energy per unit reference volume caused by the deformation after the temperature has been allowed to return to the reference value θ_R. This is in clear contrast to $\mathcal{E}_0(\boldsymbol{F})$, which measures the energy per unit reference volume after the same deformation $\boldsymbol{F}$ but before any heat flow has been allowed to take place, so that the entropy, rather than the temperature, remains at its reference zero value. It is possible to express $\mathcal{E}$ or $\tilde{\mathcal{E}}$ with respect to $\mathcal{E}_R$ rather than $\mathcal{E}_0$ by combining Equations (6.41) and (6.45) to give, after simple algebra,

$$\tilde{\mathcal{E}}(\boldsymbol{F},\theta) = \mathcal{E}_R(\boldsymbol{F}) + c_v(\theta - \theta_R). \tag{6.46}$$

It is important to note that, while both $\tilde{\mathcal{E}}(\boldsymbol{F},\theta)$ and $\mathcal{E}(\boldsymbol{F},\eta)$ represent the same physical quantity, they are different mathematical functions of their independent variables. Hence, the partial derivative of $\tilde{\mathcal{E}}(\boldsymbol{F},\theta)$ with respect to $\boldsymbol{F}$ is not the same as the partial derivative of $\mathcal{E}(\boldsymbol{F},\eta)$ with respect to $\boldsymbol{F}$. In fact, using the chain rule, it is possible to relate these derivatives as

$$\begin{aligned}\frac{\partial\tilde{\mathcal{E}}}{\partial\boldsymbol{F}} &= \frac{\partial\mathcal{E}}{\partial\boldsymbol{F}} + \frac{\partial\mathcal{E}}{\partial\eta}\frac{\partial\eta}{\partial\boldsymbol{F}}\\ &= \frac{\partial\mathcal{E}}{\partial\boldsymbol{F}} + \theta\frac{\partial\eta}{\partial\boldsymbol{F}}.\end{aligned} \tag{6.47}$$

Consequently, the first Piola–Kirchhoff stress can be expressed as

$$\begin{aligned}\boldsymbol{P} = \frac{\partial\mathcal{E}}{\partial\boldsymbol{F}} &= \frac{\partial\tilde{\mathcal{E}}}{\partial\boldsymbol{F}} - \theta\frac{\partial\eta}{\partial\boldsymbol{F}}\\ &= \frac{\partial}{\partial\boldsymbol{F}}(\tilde{\mathcal{E}} - \theta\eta).\end{aligned} \tag{6.48}$$

The term in brackets above is known as the *Helmholtz free energy* $\Psi(\boldsymbol{F},\theta)$, defined as

$$\Psi(\boldsymbol{F},\theta) = \mathcal{E}\big(\boldsymbol{F},\eta(\boldsymbol{F},\theta)\big) - \theta\eta(\boldsymbol{F},\theta). \tag{6.49}$$

Using this free energy expression, the state of stress at a given temperature can be obtained in a direct manner as

$$\boldsymbol{P}(\boldsymbol{F},\theta) = \frac{\partial \Psi}{\partial \boldsymbol{F}}. \tag{6.50}$$

In addition, observe that the derivative of $\Psi(\boldsymbol{F},\theta)$ with respect to the temperature θ gives

$$\begin{aligned}\frac{\partial \Psi}{\partial \theta} &= \frac{\partial \mathcal{E}}{\partial \eta}\frac{\partial \eta}{\partial \theta} - \theta\frac{\partial \eta}{\partial \theta} - \eta \\ &= \theta\frac{\partial \eta}{\partial \theta} - \theta\frac{\partial \eta}{\partial \theta} - \eta \\ &= -\eta. \end{aligned} \tag{6.51}$$

Consequently, a set of constitutive relationships between the pairs $\{\boldsymbol{P},\eta\}$ and $\{\boldsymbol{F},\theta\}$ is obtained as

$$\boldsymbol{P}(\boldsymbol{F},\theta) = \frac{\partial \Psi}{\partial \boldsymbol{F}}; \quad \eta(\boldsymbol{F},\theta) = -\frac{\partial \Psi}{\partial \theta}, \tag{6.52a,b}$$

which are complementary to Equations (6.14a,b), namely

$$\boldsymbol{P}(\boldsymbol{F},\eta) = \frac{\partial \mathcal{E}}{\partial \boldsymbol{F}}; \quad \theta(\boldsymbol{F},\eta) = \frac{\partial \mathcal{E}}{\partial \eta}. \tag{6.53a,b}$$

A constitutive model can thus be defined by providing either the function $\mathcal{E}(\boldsymbol{F},\eta)$ or the function $\Psi(\boldsymbol{F},\theta)$, depending on whether η or θ is used as the primary thermal variable. However, note that the Helmholtz free energy does not represent a physical variable that is conserved (like the actual internal energy), but rather a useful thermodynamic potential that allows constitutive models to be expressed more directly as a function of temperature rather than entropy.

Useful explicit expressions for Ψ often employed in the literature can be derived by substituting Equation (6.36a,b) for the entropy into Equation (6.49), which, taking into account the internal energy expression Equation (6.46) and the definition of $\mathcal{E}_R(\boldsymbol{F})$ given in Equation (6.45), gives, after some algebra,

$$\begin{aligned}\Psi(\boldsymbol{F},\theta) &= \mathcal{E}\big(\boldsymbol{F},\eta(\boldsymbol{F},\theta)\big) - \theta\eta(\boldsymbol{F},\theta) \\ &= \mathcal{E}_R(\boldsymbol{F}) + c_v(\theta-\theta_R) - \theta\left[\eta_R(\boldsymbol{F}) + c_v \ln\frac{\theta}{\theta_R}\right] \\ &= \mathcal{E}_R(\boldsymbol{F}) - \theta\eta_R(\boldsymbol{F}) + c_v\left[\Delta\theta - \theta\ln\frac{\theta}{\theta_R}\right], \end{aligned} \tag{6.54}$$

where $\Delta\theta = \theta - \theta_R$ is the temperature increment. The term $\mathcal{E}_R(\boldsymbol{F})$ can be removed in the above expression by defining the Helmholtz free energy at the reference configuration as

$$\Psi_R(\boldsymbol{F}) = \Psi(\boldsymbol{F}, \theta_R) = \mathcal{E}_R(\boldsymbol{F}) - \theta_R \eta_R(\boldsymbol{F}). \tag{6.55}$$

Substituting for $\mathcal{E}_R$ into Equation (6.54) gives a final expression for Ψ:

$$\Psi(\boldsymbol{F}, \theta) = \Psi_R(\boldsymbol{F}) - \Delta\theta \eta_R(\boldsymbol{F}) + T(\theta); \; T(\theta) = c_v \left[\Delta\theta - \theta \ln \frac{\theta}{\theta_R} \right]. \tag{6.56a,b}$$

In this expression, the first term represents the material behavior at the reference configuration temperature; the middle term accounts for the thermomechanical coupling; and the last term $T(\theta)$ is purely thermal and embodies the temperature-entropy relationships described in Section 6.4.2. For instance, the first Piola–Kirchhoff stress is now

$$\boldsymbol{P} = \boldsymbol{P}_R - \Delta\theta \frac{\partial \eta_R}{\partial \boldsymbol{F}}; \quad \boldsymbol{P}_R = \frac{\partial \Psi_R}{\partial \boldsymbol{F}}. \tag{6.57a,b}$$

In this expression, $\boldsymbol{P}_R(\boldsymbol{F})$ describes the stress-strain behavior at the reference temperature θ_R, whereas the second term provides the change in stress that results from temperature variations.

Specific constitutive models can now be defined by establishing functions $\Psi_R(\boldsymbol{F})$ and $\eta_R(\boldsymbol{F})$ instead of $\mathcal{E}_0(\boldsymbol{F})$ and $\theta_0(\boldsymbol{F})$, see Equations (6.40b,c). That is, the material is defined by its behavior at the reference temperature through $\Psi_R(\boldsymbol{F})$ and the thermomechanical coupling function $\eta_R(\boldsymbol{F})$ instead of being defined by the behavior at constant zero entropy through $\mathcal{E}_0(\boldsymbol{F})$ and the coupling function $\theta_0(\boldsymbol{F})$. There are, of course, mathematical relationships between these two pairs of functions. For instance, the relationship between $\theta_0(\boldsymbol{F})$ and $\eta_R(\boldsymbol{F})$ is given in Equation (6.37a,b) as

$$\theta_0(\boldsymbol{F}) = \theta_R e^{-\eta_R(\boldsymbol{F})/c_v} \quad \text{or} \quad \eta_R(\boldsymbol{F}) = -c_v \ln \frac{\theta_0(\boldsymbol{F})}{\theta_R}, \tag{6.58a,b}$$

and the relationship between $\mathcal{E}_0$ and Ψ_R can be obtained by combining Equations (6.45) and (6.55) to give

$$\begin{aligned} \Psi_R(\boldsymbol{F}) &= \mathcal{E}_R(\boldsymbol{F}) - \theta_R \eta_R(\boldsymbol{F}) \\ &= \mathcal{E}_0(\boldsymbol{F}) - c_v \theta_0(\boldsymbol{F}) + c_v \theta_R - \theta_R \eta_R(\boldsymbol{F}) \\ &= \mathcal{E}_0(\boldsymbol{F}) - c_v \left[\theta_0(\boldsymbol{F}) - \theta_R - \theta_R \ln \frac{\theta_0(\boldsymbol{F})}{\theta_R} \right]. \end{aligned} \tag{6.59}$$

This allows a material to be defined via either its isothermal behavior or its isentropic behavior, depending on the type of testing regime employed.

6.4.5 Entropic Elasticity

Sections 6.4.2 and 6.4.4 describe the general form of constitutive models in thermoelasticity. Consideration is now focused on one of the simplest possible models of practical significance, namely the so-called "entropic elasticity" model often used to describe thermomechanical effects in rubber materials. This is based on the assumption that during deformation processes the internal energy is entirely associated with the generation of heat in the material. Recalling the expression for the internal energy given in the previous section by Equation (6.46), entropic elasticity is therefore defined by a vanishing $\mathcal{E}_R(\boldsymbol{F})$, that is,

$$\mathcal{E}_R(\boldsymbol{F}) = 0. \tag{6.60}$$

The Helmholtz free energy at the reference temperature is now given by Equation (6.55) as

$$\Psi_R(\boldsymbol{F}) = -\theta_R \eta_R(\boldsymbol{F}), \tag{6.61}$$

and introducing this expression into Equation (6.56a,b) gives

$$\begin{aligned} \Psi(\boldsymbol{F}, \theta) &= \Psi_R(\boldsymbol{F}) - \Delta\theta \eta_R(\boldsymbol{F}) + T(\theta) \\ &= -\theta_R \eta_R(\boldsymbol{F}) + (\theta_R - \theta)\eta_R(\boldsymbol{F}) + T(\theta) \\ &= -\theta \eta_R(\boldsymbol{F}) + T(\theta). \end{aligned} \tag{6.62}$$

The final equation in this derivation justifies the name "entropic" elasticity for this model since the entire elastic behavior of the material, which is governed by the dependency of Ψ on $\boldsymbol{F}$, is determined by the entropy function at constant temperature $\eta_R(\boldsymbol{F})$. Observe also that Equations (6.62) and (6.61) can be combined to give

$$\Psi(\boldsymbol{F}, \theta) = \frac{\theta}{\theta_R} \Psi_R(\boldsymbol{F}) + T(\theta). \tag{6.63}$$

Hence the first Piola–Kirchhoff stresses are found at different temperatures by the simple relationship to the stress behavior at the reference temperature given by

$$\boldsymbol{P}(\boldsymbol{F}, \theta) = \frac{\theta}{\theta_R} \boldsymbol{P}_R(\boldsymbol{F}); \quad \boldsymbol{P}_R(\boldsymbol{F}) = \frac{\partial \Psi_R}{\partial \boldsymbol{F}}. \tag{6.64a,b}$$

Well-known specific rubber elasticity functions $\Psi_R(\boldsymbol{F})$ can be established, such as the neo-Hookean and Mooney–Rivlin models. For example, a neo-Hookean entropic elastic model is defined by

$$\Psi_R(\boldsymbol{F}) = \frac{1}{2}\mu_R(\boldsymbol{F} : \boldsymbol{F} - \boldsymbol{I}) - \mu_R \ln J + \frac{1}{2}\lambda_R (J-1)^2, \tag{6.65}$$

where μ_R and λ_R are the shear and Lamé coefficients, respectively, observed at the reference temperature θ_R.

6.4.6 Perfect Gas

Although this book is concerned with the behavior of solids, it is instructive to see that the constitutive behavior of a perfect gas can be reproduced as a simple example of entropic elasticity, where the free energy term $\Psi(\boldsymbol{F})$ is a function of $\boldsymbol{F}$ via its determinant J, that is, $\Psi_R(\boldsymbol{F}) = \Psi_R(J)$. Consequently, Equation (6.64a,b)$_a$ yields

$$\begin{aligned}\boldsymbol{P}(\boldsymbol{F}) &= \frac{\theta}{\theta_R}\frac{\partial \Psi_R}{\partial \boldsymbol{F}} \\ &= \frac{\theta}{\theta_R}\frac{\partial \Psi_R(J)}{\partial J}\frac{\partial J}{\partial \boldsymbol{F}}.\end{aligned} \tag{6.66}$$

The term $\partial J/\partial \boldsymbol{F}$ is derived from the linearization of the determinant of a tensor (see Chapter 2, NL-Statics) to give

$$\frac{\partial J}{\partial \boldsymbol{F}} = \frac{\partial \det \boldsymbol{F}}{\partial \boldsymbol{F}} = J\boldsymbol{F}^{-T}, \tag{6.67}$$

which enables the first Piola–Kirchhoff stress tensor to be expressed in terms of the pressure p as

$$\boldsymbol{P} = pJ\boldsymbol{F}^{-T}; \quad p = \frac{\theta}{\theta_R}\frac{\partial \Psi_R}{\partial J}. \tag{6.68a,b}$$

In order to complete the definition of this constitutive model, recall that for a perfect gas the pressure, volume, and temperature are related by the combined gas law as

$$\frac{pV}{\theta} = \frac{p_R V_R}{\theta_R} = \text{constant} \quad \text{or} \quad p = \frac{\theta}{\theta_R}\frac{p_R}{J}, \tag{6.69a,b}$$

where p_R and V_R are the pressure and volume of the gas at the reference state. Absolute rather than relative pressures are used above to remain compatible with the ideal gas laws. Comparing Equations (6.68a,b)$_b$ with (6.69a,b)$_b$ and integrating gives

$$\frac{\partial \Psi_R}{\partial J} = \frac{p_R}{J}; \quad \Psi_R = p_R \ln J, \tag{6.70a,b}$$

and, therefore, the complete model is defined by a Helmhotz free energy given by Equation (6.63) as

$$\Psi(J,\theta) = \frac{\theta}{\theta_R} p_R \ln J + T(\theta), \tag{6.71}$$

which is clearly a particular case of an entropic elasticity model where the Helmholtz free energy is a function of J and θ only.

EXAMPLE 6.3: Modified entropic elasticity

The entropic elasticity model described in Section 6.4.5 is rather strict and can only be applied to a few actual materials. A more useful alternative is the so-called modified entropic elasticity model, in which the energy at the reference temperature, that is, $\mathcal{E}_R(\boldsymbol{F})$, is not assumed to vanish, but rather is a function of the volume ratio J:

$$\mathcal{E}_R(\boldsymbol{F}) = \mathcal{E}_R(J).$$

The Helmholtz free energy at the reference configuration is now given by

$$\Psi_R(\boldsymbol{F}) = \mathcal{E}_R(J) - \theta_R \eta_R(\boldsymbol{F}),$$

and the full Helmholtz potential is

$$\Psi(\boldsymbol{F}, \theta) = \frac{\theta}{\theta_R}\Psi_R(\boldsymbol{F}) - \frac{\Delta\theta}{\theta_R}\mathcal{E}_R(J) + T(\theta).$$

If the term Ψ_R is split into volumetric and isochoric free energy components,

$$\Psi_R(\boldsymbol{F}) = \Psi_{R,vol}(J) + \hat{\Psi}_R(\hat{\boldsymbol{F}}),$$

it is simple to show that the deviatoric component of the stress is obtained, as in the case of entropic elasticity, as follows:

$$\boldsymbol{P}' = \frac{\theta}{\theta_R}\frac{\partial \hat{\Psi}_R(\hat{\boldsymbol{F}})}{\partial \boldsymbol{F}} = \frac{\theta}{\theta_R}\boldsymbol{P}'_R; \quad \boldsymbol{P}'_R = \frac{\partial \hat{\Psi}_R(\hat{\boldsymbol{F}})}{\partial \boldsymbol{F}},$$

whereas the pressure component is now given by

$$p = \frac{\theta}{\theta_R}p_R - \frac{\Delta\theta}{\theta_R}\frac{d\mathcal{E}_R}{dJ}; \quad p_R = \frac{d\Psi_{R,vol}}{dJ}.$$

6.4.7 Distortional-Volumetric Energy Decomposition

The models defined by the entropic elasticity assumption are very specific to rubbers and are therefore of limited use in solid mechanics. A more general formulation, valid for a wide range of materials, including most metals, relies on the assumption that the shear or distortional behavior of the material can be split from the volumetric response and, more crucially, that the thermomechanical coupling is associated only with the volumetric component of the deformation. This assumption closely matches the intuitive expectation that changes of temperature lead only to changes in volume and not to the shape distortion of the material. In addition, the assumptions enshrined in this model must be such that, if the material is completely confined so that it cannot deform or expand, then a rise in temperature will lead to an increase in pressure but not to shear stresses. Mathematically, these assumptions are expressed by means of an additive decomposition of the internal energy into volumetric and distortional components as

$$\mathcal{E}(\boldsymbol{F},\eta) = \hat{\mathcal{E}}(\hat{\boldsymbol{F}}) + U(J,\eta), \tag{6.72}$$

where the distortional component $\hat{\mathcal{E}}$ is a function of the deformation gradient $\boldsymbol{F}$ via its isochoric or distorsional component $\hat{\boldsymbol{F}}$ defined by

$$\hat{\boldsymbol{F}} = (\det \boldsymbol{F})^{-1/3}\boldsymbol{F} = J^{-1/3}\boldsymbol{F}. \tag{6.73}$$

Recall that, by construction, $\det \hat{\boldsymbol{F}} = 1$ (see Chapter 4 of NL-Statics). The fact that $\hat{\mathcal{E}}$ is not a function of the entropy implies that there is no thermomechanical coupling in relation to the deviatoric or shear behavior of the material.

In Equation (6.72) $U(J,\eta)$ represents the volumetric internal energy which is responsible for the thermomechanical coupling through its dependence upon the entropy η. In particular, recall Equations (6.40b,c), which enables the volumetric internal energy to be decomposed into a purely mechanical term and a coupled thermomechanical component as

$$U(J,\eta) = U_0(J) + c_v\theta_0(J)\left[e^{\eta/c_v} - 1\right], \tag{6.74}$$

where $U_0(J)$ denotes the internal volumetric energy relative to the entropy at the reference state. Note that the dependency of θ_0 upon $\boldsymbol{F}$ has been restricted to the volumetric component J.

The evaluation of the first Piola–Kirchhoff stress now emerges as a deviatoric-pressure decomposition given by

$$\begin{aligned}\boldsymbol{P} &= \frac{\partial \mathcal{E}}{\partial \boldsymbol{F}} = \frac{\partial \hat{\mathcal{E}}}{\partial \boldsymbol{F}} + \frac{dU}{dJ}\frac{\partial \det \boldsymbol{F}}{\partial \boldsymbol{F}} \\ &= \boldsymbol{P}' + pJ\boldsymbol{F}^{-T},\end{aligned} \tag{6.75}$$

where the deviatoric first Piola–Kirchhoff stress and pressure are

$$\boldsymbol{P}' = \frac{\partial \hat{\mathcal{E}}}{\partial \boldsymbol{F}}; \quad p = \frac{dU}{dJ}, \tag{6.76a,b}$$

and the differentiation of the determinant of the deformation gradient is evaluated in accordance with Chapter 2 of NL-Statics to give the term $J\boldsymbol{F}^{-T}$. Using Equation (6.74), the pressure can be written as

$$\begin{aligned}p(J,\eta) &= \frac{dU_0}{dJ} + c_v\frac{d\theta_0}{dJ}\left[e^{\eta/c_v} - 1\right] \\ &= p_0(J) + c_v\frac{d\theta_0}{dJ}\left[e^{\eta/c_v} - 1\right]; \quad p_0(J) = \frac{dU_0}{dJ}.\end{aligned} \tag{6.77}$$

In this expression, $p_0(J)$ describes the way in which the pressure in the solid changes as a consequence of changes in volume while the entropy remains constant, that is, before any heat flow has had an opportunity to diffuse increases or decreases in temperature that would also be a consequence of the volume change.

Using the split between distortional and volumetric components of the energy, a thermoelastic material can be fully defined through its distortional energy component $\hat{\mathcal{E}}(\hat{\boldsymbol{F}})$, its volumetric energy function at constant zero entropy $U_0(J)$, and the thermoelastic coupling function $\theta_0(J)$. This last function measures the changes in temperature due to changes in volume when the process takes place at constant zero entropy, that is, without heat flow.

The same decomposition between volumetric and distortional energy components can be established in terms of the Helmholtz free energy for processes that are either controlled, or more conveniently described, by temperature rather than entropy. For this purpose, a similar deviatoric-volumetric decomposition for Ψ can be established as

$$\Psi(\boldsymbol{F},\theta) = \hat{\Psi}(\hat{\boldsymbol{F}}) + \Psi_{vol}(J,\theta), \tag{6.78}$$

where, as in the case of the internal energy, the thermomechanical coupling takes place only in the volumtric term. The relationship between $\mathcal{E}$ and Ψ given in Equation (6.49), $\mathcal{E} = \Psi - \theta\eta$, combined with the respective volumetric-distortional decomposition of Ψ and $\mathcal{E}$, imply that

$$\hat{\mathcal{E}}(\hat{\boldsymbol{F}}) = \hat{\Psi}(\hat{\boldsymbol{F}}); \quad \Psi_{vol}(J,\theta) = U\big(J,\eta(J,\theta)\big) - \theta\eta(J,\theta). \tag{6.79a,b}$$

That is, there is no difference between the distortional internal energy and the Helmholtz free energy since the thermal effects are confined to the volumetric components. By defining $\Psi_{vol,R}$ as

$$\begin{aligned}\Psi_{vol,R}(J) &= \Psi_{vol}(J,\theta_R)\\ &= U\big(J,\eta_R(J)\big) - \theta_R\eta_R(J)\\ &= U_0(J) + c_v\big(\theta_R - \theta_0(J)\big) - \theta_R\eta_R(J),\end{aligned} \tag{6.80}$$

and repeating the derivation that led to Equation (6.56a,b) for Ψ in terms of Ψ_R,

$$\Psi(\boldsymbol{F},\theta) = \hat{\mathcal{E}}(\hat{\boldsymbol{F}}) + \Psi_{vol,R}(J) - \Delta\theta\eta_R(J) + T(\theta). \tag{6.81}$$

The first Piola–Kirchhoff stress tensor decomposition is now given by the same deviatoric component $\boldsymbol{P}' = \partial\hat{\mathcal{E}}/\partial\boldsymbol{F}$ and a pressure component, now defined by

$$\boldsymbol{P} = \boldsymbol{P}' + pJ\boldsymbol{F}^{-T}, \tag{6.82a}$$

$$\begin{aligned} p &= \frac{d\Psi_{vol,R}}{dJ} - \Delta\theta\frac{d\eta_R}{dJ} \\ &= p_R(J) - \Delta\theta\frac{d\eta_R}{dJ} \end{aligned} \tag{6.82b}$$

Hence, the material model is now given by the functions $\hat{\mathcal{E}}(\hat{\boldsymbol{F}})$, $\Psi_{vol,R}(J)$, and the coupling term $\eta_R(J)$. It is worth recalling at this juncture that the thermomechanical coupling can be expressed in terms of entropy $\eta_R(J)$ or temperature $\theta_0(J)$, and that these terms are related via Equation (6.36a,b) as

$$\theta_0(J) = \theta_R e^{-\eta_R(J)/c_v}; \quad \eta_R(J) = -c_v \ln\frac{\theta_0(J)}{\theta_R}. \tag{6.83a,b}$$

6.4.8 Mie–Grüneisen Equation of State

One of the most widely used constitutive models describing the thermal coupling in solids is the Mie–Grüeisen equation of state. This is based on observed relationships between the pressure and thermal energy obtained experimentally from a confined material. For instance, in linear elasticity, an isotropic material with a linear thermal expansion coefficient α, measured per kelvin (K^{-1}) will experience a volumetric expansion equal to 3α per unit of increased temperature. If the volume remains constant, the pressure variation due to the temperature rise would be given by the formula

$$\Delta p = -3\alpha\kappa\Delta\theta, \tag{6.84}$$

where κ is the bulk modulus; the minus sign is required as pressures are taken as positive in tension and negative in compression. If the thermal change is expressed in terms of energy using the specific heat coefficient c_v, the above relationship would become

$$\Delta p = -\frac{3\alpha\kappa}{c_v}\Delta\mathcal{E}. \tag{6.85}$$

The Mie–Grüneisen model generalizes this relationship to the nonlinear material context by expressing it as

$$J\left.\frac{dp}{d\mathcal{E}}\right|_{J=\text{constant}} = -\Gamma(J), \tag{6.86}$$

which simply means the way pressure changes with thermal energy at confined volumes ($J=$ constant) is a function only of J, not of the temperature θ (and neither is it a function of either η or $\mathcal{E}$). The function $\Gamma(J)$ is obtained through experimental measurements and is generally found to satisfy a power law relationship,

$$\Gamma(J) = \Gamma_0 J^q, \tag{6.87}$$

where Γ_0 is a positive material constant and q varies from zero for a perfect gas to unity for solid materials. Consequently, Equation (6.86) can be rewritten as

$$\left.\frac{dp}{d\mathcal{E}}\right|_{J=\text{constant}} = -\Gamma_0 J^{q-1}. \tag{6.88}$$

This relationship between pressure and energy provides a measure of the thermo-mechanical coupling, which in the sections above was generally described by the functions $\eta_R(J)$ or $\theta_0(J)$. It is therefore possible to obtain explicit expressions for either of these two coupling functions by starting from Equation (6.88). In order to achieve this, note first that changes in energy and temperature at constant volume are related by the specific heat coefficient c_v, that is, $d\mathcal{E}/d\theta|_J = c_v$. Combining this relationship with the derivative with respect to θ of the pressure expression given in Equation (6.88) gives

$$\left.\frac{dp}{d\mathcal{E}}\right|_{J=\text{constant}} = \frac{1}{c_v}\left.\frac{dp}{d\theta}\right|_{J=\text{constant}} = -\frac{1}{c_v}\frac{d\eta_R}{dJ} = -\Gamma_0 J^{q-1}, \tag{6.89}$$

from which $\eta_R(J)$ can be integrated to give

$$\int_0^{\eta_R} d\eta_R = \int_1^J c_v\Gamma_0 J^{q-1}\, dJ; \quad \eta_R(J) = c_v\Gamma_0 \frac{J^q - 1}{q}. \tag{6.90a,b}$$

For the limiting case of $q = 0$, Equation (6.90a,b)$_b$ becomes

$$\eta_R(J) = c_v\Gamma_0 \ln J, \tag{6.91}$$

which coincides with the expression derived for perfect gases in Equation (6.70a,b)$_b$ by identifying $\Gamma_0 = -p_R/c_v$. Again, the negative pressure sign is necessary as discussed previously. The variation of pressure with temperature can be evaluated using Equations (6.82b) and (6.89) to give

$$p(J,\theta) = p_R(J) - c_v\Gamma_0\Delta\theta J^{q-1}. \tag{6.92}$$

Exercises

1. Consider the Mie–Grüneisen material given by Equation (6.88) with constant $q = 1$ and let the volumetric behavior of the material at reference temperature be defined by a simple quadratic function

$$\Psi_{R,vol}(J) = \frac{1}{2}\kappa_R(J-1)^2,$$

where κ_R is the bulk modulus at temperature θ_R.

Derive the corresponding volumetric energy function $U_0(J,\eta)$. By differentiating this expression with respect to J and taking $\eta = 0$, obtain the pressure function, $p_0(J)$, at zero entropy. If κ_0 is defined as the bulk modulus at zero entropy and $J = 1$, that is,

$$\kappa_0 = \left.\frac{dp_0}{dJ}\right|_{J=1},$$

obtain a relationship between κ_R, the bulk modulus at constant reference temperature, and κ_0, the bulk modulus at constant entropy (at $J = 1$), in terms of Γ_0 and c_v.

2. Consider the same Mie–Grüneisen case as in Exercise 1 but now let the volumetric behavior of the material be defined by an energy function at constant zero entropy:

$$U_0(J) = \frac{1}{2}\kappa_0(J-1)^2.$$

Determine the corresponding behavior of the material at constant temperature by finding $\Psi_{R,vol}(J)$. If κ_R is now defined as

$$\kappa_R = \left.\frac{dp_R}{dJ}\right|_{J=1},$$

where $p_R(J)$ is the pressure at temperature θ_R, that is,

$$p_R = \frac{d\Psi_{R,vol}}{dJ},$$

show that κ_R and κ_0 are related by the same expression as in Exercise 1.

3. For the perfect gas model described in Section 6.4.6, determine the energy entropy function $U(J, \eta)$ and the corresponding pressure entropy function $p(J, \eta)$.

4. The linearization of the first Piola–Kirchhoff stress tensor with respect to changes in $\boldsymbol{F}$ and η can be written as

$$D\boldsymbol{P}[\delta\boldsymbol{F}, \delta\eta] = \boldsymbol{\mathcal{A}}_\eta : \delta\boldsymbol{F} + \boldsymbol{T}_\eta\delta\eta,$$

$$\boldsymbol{\mathcal{A}}_\eta = \frac{\partial\boldsymbol{P}}{\partial\boldsymbol{F}} = \frac{\partial^2\mathcal{E}}{\partial\boldsymbol{F}\partial\boldsymbol{F}};\ \boldsymbol{T}_\eta = \frac{\partial\boldsymbol{P}}{\partial\eta} = \frac{\partial^2\mathcal{E}}{\partial\boldsymbol{F}\partial\eta},$$

together with the linearization of the temperature function $\theta(\boldsymbol{F}, \eta)$ as

$$D\theta[\delta\boldsymbol{F}, \delta\eta] = \boldsymbol{T}_\eta : \delta\boldsymbol{F} + T_{\eta\eta}\delta\eta;\ T_{\eta\eta} = \frac{\partial\theta}{\partial\eta} = \frac{\partial^2\mathcal{E}}{\partial\eta\partial\eta}.$$

Show that, for a material with constant specific heat coefficient c_v,

$$T_{\eta\eta} = \frac{\theta}{c_v};\ \boldsymbol{T}_\eta = \frac{\theta}{\theta_0}\frac{\partial\theta_0}{\partial\boldsymbol{F}};\ \boldsymbol{\mathcal{A}}_\eta = \boldsymbol{\mathcal{A}}_0 + \frac{\theta-\theta_0}{\theta_0}\frac{\partial^2\theta_0}{\partial\boldsymbol{F}\partial\boldsymbol{F}},$$

where

$$\boldsymbol{\mathcal{A}}_0 = \frac{\partial^2 \mathcal{E}_0}{\partial \boldsymbol{F} \partial \boldsymbol{F}}.$$

Particularize these expressions for the case of a neo-Hookean Mie–Grüneisen material with $q = 1$ and

$$U_0(J) = \frac{1}{2}\kappa(J-1)^2; \quad \hat{\mathcal{E}}(\boldsymbol{F}) = \frac{1}{2}\mu(\hat{\boldsymbol{F}} : \hat{\boldsymbol{F}}).$$

5. Again consider the linearization of the first Piola–Kirchhoff stress and entropy, but now as functions of the deformation gradient $\boldsymbol{F}$ and the temperature θ via the Helmholtz free energy function as

$$DP[\delta \boldsymbol{F}, \delta\theta] = \boldsymbol{\mathcal{A}}_\theta : \delta \boldsymbol{F} + \boldsymbol{T}_\theta \delta\theta; \; \boldsymbol{\mathcal{A}}_\theta = \frac{\partial^2 \Psi}{\partial \boldsymbol{F} \partial \boldsymbol{F}}; \; \boldsymbol{T}_\theta = \frac{\partial^2 \Psi}{\partial \boldsymbol{F} \partial \theta},$$

$$-D\eta[\delta \boldsymbol{F}, \delta\theta] = \boldsymbol{T}_\theta \delta\theta + T_{\theta\theta}\delta\theta; \; T_{\theta\theta} = \frac{\partial^2 \Psi}{\partial \theta \partial \theta}.$$

Show that

$$\boldsymbol{\mathcal{A}}_\theta = \boldsymbol{\mathcal{A}}_R - \Delta\theta \frac{\partial^2 \eta_R}{\partial \boldsymbol{F} \partial \boldsymbol{F}}; \quad \boldsymbol{\mathcal{A}}_R = \frac{\partial \boldsymbol{P}_R}{\partial \boldsymbol{F}} = \frac{\partial^2 \Psi_R}{\partial \boldsymbol{F} \partial \boldsymbol{F}},$$

$$\boldsymbol{T}_\theta = -\frac{\partial \eta_R}{\partial \boldsymbol{F}}; \quad \boldsymbol{T}_{\theta\theta} = -\frac{c_v}{\theta}.$$

Derive these expressions for the particular case of entropic elasticity for which $\Psi_R = \frac{1}{2}\mu_R(\boldsymbol{F} : \boldsymbol{F} - 3) - \mu_R \ln J + \frac{1}{2}\lambda_R(J-1)^2$.

6. Show that, in general, the isentropic elasticity tensor $\boldsymbol{\mathcal{A}}_\eta$ and the isothermal elasticity tensor $\boldsymbol{\mathcal{A}}_\theta$ are related as

$$\boldsymbol{\mathcal{A}}_\eta = \boldsymbol{\mathcal{A}}_\theta + \frac{\theta}{c_v} \boldsymbol{T}_\eta \otimes \boldsymbol{T}_\eta \,; \quad \boldsymbol{\mathcal{A}}_\theta = \boldsymbol{\mathcal{A}}_\eta - \frac{c_v}{\theta} \boldsymbol{T}_\theta \otimes \boldsymbol{T}_\theta,$$

where $\boldsymbol{T}_\eta$ and $\boldsymbol{T}_\theta$ satisfy

$$\boldsymbol{T}_\eta = -\frac{c_v}{\theta} \boldsymbol{T}_\theta.$$

7. Using the relationships between internal energy $\mathcal{E}$, entropy, temperature, and the coefficient $c_v = (d\mathcal{E}/d\theta)|_J = \theta\,(d\eta/d\theta)|_J$, show that the Mie–Grüneisen coefficient $\Gamma(J)$ can be variously expressed as

$$\Gamma(J) = -J \left.\frac{dp}{d\mathcal{E}}\right|_J = \frac{J}{\theta} \frac{\partial^2 \mathcal{E}(J,\eta)}{\partial J \partial \eta} = -\left.\frac{d \ln \theta}{d \ln J}\right|_{\eta = \text{constant}}.$$

CHAPTER SEVEN

SPACE AND TIME DISCRETIZATION OF CONSERVATION LAWS IN SOLID DYNAMICS

7.1 INTRODUCTION

Industrial applications involving transient large strain deformations are typically modeled using explicit low-order displacement-based finite element formulations. This is due to their relative robustness, accuracy (especially in terms of displacements), and computational efficiency (due to their small number of integration points). The use of trilinear hexahedral elements in conjunction with a selective integration type approach (i.e. mean dilatation method) is nowadays a well-established methodology embedded in most commercial codes. Nonetheless, for challenging modeling problems such as crashworthiness and drop-impact, involving complex three-dimensional geometries subjected to large distortions, the possibility of using linear tetrahedral finite elements is seen as ideal due to the maturity and robustness of existing tetrahedral mesh generators and mesh adaptation techniques.

Unfortunately, the limited use of this finite element technology must be attributed to one or more of the following shortcomings. First, the reduced order of convergence for derived variables (i.e. second order for displacements but first order for strains and stresses) requires some form of stress recovery procedure if the latter are of interest. Second, the performance of these formulations in bending dominated scenarios can be very poor, yielding unacceptable results unless extremely fine meshes are employed across the thickness of the component. Third, numerical instabilities occur in the form of volumetric locking, shear locking, and spurious hydrostatic pressure fluctuations when large Poisson's ratios are used or in the context of isochoric plastic deformations. Fourth, from the time discretization point of view, prototypical Newmark-type methods have a tendency to introduce high frequency noise, especially in the vicinity of sharp spatial gradients, and accuracy is degraded once numerical artificial damping is employed. These schemes are thus not desirable for shock-dominated problems.

With this in mind, this chapter will present an alternative computational framework to the classical displacement-based methodology whereby the system of total Lagrangian conservation laws presented in Chapter 5 is discretized in space and time. Equal (first-order) interpolation is used for all the conservation variables seeking optimal (second-order) convergence in all variables. From the spatial discretization standpoint, this chapter introduces a Petrov–Galerkin method, where appropriately chosen work conjugate virtual fields are selected in order to guarantee the stability of the resulting scheme. From the time discretization standpoint, the system of first-order conservation laws is integrated in time via a simple second-order Runge–Kutta type explicit time integrator. The content of the chapter is divided into three generic blocks: (1) one-dimensional isothermal elasticity in terms of the one-dimensional components of the linear momentum and the deformation gradient $\{p_x, F_{xX}\}$; (2) three-dimensional isothermal elasticity in terms of the generic linear momentum and deformation gradient tensor $\{\boldsymbol{p}_R, \boldsymbol{F}\}$; (3) three-dimensional thermoelasticity, where the entropy of the system is an additional conservation variable to be solved for, namely $\{\boldsymbol{p}_R, \boldsymbol{F}, \eta\}$. The resulting set of procedures is commonly described as "mixed" in the sense that deformation, represented here by $\boldsymbol{F}$, is solved as a separate variable to geometry, represented here by $\boldsymbol{p}_R = \rho_R \boldsymbol{v}$.

7.2 A SIMPLE ONE-DIMENSIONAL SYSTEM OF CONSERVATION LAWS

We start by considering the simple case of a uniaxial deformation* taking place in a domain defined by a space interval $[0, L]$ oriented along the X axis with reference cross-sectional area A, as shown in Figure 7.1. In this case, the governing equations of the problem can be obtained from Equations (5.29a) and (5.29b), which, in the case of a single spatial dimension x in the spatial setting and X in the reference setting, become the following one-dimensional system of conservation laws:

$$\frac{\partial p_x}{\partial t} - \frac{\partial P_{xX}}{\partial X} = f_x, \tag{7.1a}$$

$$\frac{\partial F_{xX}}{\partial t} - \frac{\partial v_x}{\partial X} = 0, \tag{7.1b}$$

where v_x denotes the velocity of the motion, $p_x = \rho_R v_x$ is its corresponding linear momentum per unit undeformed volume (note the use of the material density ρ_R), f_x is a possible body force per unit of undeformed volume, and F_{xX} and P_{xX} represent the $(\cdot)_{xX}$ components of the deformation gradient and the first Piola–Kirchhoff stress tensors, respectively (oriented along the X axis). For simplicity, we

* As typical in uniaxial deformation, only the stress along the axis of deformation is considered non-negligible.

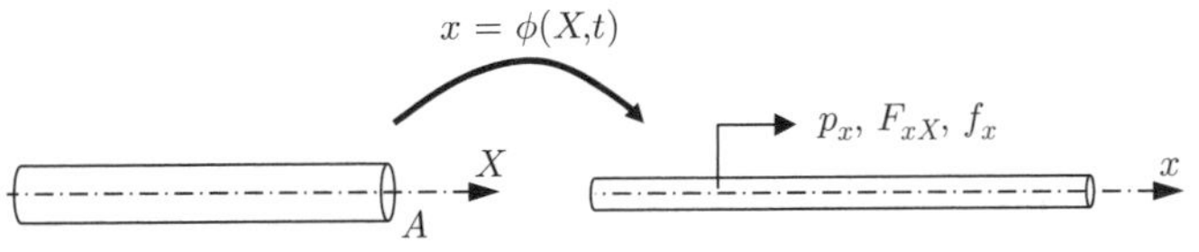

FIGURE 7.1 One-dimensional deformation.

assume that the material density ρ_R is homogeneous and thus not varying spatially. Note that in physical terms F_{xX} represents the stretch λ at a given point of the bar and AP_{xX} would give the resulting tension T.

For closure of the above system of equations, and hence to permit its resolution, in addition to appropriate initial and boundary conditions, it is necessary to supplement a constitutive law relating the stress and deformation components P_{xX} and F_{xX}. For the particular case of linear elasticity, we can introduce the simple linear relationship

$$P_{xX}(F_{xX}) = E(F_{xX} - 1), \tag{7.2}$$

where E is the Young's modulus of the material. The system of conservation laws given by Equations (7.1) can be conveniently written in a compact manner as

$$\frac{\partial \boldsymbol{\mathcal{U}}}{\partial t} + \frac{\partial \boldsymbol{\mathcal{F}}(\boldsymbol{\mathcal{U}})}{\partial X} = \boldsymbol{\mathcal{S}}, \tag{7.3}$$

where $\boldsymbol{\mathcal{U}}$ denotes the vector of conservation variables, $\boldsymbol{\mathcal{F}}(\boldsymbol{\mathcal{U}})$ represents a vector containing the flux components (dependent upon $\boldsymbol{\mathcal{U}}$), and $\boldsymbol{\mathcal{S}}$ is a vector containing the source terms, defined as

$$\boldsymbol{\mathcal{U}} = \begin{bmatrix} p_x \\ F_{xX} \end{bmatrix}; \quad \boldsymbol{\mathcal{F}}(\boldsymbol{\mathcal{U}}) = \begin{bmatrix} -P_{xX}(F_{xX}) \\ -v_x(p_x) \end{bmatrix}; \quad \boldsymbol{\mathcal{S}} = \begin{bmatrix} f_x \\ 0 \end{bmatrix}, \tag{7.4a,b,c}$$

where Equation (7.4a,b,c)$_b$ indicates the explicit dependence of P_{xX} and v_x in terms of F_{xX} and p_x, respectively. In the absence of shocks (i.e. discontinuous fields), a very convenient alternative way to rewrite the system of conservation laws in Equation (7.3) is through its so-called *quasilinear* representation, as follows:

$$\frac{\partial \boldsymbol{\mathcal{U}}}{\partial t} + \boldsymbol{\mathcal{A}}\frac{\partial \boldsymbol{\mathcal{U}}}{\partial X} = \boldsymbol{\mathcal{S}}; \quad \boldsymbol{\mathcal{A}} = \frac{\partial \boldsymbol{\mathcal{F}}(\boldsymbol{\mathcal{U}})}{\partial \boldsymbol{\mathcal{U}}}, \tag{7.5a,b}$$

where $\boldsymbol{\mathcal{A}}$ is known as the *flux Jacobian matrix*. It is important to emphasize that the above quasilinear representation can only be obtained for smooth problems, that is, when the derivative of the flux with respect to the conservation variables exists and can be computed. Referring to Equatons (7.4a,b,c)$_{a,b}$ and Equation (7.2), the matrix $\boldsymbol{\mathcal{A}}$ can be computed as

$$\boldsymbol{\mathcal{A}} = \begin{bmatrix} -\frac{\partial P_{xX}}{\partial p_x} & -\frac{\partial P_{xX}}{\partial F_{xX}} \\ -\frac{\partial v_x}{\partial p_x} & -\frac{\partial v_x}{\partial F_{xX}} \end{bmatrix} = \begin{bmatrix} 0 & -E \\ -\frac{1}{\rho_R} & 0 \end{bmatrix}, \tag{7.6}$$

where it should be noted that the diagonal terms of $\boldsymbol{\mathcal{A}}$ are zero due to the *non-explicit* dependence of P_{xX} and v_x in terms of p_x and F_{xX}, respectively. Moreover, due to the linear relationship between stress and strain, the flux Jacobian matrix is constant. This is, of course, only the case for linear elasticity. In general $\boldsymbol{\mathcal{A}}$ will be a nonlinear function of the conservation variables $\boldsymbol{\mathcal{U}}$. Example 7.1 demonstrates the existence of wave speeds for this one-dimensional problem.

EXAMPLE 7.1: Hyperbolicity and wave speeds

Here interest focuses on the possible analytical solutions of the above one-dimensional problem in the form of wave-type fields. In this case, the conservation variable vector $\boldsymbol{\mathcal{U}}$ must admit, in the absence of source terms (i.e. $\boldsymbol{\mathcal{S}} = \mathbf{0}$), the following representation:

$$\boldsymbol{\mathcal{U}}(X,t) = f(X - c_\alpha t)\bar{\boldsymbol{\mathcal{U}}}_\alpha; \quad \bar{\boldsymbol{\mathcal{U}}}_\alpha = \begin{bmatrix} \bar{p}_x^\alpha \\ \bar{F}_{xX}^\alpha \end{bmatrix},$$

where $\bar{\boldsymbol{\mathcal{U}}}_\alpha$ is a (nonzero) vector of constants and f is a scalar function dependent upon a constant c_α, known as the wave speed. This solution represents a plane wave traveling along the bar of the type discussed in Example 3.2 in Chapter 3.

In general, more than one wave-type solution can exist (in this case, it will be shown that two solutions exist), and the α subindex is used to indicate this (in this case, $\alpha = 1, 2$). Noticing that

$$\frac{\partial \boldsymbol{\mathcal{U}}}{\partial t} = -c_\alpha f'(X - c_\alpha t)\bar{\boldsymbol{\mathcal{U}}}_\alpha; \quad \frac{\partial \boldsymbol{\mathcal{U}}}{\partial X} = f'(X - c_\alpha t)\bar{\boldsymbol{\mathcal{U}}}_\alpha,$$

it is now possible to substitute these expressions into Equation (7.5a,b)$_a$, to give

$$f'(X - c_\alpha t)\,(\boldsymbol{\mathcal{A}} - c_\alpha \boldsymbol{I})\bar{\boldsymbol{\mathcal{U}}}_\alpha = \mathbf{0}; \quad \boldsymbol{I} = \begin{bmatrix} 1 & 0 \\ 0 & 1 \end{bmatrix}.$$

For the above system of equations to hold the most general case where $f'(X - c_\alpha t) \neq 0$, the following condition must then be satisfied:

$$(\boldsymbol{\mathcal{A}} - c_\alpha \boldsymbol{I})\bar{\boldsymbol{\mathcal{U}}}_\alpha = \mathbf{0},$$

which represents a standard eigenvalue/eigenvector problem for the 2×2 matrix $\boldsymbol{\mathcal{A}}$. The roots of the so-called characteristic polynomial

(continued)

Example 7.1: *(cont.)*

$p(c_{1,2}) = \det(\boldsymbol{\mathcal{A}} - c_{1,2}\boldsymbol{I})$ determine the wave speeds of the system. Naturally, for wave-type solutions to exist, the wave speeds must be real, and this, crucially, depends on the structure of $\boldsymbol{\mathcal{A}}$ and therefore on the material properties of the system. Provided the two wave speeds are real, the system of conservation laws along with its corresponding constitutive law is known as *hyperbolic*. In the case of the matrix $\boldsymbol{\mathcal{A}}$ defined in Equation (7.6), it gives

$$\begin{aligned} -E\,\bar{F}^{\alpha}_{xX} &= c_\alpha\,\bar{p}^{\alpha}_{x}, \\ -\frac{1}{\rho_R}\,\bar{p}^{\alpha}_{x} &= c_\alpha\,\bar{F}^{\alpha}_{xX}, \end{aligned}$$

which can be combined to result in

$$E\,\bar{p}^{\alpha}_{x} = c_\alpha^2\,\rho_R\,\bar{p}^{\alpha}_{x}.$$

For the nontrivial solution (i.e. $\bar{p}^{\alpha}_{x} \neq 0$) to occur, it is straightforward to show that

$$c_{1,2} = \pm c_p; \qquad c_p = \sqrt{\frac{E}{\rho_R}},$$

which lead to two wave-type solutions traveling in opposite directions (left and right) of the domain. As can be seen, *hyperbolicity* (the existence of real wave speeds) is thus guaranteed provided the material constant E is non-negative.

7.2.1 Weak Form Equations and Stabilization

Appropriate weak form statements are introduced in this section before discussing space and time discretization techniques. Contrary to the approaches presented in Chapter 4, which relied upon variational principles stemming either from a generic action integral or from the use of the principle of virtual work (power), here we introduce alternative weak statements consistently derived from the set of conservation laws given in Equations (7.1). To start with, it is convenient to rewrite the system of conservation laws (7.1) in the form of a system of residual equations as

$$\boldsymbol{\mathcal{R}} = \begin{bmatrix} \mathcal{R}_{p_x} \\ \mathcal{R}_{F_{xX}} \end{bmatrix} = \begin{bmatrix} \frac{\partial p_x}{\partial t} - \frac{\partial P_{xX}}{\partial X} - f_x \\ \frac{\partial F_{xX}}{\partial t} - \frac{\partial v_x}{\partial X} \end{bmatrix} = \mathbf{0}, \tag{7.7}$$

or, written in a compact manner,

$$\boldsymbol{\mathcal{R}} = \frac{\partial \boldsymbol{\mathcal{U}}}{\partial t} + \frac{\partial \boldsymbol{\mathcal{F}}(\boldsymbol{\mathcal{U}})}{\partial X} - \boldsymbol{\mathcal{S}} = \mathbf{0}. \tag{7.8}$$

A general Galerkin type of *weak form statement* can be obtained by first multiplying by appropriate virtual fields, then adding together and integrating over the domain the above two residual equations. These virtual fields must be, first, compatible with the existing boundary conditions and, second, work conjugates to the respective conservation variables in each equation. This will ensure that after multiplication by the residual equations they both render units of power and the summation can be consistently carried out. This is physically important as the units of the two components of the residual $\mathcal{R}$ are different.

In our case, appropriate work conjugate virtual fields are $\delta\boldsymbol{\mathcal{V}} = [\delta v_x, \delta P_{xX}]^T$, since δv_x times $\dot{p}_x$ and δP_{xX} times $\dot{F}_{xX}$ both give the rate of energy. Multiplying this vector of conjugate virtual variables, then adding together the residual expressions given by Equations (7.7), and then integrating over the material volume $V = A \times [0, L]$, results in the following weak form $\delta W(\delta\boldsymbol{\mathcal{V}},\boldsymbol{\mathcal{U}})$ statement:

$$\delta W(\delta\boldsymbol{\mathcal{V}},\boldsymbol{\mathcal{U}}) = \int_V \delta\boldsymbol{\mathcal{V}}^T \left(\frac{\partial \boldsymbol{\mathcal{U}}}{\partial t} + \frac{\partial \boldsymbol{\mathcal{F}}(\boldsymbol{\mathcal{U}})}{\partial X} - \boldsymbol{\mathcal{S}} \right) dV = 0, \tag{7.9}$$

or, rewritten equivalently in component form,

$$\int_0^L \left[\delta v_x \left(\frac{\partial p_x}{\partial t} - \frac{\partial P_{xX}}{\partial X} - f_x \right) + \delta P_{xX} \left(\frac{\partial F_{xX}}{\partial t} - \frac{\partial v_x}{\partial X} \right) \right] A dX = 0. \tag{7.10}$$

It is possible to perform integration by parts on the flux term in the above weak form $\delta W(\delta\boldsymbol{\mathcal{V}},\boldsymbol{\mathcal{U}})$ statement, to transform Equation (7.9) into

$$\int_V \delta\boldsymbol{\mathcal{V}}^T \frac{\partial \boldsymbol{\mathcal{U}}}{\partial t} \, dV = \int_V \left(\frac{\partial \delta\boldsymbol{\mathcal{V}}}{\partial X} \right)^T \boldsymbol{\mathcal{F}}(\boldsymbol{\mathcal{U}}) \, dV - \left[\delta\boldsymbol{\mathcal{V}}^T \boldsymbol{\mathcal{F}}(\boldsymbol{\mathcal{U}}) A \right]_0^L + \int_V \delta\boldsymbol{\mathcal{V}}^T \boldsymbol{\mathcal{S}} \, dV. \tag{7.11}$$

By considering independent variations δv_x and δP_{xX} applied to the individual components in Equation (7.11), separate weak form statements for the linear momentum and the geometric conservation equations emerge as

$$\int_V \delta v_x \frac{\partial p_x}{\partial t} \, dV = -\int_V \frac{\partial \delta v_x}{\partial X} P_{xX} \, dV + [\delta v_x t_B A]_0^L + \int_V \delta v_x f_x \, dV, \tag{7.12a}$$

$$\int_V \delta P_{xX} \frac{\partial F_{xX}}{\partial t} \, dV = -\int_V \frac{\partial \delta P_{xX}}{\partial X} v_x \, dV + [\delta P_{xX} v_B A]_0^L, \tag{7.12b}$$

where t_B and v_B represent possible traction or velocity boundary conditions.† The main advantage of integrating by parts as shown above is to enable the imposition

† Note that t_B and v_B must be *complementarily* applied, that is, they both cannot be applied simultaneously on the same boundary position.

of the boundary conditions. This is indeed useful for the linear momentum Equation (7.12a) as it introduces *naturally* the boundary tractions, but less so in the case of the geometric conservation law, as v_B can be imposed directly in the typical *essential* manner. In this case, Equation (7.12b) can be replaced by its counterpart before integration by parts as

$$\int_V \delta P_{xX} \frac{\partial F_{xX}}{\partial t} \, dV = \int_V \delta P_{xX} \frac{\partial v_x}{\partial X} \, dV. \tag{7.13}$$

Unfortunately, a standard (Bubnov–Galerkin) finite element spatial discretization of Equations (7.12a) and (7.12b) (or its counterpart Equation (7.13)) does not yield a stable methodology; instead, it generates spurious unphysical oscillations which can grow uncontrollably, culminating in the breakdown of the scheme. For this reason, and before proceeding to a finite element spatial discretization, in order to address this deficiency we pursue an alternative Petrov–Galerkin method, whose point of departure from the previous derivation resides on a new definition of *stabilized* work conjugate virtual fields, renamed as $\delta\boldsymbol{\mathcal{V}}_{st}$. This is achieved by introducing a stabilization term based on the flux Jacobian matrix as

$$\delta\boldsymbol{\mathcal{V}}_{st} = \delta\boldsymbol{\mathcal{V}} + \boldsymbol{\tau}\boldsymbol{\mathcal{A}}^T \frac{\partial \delta\boldsymbol{\mathcal{V}}}{\partial X}; \qquad \boldsymbol{\tau} = \begin{bmatrix} \tau_{\boldsymbol{p}} & 0 \\ 0 & \tau_{\boldsymbol{F}} \end{bmatrix}, \tag{7.14a,b}$$

where $\boldsymbol{\tau}$ is a so-called *stabilization* matrix, typically diagonal, containing user-defined (non-negative) stabilization parameters $\tau_{\boldsymbol{p}}$ and $\tau_{\boldsymbol{F}}$ (with units of time). The specific choice of the stabilized work conjugate virtual field, $\delta\boldsymbol{\mathcal{V}}_{st}$, given by Equation (7.14a,b)$_a$, will be shown in Example 7.2 to result in a stable scheme referred to as the *streamline upwind Petrov–Galerkin* (SUPG) method. The stabilized virtual field definition given by Equation (7.14a,b)$_a$ can be expanded for each of the individual components by using the explicit expression for $\boldsymbol{\mathcal{A}}$ given in Equation (7.6) to give

$$\delta v_x^{st} = \delta v_x - \frac{\tau_{\boldsymbol{p}}}{\rho_R} \frac{\partial \delta P_{xX}}{\partial X}; \qquad \delta P_{xX}^{st} = \delta P_{xX} - \tau_{\boldsymbol{F}} E \frac{\partial \delta v_x}{\partial X}. \tag{7.15a,b}$$

A *stabilized* weak form statement can now be obtained by replacing $\delta\boldsymbol{\mathcal{V}}$ by $\delta\boldsymbol{\mathcal{V}}_{st}$ (refer to Equation (7.14a,b)$_a$) into Equation (7.9) to give

$$\begin{aligned} \delta W(\delta\boldsymbol{\mathcal{V}}_{st}, \boldsymbol{\mathcal{U}}) &= \int_V \delta\boldsymbol{\mathcal{V}}_{st}^T \boldsymbol{\mathcal{R}} \, dV \\ &= \int_V \delta\boldsymbol{\mathcal{V}}^T \boldsymbol{\mathcal{R}} \, dV + \int_V \left(\boldsymbol{\tau}\boldsymbol{\mathcal{A}}^T \frac{\partial \delta\boldsymbol{\mathcal{V}}}{\partial X} \right)^T \boldsymbol{\mathcal{R}} \, dV \\ &= \delta W(\delta\boldsymbol{\mathcal{V}}, \boldsymbol{\mathcal{U}}) + \int_V \left(\frac{\partial \delta\boldsymbol{\mathcal{V}}}{\partial X} \right)^T (\boldsymbol{\mathcal{A}}\,\boldsymbol{\tau}) \boldsymbol{\mathcal{R}} \, dV \\ &= 0, \end{aligned} \tag{7.16}$$

where we have made use of Equation (7.9) and exploited the diagonal (and thus symmetric) structure of the stabilization matrix $\boldsymbol{\tau}$. As can be seen, the SUPG weak form given by Equation (7.16) is identical to the original weak form, Equation (7.9), apart from the last ($\boldsymbol{\tau}$-dependent) term. It is clear that when the stabilization matrix is zero, the stabilized SUPG weak form statement reduces to the original statement given by Equation (7.9). It is now possible to perform integration by parts in the flux term of the above stabilized weak form $\delta W(\delta\boldsymbol{\mathcal{V}}_{st},\boldsymbol{\mathcal{U}})$ statement, to transform Equation (7.16) as follows:

$$\int_V \delta\boldsymbol{\mathcal{V}}^T\frac{\partial\boldsymbol{\mathcal{U}}}{\partial t}\,dV = \int_V \left(\frac{\partial\delta\boldsymbol{\mathcal{V}}}{\partial X}\right)^T \boldsymbol{\mathcal{F}}^{st}(\boldsymbol{\mathcal{U}})\,dV - \left[\delta\boldsymbol{\mathcal{V}}^T\boldsymbol{\mathcal{F}}A\right]_0^L + \int_V \delta\boldsymbol{\mathcal{V}}^T\boldsymbol{\mathcal{S}}\,dV, \tag{7.17}$$

where $\boldsymbol{\mathcal{F}}^{st}(\boldsymbol{\mathcal{U}})$ represents a stabilized flux vector defined as

$$\boldsymbol{\mathcal{F}}^{st}(\boldsymbol{\mathcal{U}}) = \boldsymbol{\mathcal{F}}(\boldsymbol{\mathcal{U}}) + \boldsymbol{\mathcal{F}}'(\boldsymbol{\mathcal{U}}); \qquad \boldsymbol{\mathcal{F}}'(\boldsymbol{\mathcal{U}}) = \boldsymbol{\mathcal{A}}(-\boldsymbol{\tau}\boldsymbol{\mathcal{R}}). \tag{7.18a,b}$$

Note that $\boldsymbol{\mathcal{F}}(\boldsymbol{\mathcal{U}})$ represents the flux vector and $\boldsymbol{\mathcal{F}}'(\boldsymbol{\mathcal{U}})$ represents a perturbation of the flux vector defined in terms of the flux Jacobian matrix $\boldsymbol{\mathcal{A}}$ multiplied by the perturbed field $-\boldsymbol{\tau}\boldsymbol{\mathcal{R}}$. Specifically, when considering independent variations δv_x and δP_{xX}, the two individual weak forms emerging from Equation (7.17) can be formulated as

$$\int_V \delta v_x \frac{\partial p_x}{\partial t}\,dV = -\int_V \frac{\partial\delta v_x}{\partial X}P_{xX}^{st}\,dV + \left[\delta v_x t_B A\right]_0^L + \int_V \delta v_x f_x\,dV, \tag{7.19a}$$

$$\int_V \delta P_{xX}\frac{\partial F_{xX}}{\partial t}\,dV = -\int_V \frac{\partial\delta P_{xX}}{\partial X}v_x^{st}\,dV + \left[\delta P_{xX}v_B A\right]_0^L, \tag{7.19b}$$

where the terms P_{xX}^{st} and v_x^{st} are defined as

$$P_{xX}^{st} = P_{xX} + P'_{xX}; \qquad P'_{xX} = E\left(-\tau_{\boldsymbol{F}}\mathcal{R}_{F_{xX}}\right), \tag{7.20a,b}$$

$$v_x^{st} = v_x + v'_x; \qquad v'_x = \frac{1}{\rho_R}(-\tau_{\boldsymbol{p}}\mathcal{R}_{p_x}). \tag{7.21a,b}$$

Note, that the only (but crucial) difference between the unstable Bubnov–Galerkin Equation (7.11) (or (7.12)) and the newly introduced SUPG Equation (7.17) (or (7.19)) weak form statements is the use of the fields P'_{xX} and v'_x (collectively referred to as $\boldsymbol{\mathcal{F}}'(\boldsymbol{\mathcal{U}})$) in the evaluation of the stabilized fluxes P_{xX}^{st} and v_x^{st} (collectively referred to as $\boldsymbol{\mathcal{F}}^{st}(\boldsymbol{\mathcal{U}})$); see Equations (7.20a,b) and (7.21a,b).

As explained above, it is not strictly necessary to perform integration by parts for the geometrical conservation law, Equation (7.19b). In this case, if the integration by parts is not employed, we would obtain an alternative equation,

$$\int_V \delta P_{xX} \frac{\partial F_{xX}}{\partial t}\, dV = \int_V \delta P_{xX} \frac{\partial v_x}{\partial X}\, dV + \int_V \frac{\partial \delta P_{xX}}{\partial X} \frac{\tau_{\boldsymbol{p}}}{\rho_R} \mathcal{R}_{p_x}\, dV. \tag{7.22}$$

The fields $\{P'_{xX}, v'_x\}$ will be subsequently shown to be critical in order to guarantee the stability of the resulting method, and they have been carefully constructed proportional to the residuals $\{\mathcal{R}_{F_{xX}}, \mathcal{R}_{p_x}\}$; this also ensures the consistency of the method, as $\{P'_{xX}, v'_x\}$ will tend to zero as the size of the space-time discretization also tends to zero. In this sense, it transpires that these fields seem to enhance the stability of the original Bubnov–Galerkin method by *injecting back* the numerical error induced by the weak satisfaction of the residual equations. This error is more pronounced when coarse meshes and/or low-order finite element technology is employed.

It is now instructive, by recalling Equation (7.2) and taking advantage of the linear relationship between stress and strain, to rewrite Equations (7.20a,b) as

$$P^{st}_{xX} = P_{xX}(F^{st}_{xX}) = E(F^{st}_{xX} - 1); \;\; F^{st}_{xX} = F_{xX} + F'_{xX}; \;\; F'_{xX} = -\tau_{\boldsymbol{F}} \mathcal{R}_{F_{xX}}, \tag{7.23a,b,c}$$

where the stabilized stress P^{st}_{xX} is obtained through the evaluation of the constitutive law, given by Equation (7.2) in terms of the stabilized deformation gradient component F^{st}_{xX}. Observe that, whereas in Equations (7.20a,b) the stress P_{xX} is stabilized (through the addition of the term P'_{xX}), in Equations (7.23a,b,c) the deformation gradient component F_{xX} is stabilized instead (through the addition of the term F'_{xX}). This, in principle subtle, alternative way of presenting the stabilization leads to the so-called *variational multi-scale* (VMS) method, which will be preferred hereafter. Note that, although in the case of linear elasticity the SUPG and VMS methods are identical, this is not strictly the case when considering nonlinear elasticity, due to the nonlinear relationship between strain and stress. Finally, the same VMS representation can be used for the stabilized velocity flux term v^{st}_x given in Equation $(7.21\text{a,b})_a$ (due to the linear relationship between velocity and linear momentum), which can be collectively summarized for the flux vector as

$$\begin{aligned} \boldsymbol{\mathcal{F}}^{st}(\boldsymbol{\mathcal{U}}) &= \boldsymbol{\mathcal{F}}(\boldsymbol{\mathcal{U}}) + \boldsymbol{\mathcal{A}}\,(-\boldsymbol{\tau}\boldsymbol{\mathcal{R}}) \\ &= \boldsymbol{\mathcal{F}}(\boldsymbol{\mathcal{U}}) + \frac{\partial \boldsymbol{\mathcal{F}}(\boldsymbol{\mathcal{U}})}{\partial \boldsymbol{\mathcal{U}}}(-\boldsymbol{\tau}\boldsymbol{\mathcal{R}}) \\ &= \boldsymbol{\mathcal{F}}(\boldsymbol{\mathcal{U}} + \boldsymbol{\mathcal{U}}'); \qquad \boldsymbol{\mathcal{U}}' = -\boldsymbol{\tau}\boldsymbol{\mathcal{R}}. \end{aligned} \tag{7.24}$$

In summary, the weak form statements used as the starting point of the space and time discretization are those shown in Equations (7.19) (or (7.22)) and (7.23a,b,c).

EXAMPLE 7.2: Stability of the formulation

Before proceeding to the finite element discretization of the weak form statements, see Equations (7.19a) and (7.22), it is important to demonstrate the stability of the formulation, which will be accomplished through energetic considerations. We adopt the specific choice of virtual fields $\delta v_x = v_x$ and $\delta P_{xX} = P_{xX}$ in Equations (7.19a) and (7.22), respectively, to obtain

$$\int_V v_x \frac{\partial p_x}{\partial t}\, dV = -\int_V \frac{\partial v_x}{\partial X} P_{xX}^{st}\, dV + [v_x t_B A]_0^L + \int_V v_x f_x\, dV,$$

$$\int_V P_{xX} \frac{\partial F_{xX}}{\partial t}\, dV = \int_V P_{xX} \frac{\partial v_x}{\partial X}\, dV + \int_V \frac{\partial P_{xX}}{\partial X} \frac{\tau_{\boldsymbol{p}}}{\rho_R} \mathcal{R}_{p_x}\, dV.$$

Adding these two equations, recalling the definition of P_{xX}^{st} given by Equations (7.23a,b,c), and noting the following energy rate equations,

$$\frac{dK}{dt} = \int_V v_x \frac{\partial p_x}{\partial t}\, dV,$$

$$\frac{d\Pi_{\text{int}}}{dt} = \int_V P_{xX} \frac{\partial F_{xX}}{\partial t}\, dV,$$

$$\dot{W}_{\text{ext}} = [v_x t_B A]_0^L + \int_V v_x f_x\, dV,$$

in terms of the kinetic energy K, the internal elastic energy Π_{int}, and the rate of external work $\dot{W}_{\text{ext}}$, we obtain

$$\frac{dK}{dt} + \frac{d\Pi_{\text{int}}}{dt} - \dot{W}_{\text{ext}} = \int_V \frac{\partial v_x}{\partial X} \tau_{\boldsymbol{F}} E \mathcal{R}_{F_{xX}}\, dV + \int_V \frac{\partial P_{xX}}{\partial X} \frac{\tau_{\boldsymbol{p}}}{\rho_R} \mathcal{R}_{p_x}\, dV.$$

For the case of reversible elasticity and in the case of time-independent forces, the left-hand side of the above equation, which represents the rate of total energy of the system, vanishes. Thus, the term on the right-hand side symbolizes a possible dissipation introduced into the system by the addition of the stabilization terms. It is then necessary to demonstrate that this right-hand side is indeed dissipative (negative), resulting in an overall decrease of the total energy of the system. To show this, we particularize $\tau_{\boldsymbol{p}} = 0$ and adopt $\delta P_{xX} = \tau_{\boldsymbol{F}} E \frac{\partial F_{xX}}{\partial t}$ in Equation (7.22), to yield

$$\int_V \frac{\partial F_{xX}}{\partial t} \tau_{\boldsymbol{F}} E \mathcal{R}_{F_{xX}}\, dV = 0.$$

(continued)

Example 7.2: *(cont.)*

We can subtract this equation from the energy rate equation above to obtain

$$\frac{dK}{dt} + \frac{d\Pi_{\text{int}}}{dt} - \dot{W}_{\text{ext}} = -\int_V \tau_{\boldsymbol{F}} E(\mathcal{R}_{F_{xX}})^2 \, dV \leq 0,$$

which, by recalling that both E and $\tau_{\boldsymbol{F}}$ are positive, demonstrates the dissipative nature of the scheme.

7.2.2 Spatial Discretization

Following a standard finite element methodology (see Figure 7.2), the domain interval $[0, L]$ is broken into a set of non-overlapping subintervals or elements (e), of size $h_{(e)}$. The conservation variables $\boldsymbol{\mathcal{U}}$ and their virtual work conjugates $\delta\boldsymbol{\mathcal{V}}$ are discretized into nodal values $\{\boldsymbol{\mathcal{U}}_a, \delta\boldsymbol{\mathcal{V}}_a\}$ by using shape functions N_a, where $a = 1, \ldots, N$, N being the total number of nodes in the mesh. For simplicity, we will restrict ourselves to linear shape functions and two-node elements. This discretization is sufficient to guarantee second-order spatial convergence for all the fields contained in $\boldsymbol{\mathcal{U}}$.

Consider the following finite element expansion for the linear momentum p_x and its time rate:

$$p_x = \sum_{b=1}^{N} N_b p_{x,b}(t); \qquad \frac{\partial p_x}{\partial t} = \sum_{b=1}^{N} N_b \dot{p}_{x,b}(t); \qquad \dot{p}_{x,b}(t) = \frac{dp_{x,b}}{dt}. \quad (7.25\text{a,b,c})$$

Analogously, for the deformation gradient component F_{xX} and its time rate,

$$F_{xX} = \sum_{b=1}^{N} N_b F_{xX,b}(t); \qquad \frac{\partial F_{xX}}{\partial t} = \sum_{b=1}^{N} N_b \dot{F}_{xX,b}(t); \qquad \dot{F}_{xX,b}(t) = \frac{dF_{xX,b}}{dt}.$$

(7.26a,b,c)

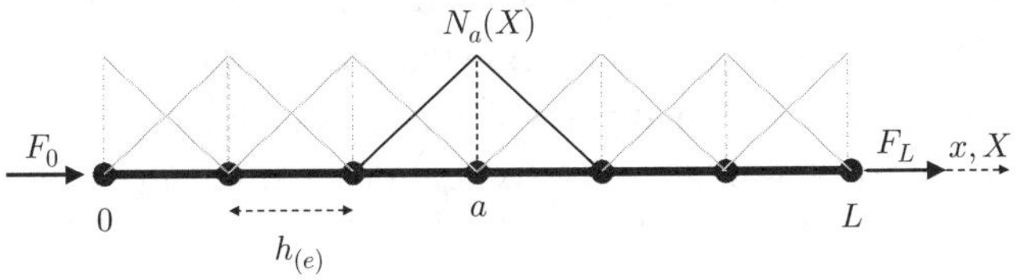

FIGURE 7.2 One-dimensional spatial discretization.

The virtual work conjugates are similarly expanded as

$$\delta v_x = \sum_{a=1}^{N} N_a \delta v_{x,a}; \qquad \delta P_{xX} = \sum_{a=1}^{N} N_a \delta P_{xX,a}, \tag{7.27a,b}$$

subject to appropriate boundary condition restrictions. Consider first the discretized weak statement for the linear momentum balance principle; see Equation (7.19a). With the purpose of obtaining the equation that corresponds to node a, we restrict δv_x in Equation (7.27a,b)$_a$ to a single nodal component $N_a \delta v_{x,a}$, before substituting into Equation (7.19a) to give, for each node $a = 1, \ldots, N$,

$$\int_V N_a \frac{\partial p_x}{\partial t}\, dV = F_0 \delta_{a1} + F_L \delta_{aN} + \int_V N_a f_x\, dV - \int_V P^{st}_{xX} \frac{\partial N_a}{\partial X}\, dV, \tag{7.28}$$

where P^{st}_{xX} is defined in Equations (7.23a,b,c), $F_0 = A\, t_0$ and $F_L = A\, t_L$ are possible boundary forces applied at the ends of the interval $[0, L]$, respectively, and δ_{a1} and δ_{aN} are standard delta Kronecker operators (e.g. $\delta_{a1} = 1$ if $a = 1$ and $\delta_{a1} = 0$ if $a \neq 1$). Note that the first three terms on the right-hand side of Equation (7.28) represent the equivalent external force acting on node a, which can be renamed as

$$F^{\boldsymbol{p}}_a = F_0 \delta_{a1} + F_L \delta_{aN} + \int_V N_a f_x\, dV. \tag{7.29}$$

This enables the reformulation of Equation (7.28) as

$$\sum_{b=1}^{N} m_{ab} \dot{p}_{x,b} = F^{\boldsymbol{p}}_a - T^{\boldsymbol{p}}_a; \qquad a = 1, \ldots, N, \tag{7.30}$$

where m_{ab} represents a consistent mass matrix type contribution and $T^{\boldsymbol{p}}_a$ represents the equivalent internal force acting on node a. These can be expanded as

$$m_{ab} = \int_V N_a N_b\, dV; \qquad T^{\boldsymbol{p}}_a = \int_V E(F^{st}_{xX} - 1) \frac{\partial N_a}{\partial X}\, dV. \tag{7.31a,b}$$

Note that, since the variable p_x is the linear momentum density, the mass-like term above does not include the density ρ_R and therefore has units of volume not mass. The evaluation of m_{ab}, $F^{\boldsymbol{p}}_a$, and $T^{\boldsymbol{p}}_a$ is carried out following the standard element by element assembly (refer to Chapter 4). Finally, the N nodal equations in Equation (7.30) can be collected into a single system of equations written in vector format as

$$\mathbf{M}\dot{\mathbf{U}}^{\boldsymbol{p}} = \mathbf{F}^{\boldsymbol{p}} - \mathbf{T}^{\boldsymbol{p}}, \tag{7.32}$$

where $\mathbf{M}$ represents a consistent mass matrix[‡] and $\mathbf{U}^{\boldsymbol{p}}$ is the vector of nodal momentum unknowns,

$$\mathbf{M} = \begin{bmatrix} m_{11} & m_{12} & \dots & m_{1N} \\ m_{21} & m_{22} & \dots & m_{2N} \\ \vdots & \vdots & & \vdots \\ m_{N1} & m_{N2} & \dots & m_{NN} \end{bmatrix}; \quad \mathbf{U}^{\boldsymbol{p}} = \begin{bmatrix} p_{x,1} \\ p_{x,2} \\ \vdots \\ p_{x,N} \end{bmatrix}, \tag{7.33a,b}$$

and $\mathbf{F}^{\boldsymbol{p}}$ and $\mathbf{T}^{\boldsymbol{p}}$ are the equivalent external and internal force vectors, which can be expanded as

$$\mathbf{F}^{\boldsymbol{p}} = \begin{bmatrix} F_1^{\boldsymbol{p}} \\ F_2^{\boldsymbol{p}} \\ \vdots \\ F_N^{\boldsymbol{p}} \end{bmatrix}; \quad \mathbf{T}^{\boldsymbol{p}} = \begin{bmatrix} T_1^{\boldsymbol{p}} \\ T_2^{\boldsymbol{p}} \\ \vdots \\ T_N^{\boldsymbol{p}} \end{bmatrix}. \tag{7.34a,b}$$

In most cases (for computational convenience), the consistent mass matrix is substituted by its lumped counterpart $\mathbf{M}^L$, defined as

$$\mathbf{M}^L = \begin{bmatrix} m_{11}^L & 0 & \dots & 0 \\ 0 & m_{22}^L & \dots & 0 \\ \vdots & \vdots & & \vdots \\ 0 & 0 & \dots & m_{NN} \end{bmatrix}; \quad m_{aa}^L = \sum_{b=1}^{N} m_{ab} = \int_V N_a \, dV. \tag{7.35a,b}$$

Attention is now focused on the discretized weak statement for the geometric conservation balance principle; see Equation (7.19b). In order to find the equation that corresponds to node a, we restrict δP_{xX} in Equation $(7.27\text{a,b})_b$ to a single nodal component $N_a \delta P_{xX,a}$ before substituting into Equation (7.19b) to give

$$\int_V N_a \frac{\partial F_{xX}}{\partial t} \, dV = A\left(v_0 \delta_{a1} + v_L \delta_{aN}\right) - \int_V v_x^{st} \frac{\partial N_a}{\partial X} \, dV, \tag{7.36}$$

where v_x^{st} is defined in Equation $(7.21\text{a,b})_a$ and v_0 and v_L are possible boundary velocities applied at end of the interval $[0, L]$. For consistency with the above derivation for the linear momentum balance principle, redefine the first two terms on the right-hand side of Equation (7.36) as $F_a^{\boldsymbol{F}} = A\left(v_0 \delta_{a1} + v_L \delta_{aN}\right)$ and the last term as $T_a^{\boldsymbol{F}} = -\int_V v_x^{st} \frac{\partial N_a}{\partial X} \, dV$. Using the finite element expansion Equation $(7.26\text{a,b,c})_{b,c}$, for the deformation gradient tensor, Equation (7.36) gives

$$\sum_{b=1}^{N} m_{ab} \dot{F}_{xX,b} = F_a^{\boldsymbol{F}} - T_a^{\boldsymbol{F}}; \quad a = 1, \dots, N, \tag{7.37}$$

‡ Note, that although $\mathbf{M}$ has been called the mass matrix, its entries m_{ab} in Equation $(7.31\text{a,b})_a$ do not have units of mass, but volume.

which can be gathered into a vector format as

$$\mathbf{M}\dot{\mathbf{U}}^{\boldsymbol{F}} = \mathbf{F}^{\boldsymbol{F}} - \mathbf{T}^{\boldsymbol{F}}, \tag{7.38}$$

with identical terms to those in Equation (7.32), but where we have replaced the upper index $\boldsymbol{p}$ by $\boldsymbol{F}$, namely

$$\mathbf{U}^{\boldsymbol{F}} = \begin{bmatrix} F_{xX,1} \\ F_{xX,2} \\ \vdots \\ F_{xX,N} \end{bmatrix}; \quad \mathbf{F}^{\boldsymbol{F}} = \begin{bmatrix} F_1^{\boldsymbol{F}} \\ F_2^{\boldsymbol{F}} \\ \vdots \\ F_N^{\boldsymbol{F}} \end{bmatrix}; \quad \mathbf{T}^{\boldsymbol{F}} = \begin{bmatrix} T_1^{\boldsymbol{F}} \\ T_2^{\boldsymbol{F}} \\ \vdots \\ T_N^{\boldsymbol{F}} \end{bmatrix}. \tag{7.39a,b,c}$$

Equations (7.32) and (7.38) represent a system of (semi-discrete) first-order ordinary differential equations in time which can be solved through a suitable time integrator.

7.2.3 Time Discretization

This section introduces a time integration scheme for the solution of the semi-discrete systems of Equations (7.32) and (7.38). It is instructive first to rewrite the above systems as a generic system,

$$\mathbf{M}\dot{\mathbf{U}} = \mathbf{S}(\mathbf{U}, \dot{\mathbf{U}}, t); \qquad \mathbf{S}(\mathbf{U}, \dot{\mathbf{U}}, t) = \mathbf{F}(t) - \mathbf{T}(\mathbf{U}, \dot{\mathbf{U}}), \tag{7.40a,b}$$

where $\mathbf{U}$ is a generic vector of unknowns and the vector $\mathbf{S}$ represents the *out-of-balance* vector between external and internal contributions. In the above equation, it is emphasized that the external vector contribution $\mathbf{F}$ can potentially depend explicitly on time, due to time-varying body forces and/or boundary tractions/velocities, while the internal vector contribution depends on the unknown vector $\mathbf{U}$ and its time rate $\dot{\mathbf{U}}$, through the stabilized unknown fields $F_{xX}^{st} = F_{xX} - \tau_{\boldsymbol{F}} \mathcal{R}_{F_{xX}}$ (refer to Equation (7.23a,b,c)$_b$) and $v_x^{st} = \frac{1}{\rho_R}(p_x - \tau_{\boldsymbol{p}} \mathcal{R}_{p_x})$ (refer to Equation (7.21a,b)$_a$). Note that the dependency with respect to $\dot{\mathbf{U}}$ is due to the residuals $\mathcal{R}_{F_{xX}}$ and $\mathcal{R}_{p_x}$ appearing in the stabilization terms.

Equation (7.40a,b)$_a$ represents a system of first-order differential equations in time. For its time integration, we split the time interval $[0, T]$ into N_t time subintervals and seek time-discrete solutions $\mathbf{U}_n$ (corresponding to time step t_n) with $n = 0, \ldots, N_t - 1$. Having shown that the system is hyperbolic, namely that it admits traveling waves propagating in space-time, it is possible to advocate for an explicit time integrator whose stability will be controlled by the wave speed c_p. For ease of understanding, a simple one-step two-stage (second order in time) Runge–Kutta (RK) method can be adopted which permits the advancement in time of the solution from time step t_n to time step t_{n+1}, with $\Delta t = t_{n+1} - t_n$, which can be formulated as

$$\mathbf{U}_{[0]} = \mathbf{U}_n, \tag{7.41a}$$

$$\mathbf{M}\,\dot{\mathbf{U}}_{[0]} = \mathbf{S}\left(\mathbf{U}_{[0]}, \dot{\mathbf{U}}_{[0]}, t_n\right), \tag{7.41b}$$

$$\mathbf{U}_{[1]} = \mathbf{U}_{[0]} + \Delta t\,\dot{\mathbf{U}}_{[0]}, \tag{7.41c}$$

$$\mathbf{M}\,\dot{\mathbf{U}}_{[1]} = \mathbf{S}\left(\mathbf{U}_{[1]}, \dot{\mathbf{U}}_{[1]}, t_{n+1}\right), \tag{7.41d}$$

$$\mathbf{U}_{n+1} = \mathbf{U}_n + \frac{1}{2}\Delta t\left(\dot{\mathbf{U}}_{[0]} + \dot{\mathbf{U}}_{[1]}\right), \tag{7.41e}$$

where the subindex $[I] = [0, 1, 2]$ is used to denote the stages of the time integrator and $\mathbf{U}_{[I]}$ denotes their corresponding intermediate stage solutions. Defining the solution stage $\mathbf{U}_{[2]}$ as

$$\mathbf{U}_{[2]} = \mathbf{U}_{[1]} + \Delta t\,\dot{\mathbf{U}}_{[1]}, \tag{7.42}$$

and substituting Equation (7.42) into Equations (7.41), results in an alternative representation of the RK time integrator as follows:

$$\mathbf{U}_{[0]} = \mathbf{U}_n, \tag{7.43a}$$

$$\mathbf{M}\left(\frac{\mathbf{U}_{[1]} - \mathbf{U}_{[0]}}{\Delta t}\right) = \mathbf{S}\left(\mathbf{U}_{[0]}, \frac{\mathbf{U}_{[1]} - \mathbf{U}_{[0]}}{\Delta t}, t_n\right), \tag{7.43b}$$

$$\mathbf{M}\left(\frac{\mathbf{U}_{[2]} - \mathbf{U}_{[1]}}{\Delta t}\right) = \mathbf{S}\left(\mathbf{U}_{[1]}, \frac{\mathbf{U}_{[2]} - \mathbf{U}_{[1]}}{\Delta t}, t_{n+1}\right), \tag{7.43c}$$

$$\mathbf{U}_{n+1} = \frac{1}{2}\left(\mathbf{U}_{[0]} + \mathbf{U}_{[2]}\right). \tag{7.43d}$$

For relatively short to medium term simulations,§ it is typical to replace the consistent mass matrix $\mathbf{M}$ by its diagonally lumped counterpart $\mathbf{M}^L$, as this enhances the computational speed of the scheme while retaining the (second) order of convergence in time of the scheme. As can be seen, each of above two stages is *implicit* due to the presence of the term $\mathbf{U}_{[I]}$ on the right-hand side of Equations (7.43b) and (7.43c), where, for simplicity, a typical fixed-point iteration can be used to converge to the stage solution $\mathbf{U}_{[I]}$ as follows:

§ Very long term simulations might require the use of more sophisticated implicit time integrators in order to prevent possible dispersion errors, which would end up undermining the quality of the final transient simulation.

$$\mathbf{M}^L \left(\frac{\mathbf{U}^{(k+1)}_{[I]} - \mathbf{U}_{[I-1]}}{\Delta t} \right) = \mathbf{S} \left(\mathbf{U}_{[I-1]}, \frac{\mathbf{U}^{(k)}_{[I]} - \mathbf{U}_{[I-1]}}{\Delta t}, t_{n+I-1} \right); \; I = 1, 2, \tag{7.44}$$

where $k = 1, 2, \ldots$ represents a generic fixed-point iteration with $\mathbf{U}^{(0)}_{[I]} = \mathbf{U}_{[I-1]}$.

Fortunately, the costly need of this fixed-point iteration algorithm at each stage of the time integrator can be circumvented by selecting the stabilization parameter τ_p to be zero. In this special case, the right-hand side of Equation (7.38) does not depend upon the residual term $\mathcal{R}_{p_x}$, and v_x^{st} simply reduces to v_x (see Equation (7.21a,b)$_a$). This enables the Ith Runge–Kutta stage equation for the unknown vector field $\mathbf{U}^{\boldsymbol{F}}$ to be written as

$$\mathbf{M}^L \left(\frac{\mathbf{U}^{\boldsymbol{F}}_{[I]} - \mathbf{U}^{\boldsymbol{F}}_{[I-1]}}{\Delta t} \right) = \mathbf{S} \left(\mathbf{U}^{\boldsymbol{p}}_{[I-1]}, t_{n+I-1} \right); \; I = 1, 2, \tag{7.45}$$

which, as can be seen, can be advanced *explicitly* from stage $I - 1$ to stage I. Subsequently, the Ith Runge–Kutta stage equation for Equation (7.32) arises as

$$\mathbf{M}^L \left(\frac{\mathbf{U}^{\boldsymbol{p}}_{[I]} - \mathbf{U}^{\boldsymbol{p}}_{[I-1]}}{\Delta t} \right) = \mathbf{S} \left(\mathbf{U}^{\boldsymbol{F}}_{[I-1]}, \frac{\mathbf{U}^{\boldsymbol{F}}_{[I]} - \mathbf{U}^{\boldsymbol{F}}_{[I-1]}}{\Delta t}, t_{n+I-1} \right); \; I = 1, 2, \tag{7.46}$$

which can also be advanced explicitly, as a result of the *staggered* solution process.

A final important point to highlight is the need to control the size of the chosen time step Δt in order to prevent the instability of the discrete system of equations and its possible breakdown. As previously discussed (refer to Example 2.3 in Chapter 2), this requires knowledge of the wave speed as well as the size of the smallest element in the mesh. This can be formally written as

$$\Delta t = \alpha_{CFL} \frac{h}{c_p}; \qquad h = \min_{(e)} h_{(e)}, \tag{7.47a,b}$$

where c_p was defined in Example 7.1, $h_{(e)}$ is the size of a generic element in the mesh, and α_{CFL} is known as the Courant–Friedrichs–Lewy constant, which depends on the specific time integrator and typically takes values in the interval $[0, 1]$. Clearly, the finer the discretization, the smaller the time step. Similarly, the stiffer and/or lighter the solid, the smaller the time step; see Equations (7.47a,b). For completeness, a flowchart is presented in Box 7.1 summarizing the overall methodology. In addition, a simple MATLAB program is presented for the analysis of a simple one-dimensional problem (see Figure 7.1), where the left end is kept fixed and the right end is subjected to an external time-varying load (refer to Box 7.3 at the end of the chapter).

BOX 7.1: Conservation laws – one-dimensional algorithm

- INPUT geometry, material properties, and solution parameters
- INITIALIZE $\mathbf{U}^p = \mathbf{U}_0^p$, $\mathbf{U}^F = \mathbf{U}_0^F$
- FIND mass matrix $\mathbf{M}$ (or $\mathbf{M}^L$) (7.35a,b)
- WHILE $t < T$ (time steps)
  - FIND Δt (7.47a,b)
  - SET $\mathbf{U}_{[0]}^p = \mathbf{U}_n^p$, $\mathbf{U}_{[0]}^F = \mathbf{U}_n^F$ (7.43a)
  - FOR I=1,2 (Runge–Kutta stages)
    - SET $\mathbf{F}^F = \mathbf{F}^F(t_{n+I-1})$ (7.37)
    - FIND $\mathbf{T}^F$ (7.21a,b), (7.37)
    - SOLVE for $\mathbf{U}_{[I]}^F$ (7.45)
    - SET $\mathbf{F}^p = \mathbf{F}^p(t_{n+I-1})$ (7.30)
    - FIND $\mathbf{T}^p$ (7.23a,b,c), (7.30)
    - SOLVE for $\mathbf{U}_{[I]}^p$ (7.46)
  - ENDLOOP
  - SOLVE for $\mathbf{U}_{n+1}^p$, $\mathbf{U}_{n+1}^F$ (7.43c)
- ENDLOOP

7.2.4 The One-Dimensional Example

In order to illustrate the theory presented for the one-dimensional case, and before proceeding to the more general three-dimensional case, we present a simple example accompanied by a MATLAB code included at the end of this chapter. Consider $A = 1$, $L = 10$, and $E = 1$, and that the left end of the domain is kept fixed while the right end is free to oscillate and is subjected to a time-varying load defined as $P = P_0 \sin(\omega t)$, where $P_0 = 10^{-3}$. The MATLAB program displays the input data used in the calculation, followed by a standard initialization process. The program then loops over the time steps and accesses the two-stage staggered Runge–Kutta time integrator to advance the unknown variables over time. Figures 7.3 and 7.4 show the time history of the velocity and stress at three different locations (left end, right end, and mid-point), for two different values of the forcing frequency ω, namely $\omega = 0.1$ and $\omega = 0.4$, respectively. It is interesting to observe how the wave travels from right to left until it is reflected back toward the right due to the left fixed

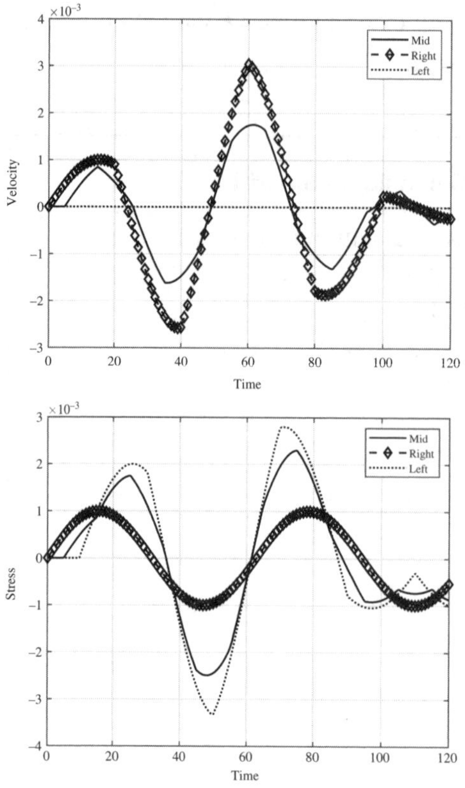

FIGURE 7.3 One-dimensional problem ($\omega = 0.1$). Time history of velocity and stress and different locations: left end, right end, and mid-point.

boundary condition. Different forcing frequencies give rise to a different system response.

7.3 A GENERAL ISOTHERMAL SYSTEM OF CONSERVATION LAWS

The solution process for the two conservation laws presented in Section 7.2 for the one-dimensional case is now generalized to three dimensions. Consider the deformation process of an isothermal deformable solid with initial volume V; we seek its simulation via the system of total Lagrangian conservation laws presented in Chapter 5 (refer to Equations (5.29a) and (5.29b)), expressed in terms of the velocity of the motion $\boldsymbol{v}$, the linear momentum per unit undeformed volume $\boldsymbol{p}_R = \rho_R \boldsymbol{v}$ (where ρ_R is the density in the undeformed configuration), the deformation gradient $\boldsymbol{F}$, the first Piola–Kirchhoff stress tensor $\boldsymbol{P}$, and a possible body force

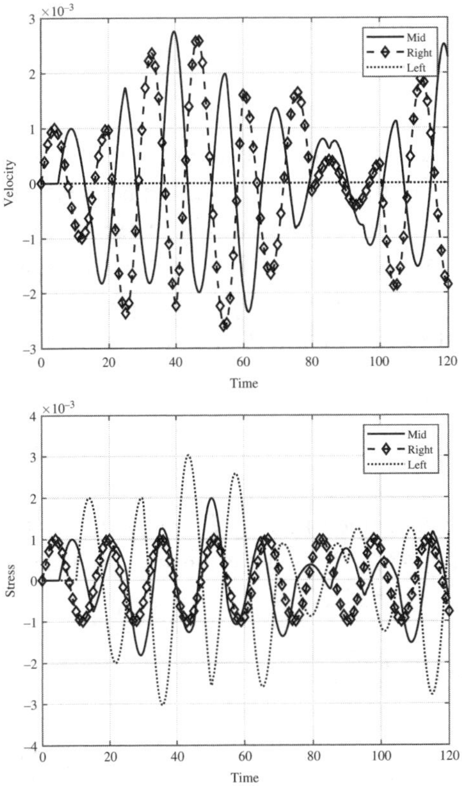

FIGURE 7.4 One-dimensional problem ($\omega = 0.4$). Time history of velocity and stress and different locations: left end, right end, mid-point.

per unit undeformed volume $\boldsymbol{f}_R$. For simplicity, the density ρ_R is assumed to be homogeneous throughout the domain. These equations are presented here again for convenience:

$$\frac{\partial \boldsymbol{p}_R}{\partial t} - \text{DIV}\boldsymbol{P} = \boldsymbol{f}_R; \tag{7.48a}$$

$$\frac{\partial \boldsymbol{F}}{\partial t} - \text{DIV}(\boldsymbol{v} \otimes \boldsymbol{I}) = \boldsymbol{0}. \tag{7.48b}$$

To complete the system of conservation laws given in Equations (7.48), such that the flux contributions are expressed in terms of the conservation variables, it is necessary to provide an additional (constitutive) relationship between the first Piola–Kirchhoff stress tensor $\boldsymbol{P}$ and the deformation gradient $\boldsymbol{F}$. In what follows a simple isothermal reversible nonlinear elasticity model is selected, which, in this case, is given in terms of a compressible neo-Hookean model and is defined through the following strain energy density:

$$\Psi(\boldsymbol{F}) = \frac{\mu}{2}(II_{\boldsymbol{F}} - 3) + f(J);\;\; f(J) = -\mu \ln J + \frac{\lambda}{2}(\ln J)^2;\;\; J = \det \boldsymbol{F}, \tag{7.49}$$

where λ and μ are material parameters with a similar interpretation to the classical Lamé parameters of linear elasticity, and $II_{\boldsymbol{F}} = \boldsymbol{F} : \boldsymbol{F}$ denotes the second invariant of $\boldsymbol{F}$. Although above constitutive law can be written alternatively in terms of the right Cauchy–Green strain tensor $\boldsymbol{C}$ (see NL-Statics, Section 6.4.3), it is more convenient, from the algebraic standpoint, to adopt the representation given above in terms of $\boldsymbol{F}$ in what follows. Thus, the first Piola–Kirchhoff stress tensor can now be obtained as

$$\boldsymbol{P}(\boldsymbol{F}) = \frac{\partial \Psi(\boldsymbol{F})}{\partial \boldsymbol{F}} = \mu \boldsymbol{F} + (\lambda \ln J - \mu)\boldsymbol{F}^{-T}. \tag{7.50}$$

Similarly, the fourth-order elasticity tensor defined here as $\boldsymbol{\mathcal{C}} = \frac{\partial \boldsymbol{P}}{\partial \boldsymbol{F}} = \frac{\partial^2 \Psi}{\partial \boldsymbol{F} \partial \boldsymbol{F}}$ and obtained as the derivative of the first Piola–Kirchhoff stress tensor with respect to the deformation gradient tensor,¶ can be obtained as

$$\begin{aligned}\boldsymbol{\mathcal{C}} &= \mu \frac{\partial \boldsymbol{F}}{\partial \boldsymbol{F}} + \frac{\lambda}{J}\boldsymbol{F}^{-T} \otimes \frac{\partial J}{\partial \boldsymbol{F}} - (\mu - \lambda \ln J)\frac{\partial \boldsymbol{F}^{-T}}{\partial \boldsymbol{F}} \\ &= \mu \boldsymbol{\mathcal{I}} + \lambda \boldsymbol{F}^{-T} \otimes \boldsymbol{F}^{-T} + (\mu - \lambda \ln J)\boldsymbol{\mathcal{H}},\end{aligned} \tag{7.51}$$

where $\boldsymbol{\mathcal{I}} = \frac{\partial \boldsymbol{F}}{\partial \boldsymbol{F}}$ is the fourth-order identity tensor (NL-Statics, Equation (2.78a)) and $\boldsymbol{\mathcal{H}} = -\frac{\partial \boldsymbol{F}^{-T}}{\partial \boldsymbol{F}}$ is a fourth-order tensor (see Example 7.3), both defined in indicial manner as

$$[\boldsymbol{\mathcal{I}}]_{iIjJ} = \delta_{ij}\delta_{IJ};\quad [\boldsymbol{\mathcal{H}}]_{iIjJ} = \left[\boldsymbol{F}^{-1}\right]_{Ji}\left[\boldsymbol{F}^{-1}\right]_{Ij}. \tag{7.52a,b}$$

For the special case of *linear elasticity* (see NL-Statics, Example 2.8), where the deformations can be considered small, the stress tensor $\boldsymbol{P}$ is related to the deformation gradient tensor $\boldsymbol{F}$ as

$$\begin{aligned}\boldsymbol{P} = \boldsymbol{P}_{lin}(\boldsymbol{F}) &= \boldsymbol{\mathcal{C}}_{lin} : (\boldsymbol{F} - \boldsymbol{I}) \\ &= \boldsymbol{\mathcal{C}}|_{\boldsymbol{F}=\boldsymbol{I}} : (\boldsymbol{F} - \boldsymbol{I}) \\ &= \left(\lambda \boldsymbol{I} \otimes \boldsymbol{I} + \mu(\boldsymbol{\mathcal{I}} + \tilde{\boldsymbol{\mathcal{I}}})\right) : (\boldsymbol{F} - \boldsymbol{I}),\end{aligned} \tag{7.53}$$

where $\boldsymbol{\mathcal{C}}|_{\boldsymbol{F}=\boldsymbol{I}}$ represents the evaluation of the constitutive tensor Equation (7.51) in the origin of deformations, namely $\boldsymbol{F} = \boldsymbol{I}$, and where $\tilde{\boldsymbol{\mathcal{I}}}$ is a fourth-order isotropic

¶ Although the same letter $\boldsymbol{\mathcal{C}}$ is used here, this fourth-order tensor should not be confused with the fourth-order elasticity tensor obtained as the derivative of the second Piola–Kirchhoff stress tensor with respect to the right Cauchy–Green strain tensor, namely $2\partial \boldsymbol{S}/\partial \boldsymbol{C}$ (refer to NL-Statics, Section 6.3.1).

tensor defined as $[\tilde{\mathcal{I}}]_{iIjJ} = \delta_{Ji}\delta_{Ij}$ (refer to NL-Statics, Equation (2.81)). In this case, the linear elasticity constitutive tensor is constant and of value $\mathcal{C}_{lin} = \lambda \boldsymbol{I} \otimes \boldsymbol{I} + \mu(\mathcal{I} + \tilde{\mathcal{I}})$. Example 7.4 demonstrates the existence of wave speeds for the general isothermal system of conservation laws.

EXAMPLE 7.3: Computation of fourth-order tensor $\mathcal{H} = -\frac{\partial \boldsymbol{F}^{-T}}{\partial \boldsymbol{F}}$

Begin by computing the directional derivative of $\boldsymbol{F}\boldsymbol{F}^{-1} = \boldsymbol{I}$, to yield

$$\begin{aligned} D(\boldsymbol{F}\boldsymbol{F}^{-1})[\boldsymbol{u}] &= D\boldsymbol{F}[\boldsymbol{u}]\boldsymbol{F}^{-1} + \boldsymbol{F}D\boldsymbol{F}^{-1}[\boldsymbol{u}] \\ &= (\boldsymbol{\nabla}_0\boldsymbol{u})\boldsymbol{F}^{-1} + \boldsymbol{F}D\boldsymbol{F}^{-1}[\boldsymbol{u}] = \boldsymbol{0}. \end{aligned}$$

By simple transposition and computing the transpose,

$$D\boldsymbol{F}^{-T}[\boldsymbol{u}] = -\boldsymbol{F}^{-T}(\boldsymbol{\nabla}_0\boldsymbol{u})^T\boldsymbol{F}^{-T},$$

and noticing that

$$D\boldsymbol{F}^{-T}[\boldsymbol{u}] = \frac{\partial \boldsymbol{F}^{-T}}{\partial \boldsymbol{F}} : \boldsymbol{\nabla}_0\boldsymbol{u},$$

the identification can be deduced, namely that

$$\left[\frac{\partial \boldsymbol{F}^{-T}}{\partial \boldsymbol{F}}\right]_{iIjJ} [\boldsymbol{\nabla}_0\boldsymbol{u}]_{jJ} = -[\boldsymbol{F}^{-1}]_{Ji}\,[\boldsymbol{\nabla}_0\boldsymbol{u}]_{jJ}\,[\boldsymbol{F}^{-1}]_{Ij},$$

which gives

$$[\mathcal{H}]_{iIjJ} = -\left[\frac{\partial \boldsymbol{F}^{-T}}{\partial \boldsymbol{F}}\right]_{iIjJ} = [\boldsymbol{F}^{-1}]_{Ji}[\boldsymbol{F}^{-1}]_{Ij}.$$

EXAMPLE 7.4: Hyperbolicity for isothermal elasticity

In the absence of source terms ($\boldsymbol{f}_R = \boldsymbol{0}$), possible (plane) wave-type solution fields can be formulated as

$$\boldsymbol{p}_R(\boldsymbol{X},t) = f(\boldsymbol{X}\cdot\boldsymbol{\nu} - c_\alpha t)\bar{\boldsymbol{p}}^\alpha; \quad \boldsymbol{F}(\boldsymbol{X},t) = f(\boldsymbol{X}\cdot\boldsymbol{\nu} - c_\alpha t)\bar{\boldsymbol{F}}^\alpha,$$

where $\bar{\boldsymbol{p}}_R^\alpha$ and $\bar{\boldsymbol{F}}^\alpha$ denote nonzero constant fields and f is a scalar function dependent upon the wave speed c_α and an arbitrary material unit normal vector $\boldsymbol{\nu}$, where $\alpha = 1, 2, \ldots$ represents the number of possible plane wave-type

(continued)

Example 7.4: *(cont.)*

solutions. It is then possible to compute the time derivative of the above two terms as

$$\frac{\partial \boldsymbol{p}_R}{\partial t} = -c_\alpha f'(\boldsymbol{X}\cdot\boldsymbol{\nu} - c_\alpha t)\bar{\boldsymbol{p}}_R^\alpha,$$

$$\frac{\partial \boldsymbol{F}}{\partial t} = -c_\alpha f'(\boldsymbol{X}\cdot\boldsymbol{\nu} - c_\alpha t)\bar{\boldsymbol{F}}^\alpha.$$

Similarly, making use of the chain rule (where the flux variables are expressed in terms of the conservation variables), we can obtain

$$\text{DIV}\boldsymbol{P} = f'(\boldsymbol{X}\cdot\boldsymbol{\nu} - c_\alpha t)\left(\boldsymbol{\mathcal{C}}\colon \bar{\boldsymbol{F}}^\alpha\right)\boldsymbol{\nu},$$

$$\text{DIV}(\boldsymbol{v}\otimes\boldsymbol{I}) = \frac{1}{\rho_R} f'(\boldsymbol{X}\cdot\boldsymbol{\nu} - c_\alpha t)\,\bar{\boldsymbol{p}}_R^\alpha\otimes\boldsymbol{\nu},$$

which, after substitution into the system of conservation laws given by Equations (7.48) and considering the general case where $f'(\boldsymbol{X}\cdot\boldsymbol{\nu} - c_\alpha t) \neq 0$, gives

$$-\left(\boldsymbol{\mathcal{C}}\colon \bar{\boldsymbol{F}}^\alpha\right)\boldsymbol{\nu} = c_\alpha\,\bar{\boldsymbol{p}}_R^\alpha,$$

$$-\frac{1}{\rho_R}\,\bar{\boldsymbol{p}}_R^\alpha\otimes\boldsymbol{\nu} = c_\alpha\,\bar{\boldsymbol{F}}^\alpha.$$

The above two equations can be combined to result in the following eigenvalue problem:

$$\boldsymbol{\mathcal{C}}_{\boldsymbol{\nu\nu}}\,\bar{\boldsymbol{p}}_R^\alpha = c_\alpha^2\,\rho_R\,\bar{\boldsymbol{p}}_R^\alpha;\qquad [\boldsymbol{\mathcal{C}}_{\boldsymbol{\nu\nu}}]_{ij} = \sum_{I,J=1}^{3}\mathcal{C}_{iIjJ}\nu_I\nu_J,$$

where $\boldsymbol{\mathcal{C}}_{\boldsymbol{\nu\nu}}$ is a (symmetric) second-order tensor, known as the *acoustic tensor*. Similarly to Example 7.1, the solution to the above eigenvalue problem allows identification of the wave speeds and establishes conditions on the material properties to ensure the hyperbolicity of the system.

EXAMPLE 7.5: Wave speeds for the compressible neo-Hookean model

For the constitutive tensor defined in Equation (7.51) corresponding to a compressible neo-Hookean material, the acoustic tensor takes the following expression:

$$\boldsymbol{\mathcal{C}}_{\boldsymbol{\nu\nu}} = \mu\boldsymbol{I} + (\lambda + \mu - \lambda\ln J)(\boldsymbol{F}^{-T}\boldsymbol{\nu})\otimes(\boldsymbol{F}^{-T}\boldsymbol{\nu}).$$

(continued)

Example 7.5: *(cont.)*

In order to obtain closed-form expressions of the eigenvalues and eigenvectors of the above problem, we first introduce the following unit normal vector:

$$\boldsymbol{\xi} = \frac{\boldsymbol{F}^{-T}\boldsymbol{\nu}}{\Lambda_{\boldsymbol{\nu}}}; \qquad \Lambda_{\boldsymbol{\nu}} = ||\boldsymbol{F}^{-T}\boldsymbol{\nu}||.$$

After introducing two arbitrary unit normal vectors $\boldsymbol{t}_1$ and $\boldsymbol{t}_2$ such that the triad $\{\boldsymbol{\xi}, \boldsymbol{t}_1, \boldsymbol{t}_2\}$ is orthonormal, and expanding the identity tensor as $\boldsymbol{I} = \boldsymbol{\xi} \otimes \boldsymbol{\xi} + \boldsymbol{t}_1 \otimes \boldsymbol{t}_1 + \boldsymbol{t}_2 \otimes \boldsymbol{t}_2$, we can rewrite the acoustic tensor as

$$\boldsymbol{\mathcal{C}}_{\boldsymbol{\nu}\boldsymbol{\nu}} = \mu \boldsymbol{t}_1 \otimes \boldsymbol{t}_1 + \mu \boldsymbol{t}_2 \otimes \boldsymbol{t}_2 + (\mu + (\lambda + \mu - \lambda \ln J)\Lambda_{\boldsymbol{\nu}}^2)\, \boldsymbol{\xi} \otimes \boldsymbol{\xi}.$$

This dyadic representation of the acoustic tensor $\boldsymbol{\mathcal{C}}_{\boldsymbol{\nu}\boldsymbol{\nu}}$ permits immediate identification of the eigenvalues and eigenvectors. Indeed, we can, by inspection, realize that the triad of vectors $\{\boldsymbol{\xi}, \boldsymbol{t}_1, \boldsymbol{t}_2\}$ are actually the eigenvectors of the problem and, by choosing recursively $\bar{\boldsymbol{p}}_R^{1,2} = \boldsymbol{\xi}$, $\bar{\boldsymbol{p}}_R^{3,4} = \boldsymbol{t}_1$, and $\bar{\boldsymbol{p}}_R^{5,6} = \boldsymbol{t}_2$, we can obtain the so-called *p-wave* speeds as

$$c_{1,2} = \pm c_p; \qquad c_p = \sqrt{\frac{\mu + (\lambda + \mu - \lambda \ln J)\Lambda_{\boldsymbol{\nu}}^2}{\rho_R}},$$

and the so-called *s-wave* speeds as

$$c_{3,4} = c_{5,6} = \pm c_s; \qquad c_s = \sqrt{\frac{\mu}{\rho_R}}.$$

Naturally, to ensure the *hyperbolicity* of the system of conservation laws, the above *p,s-wave* speeds must be real, and this can only be guaranteed if the following two conditions hold:

$$\mu \geq 0; \qquad \lambda \geq \mu \frac{1 + \Lambda_{\boldsymbol{\nu}}^2}{\Lambda_{\boldsymbol{\nu}}^2(\ln J - 1)};$$

these impose physical restriction on the values that the material parameters λ and μ can adopt. For the simple case of linear elasticity, when deformations are small, namely $\ln J \approx 0$ and $\Lambda_{\boldsymbol{\nu}} \approx 1$, the *s-wave* speed adopts the same expression as above while the *p-wave* speed reduces to $c_p \approx \sqrt{\frac{\lambda+2\mu}{\rho}}$.

7.3.1 Three-Dimensional Weak Form Equations and Stabilization

This section aims to generalize the methodology presented in Section 7.2.1 to three dimensions. Thus, we start by defining the residual equations

$$\boldsymbol{\mathcal{R}} = \boldsymbol{0}; \qquad \boldsymbol{\mathcal{R}} = \begin{bmatrix} \boldsymbol{\mathcal{R}}_{\boldsymbol{p}} \\ \boldsymbol{\mathcal{R}}_{\boldsymbol{F}} \end{bmatrix}, \tag{7.54a,b}$$

where the momentum residual and the deformation gradient residual are given by the set of conservation laws in the total Lagrangian setting derived in Sections 5.3.2 and 5.3.4, respectively, as

$$\boldsymbol{\mathcal{R}}_{\boldsymbol{p}} = \frac{\partial \boldsymbol{p}}{\partial t} - \text{DIV}\boldsymbol{P} - \boldsymbol{f}_R; \qquad \boldsymbol{\mathcal{R}}_{\boldsymbol{F}} = \frac{\partial \boldsymbol{F}}{\partial t} - \boldsymbol{\nabla}_0 \boldsymbol{v}. \tag{7.55a,b}$$

A general *weak form statement* can be obtained by multiplying these residual equations by the appropriate (and compatible with the boundary conditions) work conjugate virtual fields $\delta\boldsymbol{\mathcal{V}}$, defined as

$$\delta\boldsymbol{\mathcal{V}} = \begin{bmatrix} \delta\boldsymbol{v} \\ \delta\boldsymbol{P} \end{bmatrix}, \tag{7.56}$$

in terms of virtual velocities $\delta\boldsymbol{v}$ and stresses $\delta\boldsymbol{P}$, leading to a mixed form of the *principle of virtual power* formulated as

$$\delta W(\delta\boldsymbol{\mathcal{V}}, \boldsymbol{\mathcal{U}}) = \int_V (\delta\boldsymbol{v}\cdot\boldsymbol{\mathcal{R}}_{\boldsymbol{p}} + \delta\boldsymbol{P} : \boldsymbol{\mathcal{R}}_{\boldsymbol{F}})\, dV = 0. \tag{7.57}$$

Note that it is important that the chosen virtual fields $\delta\boldsymbol{v}$ and $\delta\boldsymbol{P}$ are work conjugates to their respective residuals $\boldsymbol{\mathcal{R}}_{\boldsymbol{p}}$ and $\boldsymbol{\mathcal{R}}_{\boldsymbol{F}}$ in order to ensure that the terms within the integrand of Equation (7.57) are consistent in terms of units (i.e. power). Independent weak form statements can be obtained, considering individual virtual variations $\delta\boldsymbol{v}$ and $\delta\boldsymbol{P}$, which, after making use of the Gauss theorem for the flux terms, are given by

$$\int_V \delta\boldsymbol{v}\cdot\frac{\partial \boldsymbol{p}_R}{\partial t}\, dV = -\int_V \boldsymbol{\nabla}_0\delta\boldsymbol{v} : \boldsymbol{P}\, dV + \int_{\partial V} \delta\boldsymbol{v}\cdot\boldsymbol{t}_B\, dA + \int_V \delta\boldsymbol{v}\cdot\boldsymbol{f}_R\, dV, \tag{7.58a}$$

$$\int_V \delta\boldsymbol{P} : \frac{\partial \boldsymbol{F}}{\partial t}\, dV = -\int_V \text{DIV}\delta\boldsymbol{P}\cdot\boldsymbol{v}\, dV + \int_{\partial V} \delta\boldsymbol{P} : (\boldsymbol{v}_B \otimes \boldsymbol{N})\, dA, \tag{7.58b}$$

where $\boldsymbol{N}$ represents the unit outward normal vector on the boundary ∂V of the domain and $\boldsymbol{t}_B$ and $\boldsymbol{v}_B$ are possible traction and velocity boundary conditions acting on complementary parts of ∂V. It is also possible to present Equation (7.58b) without making use of the Gauss theorem as

$$\int_V \delta\boldsymbol{P} : \frac{\partial \boldsymbol{F}}{\partial t}\, dV = \int_V \delta\boldsymbol{P} : \boldsymbol{\nabla}_0\boldsymbol{v}\, dV. \tag{7.59}$$

In Equation (7.59), the velocity boundary conditions $\boldsymbol{v}_B$ are imposed *essentially*, instead of *naturally* as in Equation $(7.58)_b$. As anticipated in Section 7.2.1, a standard Bubnov–Galerkin formulation of the above weak form statements can potentially lead to spurious oscillations and instability of the final scheme. Consequently the following SUPG stabilized weak form statements are proposed (refer to Equations (7.19), (7.20a,b), and (7.21a,b)) as

$$\int_V \delta \boldsymbol{v} \cdot \frac{\partial \boldsymbol{p}_R}{\partial t}\, dV = -\int_V \boldsymbol{\nabla}_0 \delta \boldsymbol{v} \colon \boldsymbol{P}^{st}\, dV + \int_{\partial V} \delta \boldsymbol{v} \cdot \boldsymbol{t}_B\, dA + \int_V \delta \boldsymbol{v} \cdot \boldsymbol{f}_R\, dV, \tag{7.60a}$$

$$\int_V \delta \boldsymbol{P} \colon \frac{\partial \boldsymbol{F}}{\partial t}\, dV = -\int_V \text{DIV}\delta \boldsymbol{P} \cdot \boldsymbol{v}^{st}\, dV + \int_{\partial V} \delta \boldsymbol{P} \colon (\boldsymbol{v}_B \otimes \boldsymbol{N})\, dA, \tag{7.60b}$$

where the stabilized terms $\boldsymbol{P}^{st}$ and $\boldsymbol{v}^{st}$ in Equations (7.60a) and (7.60b), respectively, are defined as

$$\boldsymbol{P}^{st} = \boldsymbol{P} + \boldsymbol{P}'; \qquad \boldsymbol{P}' = \boldsymbol{\mathcal{C}} \colon (-\tau_{\boldsymbol{F}} \boldsymbol{\mathcal{R}}_{\boldsymbol{F}}), \tag{7.61a,b}$$

$$\boldsymbol{v}^{st} = \boldsymbol{v} + \boldsymbol{v}'; \qquad \boldsymbol{v}' = \frac{1}{\rho_R}(-\tau_{\boldsymbol{p}} \boldsymbol{\mathcal{R}}_{\boldsymbol{p}}), \tag{7.62a,b}$$

formulated in terms of the fourth-order constitutive tensor $\boldsymbol{\mathcal{C}}$ and two non-negative stabilization parameters $\tau_{\boldsymbol{p}}$ and $\tau_{\boldsymbol{F}}$. Equations (7.60)–(7.61a,b) constitute a straight-forward extension from the one-dimensional case given in Equations (7.20a,b)–(7.21a,b) to the three-dimensional case. Clearly, the difference between the weak form statements in Equations (7.58) and (7.60) relates to the use of the fields $\boldsymbol{P}'$ and $\boldsymbol{v}'$ in the evaluation of $\boldsymbol{P}^{st}$ and $\boldsymbol{v}^{st}$, respectively. For a linear elasticity model, where the relationship between $\boldsymbol{P}$ and $\boldsymbol{F}$ is linear (see Equation (7.53)) it is possible to rewrite Equation (7.61a,b) as

$$\boldsymbol{P}^{st} = \boldsymbol{\mathcal{C}}_{lin} \colon (\boldsymbol{F}^{st} - \boldsymbol{I}); \qquad \boldsymbol{F}^{st} = \boldsymbol{F} + \boldsymbol{F}'; \qquad \boldsymbol{F}' = -\tau_{\boldsymbol{F}} \boldsymbol{\mathcal{R}}_{\boldsymbol{F}}. \tag{7.63a,b,c}$$

In the more general nonlinear case, after making use of the concept of directional derivative, namely $\boldsymbol{P}(\boldsymbol{F} + \boldsymbol{F}') - \boldsymbol{P}(\boldsymbol{F}) \approx D\boldsymbol{P}(\boldsymbol{F})[\boldsymbol{F}'] = \boldsymbol{\mathcal{C}} \colon \boldsymbol{F}'$, it is possible to approximate $\boldsymbol{P}^{st}$ in Equation (7.61a,b) as

$$\boldsymbol{P}^{st} \approx \boldsymbol{P}(\boldsymbol{F}^{st}); \qquad \boldsymbol{F}^{st} = \boldsymbol{F} + \boldsymbol{F}'; \qquad \boldsymbol{F}' = -\tau_{\boldsymbol{F}} \boldsymbol{\mathcal{R}}_{\boldsymbol{F}}. \tag{7.64a,b,c}$$

It is the stress tensor $\boldsymbol{P}$ that is stabilized in Equation (7.61a,b), whereas in Equations (7.64a,b,c) the deformation gradient tensor $\boldsymbol{F}$ is stabilized. The second approach leads to the variational multi-scale (VMS) method, which is preferred for discretization purposes. From the implementation point of view, the VMS approach is more advantageous as it avoids the need to compute explicitly the constitutive tensor $\boldsymbol{\mathcal{C}}$ for the computation of the stabilization term. Note that it is not strictly necessary to use the Gauss theorem on the geometric conservation law. Thus, alternatively, Equation (7.60b) would transform into

$$\int_V \delta\boldsymbol{P} : \frac{\partial \boldsymbol{F}}{\partial t}\, dV = \int_V \delta\boldsymbol{P} : \boldsymbol{\nabla}_0 \boldsymbol{v}\, dV + \int_V \mathrm{DIV}\delta\boldsymbol{P} \cdot \frac{\tau_{\boldsymbol{p}}}{\rho_R} \boldsymbol{\mathcal{R}}_{\boldsymbol{p}}\, dV. \tag{7.65}$$

The combined weak form statements used as a starting point for the space and time discretization are those in Equations (7.60a) and (7.60b) (or (7.65)) and (7.64a,b,c). Finally, insofar as the deformation gradient $\boldsymbol{F}$ is treated as an unknown variable, Equation (7.64a,b,c)$_c$ can be further enhanced by adding a time integrated stabilization term which controls not only the difference between the rate of the deformation $\partial \boldsymbol{F}/\partial t$ and the material gradient of the velocities $\boldsymbol{\nabla}_0 \boldsymbol{v}$ given by $\boldsymbol{\mathcal{R}}_{\boldsymbol{F}}$ in Equation (7.64a,b,c)$_c$, but also the true difference between $\boldsymbol{F}$ and the material gradient of the current deformed geometry $\boldsymbol{\nabla}_0 \boldsymbol{x}$. This can be written as

$$\boldsymbol{F}^{st} = \boldsymbol{F} - \tau_{\boldsymbol{F}} \boldsymbol{\mathcal{R}}_{\boldsymbol{F}} - \alpha\,(\boldsymbol{F} - \boldsymbol{\nabla}_0 \boldsymbol{x}), \tag{7.66}$$

where α is a small dimensionless positive stabilization parameter, typically in the range $[0, 0.5]$. Note that for an approximate discretized solution using the mixed formulation proposed in this chapter, the term within parentheses is nonzero in general as the unknown $\boldsymbol{F}$ is not computed through the material gradient of the geometry, as is customary in displacement-based formulations.

7.3.2 Three-Dimensional Spatial Discretization

Following a standard finite element methodology, the reference domain V is broken down into a set of non-overlapping finite elements (e), and the conservation variables $\boldsymbol{\mathcal{U}}$ and their virtual work conjugates $\delta\boldsymbol{\mathcal{V}}$ are discretized into nodal values $\{\boldsymbol{\mathcal{U}}_a, \delta\boldsymbol{\mathcal{V}}_a\}$ and suitable shape functions N_a, where $a = 1, \ldots, N$, N being the total number of nodes in the mesh. In our case, we will restrict the interpolation to linear shape functions and four-node tetrahedral elements. This discretization is sufficient to guarantee second-order spatial convergence for all the fields contained in $\boldsymbol{\mathcal{U}}$.

Consider the following finite element expansion for the linear momentum $\boldsymbol{p}_R$ and its time rate:

$$\boldsymbol{p}_R = \sum_{b=1}^{N} N_b \boldsymbol{p}_{R,b}(t); \qquad \frac{\partial \boldsymbol{p}_R}{\partial t} = \sum_{b=1}^{N} N_b \dot{\boldsymbol{p}}_{R,b}(t); \qquad \dot{\boldsymbol{p}}_{R,b}(t) = \frac{d\boldsymbol{p}_{R,b}}{dt}. \tag{7.67a,b,c}$$

In a similar manner, consider the finite element expansion for the deformation gradient tensor and its time rate as

$$\boldsymbol{F} = \sum_{b=1}^{N} N_b \boldsymbol{F}_b(t); \qquad \frac{\partial \boldsymbol{F}}{\partial t} = \sum_{b=1}^{N} N_b \dot{\boldsymbol{F}}_b(t); \qquad \dot{\boldsymbol{F}}_b(t) = \frac{d\boldsymbol{F}_b}{dt}. \tag{7.68a,b,c}$$

The virtual work conjugates are expanded using the same approach as

$$\delta \boldsymbol{v} = \sum_{a=1}^{N} N_a \delta \boldsymbol{v}_a; \qquad \delta \boldsymbol{P} = \sum_{a=1}^{N} N_a \delta \boldsymbol{P}_a, \tag{7.69a,b}$$

which are subject to relevant boundary condition restrictions. Figure 7.5 displays the finite element discretization employed in a tetrahedral finite element. Considering first the discretized weak statement for the linear momentum balance principle, we particularize $\delta \boldsymbol{v}$ as $\delta \boldsymbol{v} = N_a \delta \boldsymbol{v}_a$ and substitute into Equation (7.60a), resulting in the following discrete weak form statement for each node $a = 1, \ldots, N$:

$$\int_V N_a \frac{\partial \boldsymbol{p}_R}{\partial t}\, dV = \int_{\partial V} N_a \boldsymbol{t}_B \, dA + \int_V N_a \boldsymbol{f}_R \, dV - \int_V \boldsymbol{P}^{st} \boldsymbol{\nabla}_0 N_a \, dV. \tag{7.70}$$

The first two terms on the right-hand side of Equation (7.70) represent the equivalent external force vector acting on node a, which we can rename as $\boldsymbol{F}_a^{\boldsymbol{p}}$, not to be confused with the deformation gradient $\boldsymbol{F}_a$ at node a. The last term on the right-hand side of Equation (7.70) is the internal force vector acting on node a, which we can denote as $\boldsymbol{T}_a^{\boldsymbol{p}}$, namely

$$\boldsymbol{F}_a^{\boldsymbol{p}} = \int_{\partial V} N_a \boldsymbol{t}_B \, dA + \int_V N_a \boldsymbol{f}_R \, dV; \qquad \boldsymbol{T}_a^{\boldsymbol{p}} = \int_V \boldsymbol{P}^{st} \boldsymbol{\nabla}_0 N_a \, dV. \tag{7.71a,b}$$

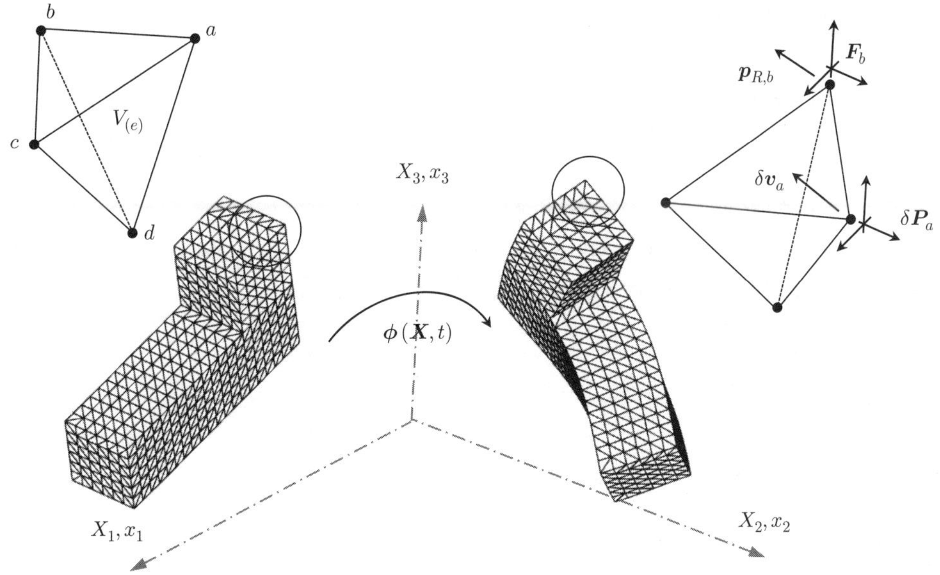

FIGURE 7.5 Three-dimensional spatial discretization – arrangement of real fields $\{\boldsymbol{p}_R, \boldsymbol{F}\}$ and work conjugate virtual fields $\{\delta \boldsymbol{v}, \delta \boldsymbol{P}\}$ in a four-noded tetrahedral finite element.

Thus, we can reformulate Equation (7.70) as

$$\sum_{b=1}^{N} m_{ab}\dot{\boldsymbol{p}}_{R,b} = \boldsymbol{F}_a^{\boldsymbol{p}} - \boldsymbol{T}_a^{\boldsymbol{p}}; \qquad a = 1,\ldots,N, \tag{7.72}$$

where m_{ab} represents the consistent mass like matrix contribution (refer to Equation (7.31a,b)$_a$) and $\boldsymbol{T}_a^{\boldsymbol{p}}$ represents the equivalent internal force vector acting on node a. Note that the evaluation of m_{ab}, $\boldsymbol{F}_a^{\boldsymbol{p}}$, and $\boldsymbol{T}_a^{\boldsymbol{p}}$ is carried out following the standard element by element assembly. Finally, the N nodal equations in Equation (7.72) can be collected into a single system of equations, written in vector format as

$$\mathbf{M}\dot{\mathbf{U}}^{\boldsymbol{p}} = \mathbf{F}^{\boldsymbol{p}} - \mathbf{T}^{\boldsymbol{p}}, \tag{7.73}$$

where $\mathbf{M}$ represents a consistent mass like matrix[‖] and $\mathbf{U}^{\boldsymbol{p}}$ is the vector of nodal unknowns,

$$\mathbf{M} = \begin{bmatrix} m_{11}\boldsymbol{I}_{3\times3} & m_{12}\boldsymbol{I}_{3\times3} & \cdots & m_{1N}\boldsymbol{I}_{3\times3} \\ m_{21}\boldsymbol{I}_{3\times3} & m_{22}\boldsymbol{I}_{3\times3} & \cdots & m_{2N}\boldsymbol{I}_{3\times3} \\ \vdots & \vdots & & \vdots \\ m_{N1}\boldsymbol{I}_{3\times3} & m_{N2}\boldsymbol{I}_{3\times3} & \cdots & m_{NN}\boldsymbol{I}_{3\times3} \end{bmatrix}; \quad \mathbf{U}^{\boldsymbol{p}} = \begin{bmatrix} \boldsymbol{p}_{R,1} \\ \boldsymbol{p}_{R,2} \\ \vdots \\ \boldsymbol{p}_{R,N} \end{bmatrix}, \tag{7.74a,b}$$

with $\boldsymbol{I}_{3\times3}$ the second-order identity tensor. Similarly, $\mathbf{F}^{\boldsymbol{p}}$ and $\mathbf{T}^{\boldsymbol{p}}$ are the equivalent external and internal force vectors, which can be expanded as

$$\mathbf{F}^{\boldsymbol{p}} = \begin{bmatrix} \boldsymbol{F}_1^{\boldsymbol{p}} \\ \boldsymbol{F}_2^{\boldsymbol{p}} \\ \vdots \\ \boldsymbol{F}_N^{\boldsymbol{p}} \end{bmatrix}; \quad \mathbf{T}^{\boldsymbol{p}} = \begin{bmatrix} \boldsymbol{T}_1^{\boldsymbol{p}} \\ \boldsymbol{T}_2^{\boldsymbol{p}} \\ \vdots \\ \boldsymbol{T}_N^{\boldsymbol{p}} \end{bmatrix}. \tag{7.75a,b}$$

Typically, for computational convenience, the above consistent mass matrix $\mathbf{M}$ can be replaced by its lumped counterpart $\mathbf{M}^L$, defined as

$$\mathbf{M}^L = \begin{bmatrix} m_{11}^L\boldsymbol{I}_{3\times3} & \mathbf{0}_{3\times3} & \cdots & \mathbf{0}_{3\times3} \\ \mathbf{0}_{3\times3} & m_{22}^L\boldsymbol{I}_{3\times3} & \cdots & \mathbf{0}_{3\times3} \\ \vdots & \vdots & & \vdots \\ \mathbf{0}_{3\times3} & \mathbf{0}_{3\times3} & \cdots & m_{NN}^L\boldsymbol{I}_{3\times3} \end{bmatrix};$$

$$m_{aa}^L = \sum_{b=1}^{N} m_{ab} = \int_V N_a \, dV. \tag{7.76a,b}$$

‖ As in the one-dimensional case, note that $\mathbf{M}$ is not strictly a mass matrix as its units are not of mass but of volume.

Attention now turns to the discretized weak statement for the geometric conservation law given by Equation (7.60b) and particularizing $\delta \boldsymbol{P}$ as $\delta \boldsymbol{P} = N_a \delta \boldsymbol{P}_a$, which, upon substitution into Equation (7.60b), gives the following nodal weak form statement for each node a:

$$\int_V N_a \frac{\partial \boldsymbol{F}}{\partial t}\, dV = \int_{\partial V} N_a \left(\boldsymbol{v}_B \otimes \boldsymbol{N}\right)\, dA - \int_V \boldsymbol{v}^{st} \otimes \boldsymbol{\nabla}_0 N_a\, dV. \tag{7.77}$$

Using the finite element discretization for the deformation gradient tensor given by Equation (7.68a,b,c) and substituting into Equation (7.77) gives

$$\sum_{b=1}^{N} m_{ab} \dot{\boldsymbol{F}}_b = \boldsymbol{F}_a^{\boldsymbol{F}} - \boldsymbol{T}_a^{\boldsymbol{F}}; \qquad a = 1, \ldots, N, \tag{7.78}$$

where

$$\boldsymbol{F}_a^{\boldsymbol{F}} = \int_{\partial V} N_a \left(\boldsymbol{v}_B \otimes \boldsymbol{N}\right)\, dA; \qquad \boldsymbol{T}_a^{\boldsymbol{F}} = \int_V \boldsymbol{v}^{st} \otimes \boldsymbol{\nabla}_0 N_a\, dV. \tag{7.79a,b}$$

Finally, grouping Equation (7.78) into a vector format gives

$$\mathbf{M}\dot{\mathbf{U}}^{\boldsymbol{F}} = \mathbf{F}^{\boldsymbol{F}} - \mathbf{T}^{\boldsymbol{F}}, \tag{7.80}$$

where the mass like matrix $\mathbf{M}$ can be replaced by its lumped counterpart $\mathbf{M}^L$ and with identical terms to those in Equation (7.73), except that the upper index $\boldsymbol{p}$ is replaced with $\boldsymbol{F}$, namely

$$\mathbf{U}^{\boldsymbol{F}} = \begin{bmatrix} \boldsymbol{F}_1 \\ \boldsymbol{F}_2 \\ \vdots \\ \boldsymbol{F}_N \end{bmatrix}; \quad \mathbf{F}^{\boldsymbol{F}} = \begin{bmatrix} \boldsymbol{F}_1^{\boldsymbol{F}} \\ \boldsymbol{F}_2^{\boldsymbol{F}} \\ \vdots \\ \boldsymbol{F}_N^{\boldsymbol{F}} \end{bmatrix}; \quad \mathbf{T}^{\boldsymbol{F}} = \begin{bmatrix} \boldsymbol{T}_1^{\boldsymbol{F}} \\ \boldsymbol{T}_2^{\boldsymbol{F}} \\ \vdots \\ \boldsymbol{T}_N^{\boldsymbol{F}} \end{bmatrix}. \tag{7.81a,b,c}$$

Equations (7.73) and (7.80) represent a system of (semi-discrete) first-order ordinary differential equations in time which can be solved through a suitable time integrator. As can be seen, these two equations represent a generalization of Equations (7.32) and (7.38) from the one-dimensional to the three-dimensional case.

7.3.3 Time Discretization

The systems of semi-discrete conservation Equations (7.73) and (7.80) can be rewritten together as

$$\mathbf{M}\dot{\mathbf{U}}^{\boldsymbol{p}} = \mathbf{F}^{\boldsymbol{p}}(t) - \mathbf{T}^{\boldsymbol{p}}(\mathbf{U}^{\boldsymbol{F}}, \dot{\mathbf{U}}^{\boldsymbol{F}}, \mathbf{x}), \tag{7.82a}$$

$$\mathbf{M}\dot{\mathbf{U}}^{\boldsymbol{F}} = \mathbf{F}^{\boldsymbol{F}}(t) - \mathbf{T}^{\boldsymbol{F}}(\mathbf{U}^{\boldsymbol{p}}, \dot{\mathbf{U}}^{\boldsymbol{p}}), \tag{7.82b}$$

along with the geometry update

$$\dot{\mathbf{x}} = \frac{1}{\rho_R}\mathbf{U}^{\boldsymbol{p}}, \tag{7.83}$$

where the explicit dependence of the external and internal contributions on the right-hand sides of above equations has been included. As it is presented, the external contributions $\mathbf{F}^{\boldsymbol{p}}$ and $\mathbf{F}^{\boldsymbol{F}}$ can depend on time due to time-varying body forces and/or boundary tractions/velocities. The internal contribution $\mathbf{T}^{\boldsymbol{F}}$ depends on the unknown variables $\mathbf{U}^{\boldsymbol{p}}$ and their time rates $\dot{\mathbf{U}}^{\boldsymbol{p}}$ through the stabilized field $\boldsymbol{v}^{st}$ (see Equation (7.62a,b)). As for the internal contribution $\mathbf{T}^{\boldsymbol{p}}$, this depends on the unknown variables $\mathbf{U}^{\boldsymbol{F}}$, their time rates $\dot{\mathbf{U}}^{\boldsymbol{F}}$, and the vector of nodal coordinates $\mathbf{x}$ through the stabilized deformation gradient field $\boldsymbol{F}^{st}$ (see Equation (7.66)).

We now split the time interval $[0, T]$ into N_t time subintervals and seek time-discrete solutions $\mathbf{U}_n$ (and $\mathbf{x}_n$) (corresponding to time step t_n) with $n = 0, \ldots, N_t - 1$ via a simple one-step two-stage (second order in time) Runge–Kutta (RK) method, which permits the advancement in time of the solution from time step t_n to t_{n+1}, with $\Delta t = t_{n+1} - t_n$. Unfortunately, given that the time rates of the unknown variables feature on the right-hand side of Equations (7.82), this prevents the possibility of straightforwardly employing the explicit time integrator and would require the use of a fixed-point type iteration algorithm at each stage, as previously explained in Section 7.2.3.

Fortunately, the costly need of this fixed-point iteration algorithm at each stage of the time integrator can be circumvented by selecting the stabilization parameter $\tau_{\boldsymbol{p}}$ to be zero. In this special case, the right-hand side of Equation (7.82b) stops depending on $\dot{\mathbf{U}}^{\boldsymbol{p}}$ as it does not depend upon the residual term $\boldsymbol{\mathcal{R}}_{\boldsymbol{p}}$ ($\boldsymbol{v}^{st}$ simply reduces to $\boldsymbol{v}$). This enables the Ith Runge–Kutta stage equation for the unknown vector field $\mathbf{U}^{\boldsymbol{F}}$ to be written as

$$\mathbf{M}^L\left(\frac{\mathbf{U}^{\boldsymbol{F}}_{[I]} - \mathbf{U}^{\boldsymbol{F}}_{[I-1]}}{\Delta t}\right) = \mathbf{F}^{\boldsymbol{F}}(t_{n+I-1}) - \mathbf{T}^{\boldsymbol{F}}(\mathbf{U}^{\boldsymbol{p}}_{[I-1]}); \quad I = 1, 2, \tag{7.84}$$

which, as can be seen, can be advanced explicitly from stage $I - 1$ to stage I. The values at $[0]$ are assumed to coincide with the values at the previous time step n. In the above equation, the consistent mass matrix has been replaced by its diagonally lumped counterpart $\mathbf{M}^L$ as this enhances the computational speed of the scheme while retaining the (second) order of convergence in time. Subsequently, the Ith Runge–Kutta stage equation for the advancement of $\mathbf{U}^{\boldsymbol{p}}$ arises as

$$\begin{aligned}\mathbf{M}^L\left(\frac{\mathbf{U}^{\boldsymbol{p}}_{[I]} - \mathbf{U}^{\boldsymbol{p}}_{[I-1]}}{\Delta t}\right) &= \mathbf{F}^{\boldsymbol{p}}(t_{n+I-1}) - \mathbf{T}^{\boldsymbol{p}}\left(\mathbf{U}^{\boldsymbol{F}}_{[I-1]}, \frac{\mathbf{U}^{\boldsymbol{F}}_{[I]} - \mathbf{U}^{\boldsymbol{F}}_{[I-1]}}{\Delta t}, \mathbf{x}_{[I-1]}\right);\\ I &= 1, 2,\end{aligned} \tag{7.85}$$

which can also be advanced explicitly, as a result of the staggered solution process. Finally, the geometry can be updated as

$$\left(\frac{\mathbf{x}_{[I]} - \mathbf{x}_{[I-1]}}{\Delta t}\right) = \frac{1}{\rho_R}\mathbf{U}^p_{[I-1]}; \qquad I = 1, 2. \tag{7.86}$$

The final evolution of the set of variables at step $n+1$ is obtained via Equation (7.43d) as

$$\mathbf{U}_{n+1} = \frac{1}{2}\left(\mathbf{U}_{[0]} + \mathbf{U}_{[2]}\right), \tag{7.87}$$

and the geometry as

$$\mathbf{x}_{n+1} = \frac{1}{2}\left(\mathbf{x}_{[0]} + \mathbf{x}_{[2]}\right). \tag{7.88}$$

For time stability, it is important to control the size of the chosen time step Δt. A similar time stability criterion to that shown in Equation (7.47a,b) is used,

$$\Delta t = \alpha_{CFL}\frac{h}{c_{max}}; \qquad c_{max} = c_p; \qquad h = \min_{(e)} h_{(e)}, \tag{7.89a,b,c}$$

where $h_{(e)}$ is the size of a generic element in the mesh and α_{CFL} is the Courant–Friedrichs–Lewy constant, which depends on the specific time integrator and, typically, takes values in the interval $[0, 1]$. In this case, the maximum wave speed is that of the *p-wave* speed, which was computed in Example 7.5. A flowchart is presented in Box 7.2 summarizing the overall algorithm.

BOX 7.2: Conservation laws – three-dimensional algorithm

- INPUT geometry, material properties, and solution parameters
- INITIALIZE $\mathbf{U}^p = \mathbf{U}^p_0$, $\mathbf{U}^F = \mathbf{U}^F_0$, $\mathbf{x} = \mathbf{X}$
- FIND mass matrix $\mathbf{M}$ (or $\mathbf{M}^L$) (7.76a,b)
- WHILE $t < T$ (time steps)
  - FIND Δt (7.89a,b,c)
  - SET $\mathbf{U}^p_{[0]} = \mathbf{U}^p_n$, $\mathbf{U}^F_{[0]} = \mathbf{U}^F_n$, $\mathbf{x}_{[0]} = \mathbf{x}_n$ (7.43a)
  - FOR I=1,2 (Runge–Kutta stages)
    - SET $\mathbf{F}^F = \mathbf{F}^F(t_{n+I-1})$ (7.79a,b)$_a$
    - FIND $\mathbf{T}^F$ (7.62a,b), (7.79a,b)$_b$
    - SOLVE for $\mathbf{U}^F_{[I]}$ (7.84)
    - SET $\mathbf{F}^p = \mathbf{F}^p(t_{n+I-1})$ (7.72)
    - FIND $\mathbf{T}^p$ (7.64a,b,c), (7.72)
    - SOLVE for $\mathbf{U}^p_{[I]}$, $\mathbf{x}_{[I]}$ (7.85), (7.86)
  - ENDLOOP

(continued)

Box 7.2: *(cont.)*

- SOLVE for $\mathbf{U}^{p}_{n+1}$, $\mathbf{U}^{F}_{n+1}$, $\mathbf{x}_{n+1}$ (7.43c)
- ENDLOOP

The time integration algorithm, as presented, does not necessarily preserve the global angular momentum (it only satisfies the conservation of global linear momentum by construction). Although this is not strictly critical for the majority of problems of interest, it is possible to satisfy this requirement. In order to preserve global angular momentum, it is necessary to introduce a suitable modification of the time integration algorithm by identifying suitable kinematic constraints. This is shown in Examples 7.6 and 7.7.

EXAMPLE 7.6: Conservation of angular momentum

The time-discrete variation of angular momentum from a given time step t_n to the next time step t_{n+1} can be written as

$$\boldsymbol{A}_{n+1} - \boldsymbol{A}_n = \sum_{a,b=1}^{N} M_{ab}(\boldsymbol{x}_{a,n+1} \times \boldsymbol{p}_{b,n+1}) - \sum_{a,b=1}^{N} M_{ab}(\boldsymbol{x}_{a,n} \times \boldsymbol{p}_{b,n}).$$

The Runge–Kutta update given by Equations (7.41a)–(7.41e) for the linear momentum at node b is

$$\boldsymbol{p}_{b,n+1} = \boldsymbol{p}_{b,n} + \frac{\Delta t}{2}\left(\dot{\boldsymbol{p}}_{b,[0]} + \dot{\boldsymbol{p}}_{b,[1]}\right).$$

Substitution of the linear momentum update gives

$$\begin{aligned}\boldsymbol{A}_{n+1} - \boldsymbol{A}_n = &\sum_{a,b=1}^{N} M_{ab}\left(\boldsymbol{x}_{a,n+1} \times \left(\boldsymbol{p}_{n,b} + \frac{\Delta t}{2}\dot{\boldsymbol{p}}_{b,[0]}\right)\right)\\ &+ \sum_{a,b=1}^{N} M_{ab}\left(\boldsymbol{x}_{a,n+1} \times \frac{\Delta t}{2}\dot{\boldsymbol{p}}_{b,[1]}\right)\\ &- \sum_{a,b=1}^{N} M_{ab}\left(\boldsymbol{x}_{a,n} \times \boldsymbol{p}_{b,n}\right).\end{aligned}$$

Similarly, the Runge–Kutta update given by Equations (7.41a)–(7.41e) for the geometry at node a can be formulated as

$$\boldsymbol{x}_{a,n+1} = \boldsymbol{x}_{a,n} + \frac{\Delta t}{2}\left(\dot{\boldsymbol{x}}_{a,[0]} + \dot{\boldsymbol{x}}_{a,[1]}\right) = \boldsymbol{x}_{a,n} + \frac{\Delta t}{2\rho_R}\left(\boldsymbol{p}_{a,n} + \boldsymbol{p}_{a,[1]}\right)$$

(continued)

Example 7.6: *(cont.)*

$$= \boldsymbol{x}_{a,n} + \frac{\Delta t}{2\rho_R}\left(\boldsymbol{p}_{a,n} + \left(\boldsymbol{p}_{a,n} + \Delta t \dot{\boldsymbol{p}}_{a,[0]}\right)\right)$$
$$= \boldsymbol{x}_{a,n} + \frac{\Delta t}{\rho_R}\boldsymbol{p}_{a,n} + \frac{\Delta t^2}{2\rho_R}\dot{\boldsymbol{p}}_{a,[0]}.$$

Substituting the geometry update into the first term on the right-hand side of the angular momentum update, then expanding and simplifying appropriately, results in

$$\boldsymbol{A}_{n+1} - \boldsymbol{A}_n = \sum_{a,b=1}^{N} \boldsymbol{M}_{ab}\left(\boldsymbol{x}_{a,n} \times \frac{\Delta t}{2}\dot{\boldsymbol{p}}_{b,[0]}\right) + \sum_{a,b=1}^{N} \boldsymbol{M}_{ab}\left(\boldsymbol{x}_{a,n+1} \times \frac{\Delta t}{2}\dot{\boldsymbol{p}}_{b,[1]}\right).$$

Thus, in order to ensure the discrete satisfaction of angular momentum, it suffices to show that both terms on the right-hand side of the preceding equation cancel, that is,

$$\sum_{a,b=1}^{N} \boldsymbol{M}_{ab}\left(\boldsymbol{x}_{a,n} \times \dot{\boldsymbol{p}}_{b,[0]}\right) = \boldsymbol{0}; \qquad \sum_{a,b=1}^{N} \boldsymbol{M}_{ab}\left(\boldsymbol{x}_{a,n+1} \times \dot{\boldsymbol{p}}_{b,[1]}\right) = \boldsymbol{0},$$

which imposes constraints on $\dot{\boldsymbol{p}}_{b,[0]}$ and $\dot{\boldsymbol{p}}_{b,[1]}$, to be fulfilled at each of the two Runge–Kutta stages.

EXAMPLE 7.7: A projection procedure for angular momentum conservation

In order to ensure conservation of both linear and angular momenta simultaneously, the discrete nodal values $\dot{\boldsymbol{p}}_{b,[0]}$ (featuring in the first Runge–Kutta stage) and $\dot{\boldsymbol{p}}_{b,[1]}$ (featuring in the second Runge–Kutta stage) can be modified, resulting in the new corrected values $\dot{\boldsymbol{p}}^c_{b,[0]}$ and $\dot{\boldsymbol{p}}^c_{b,[1]}$, respectively. This correction can be carried out through a least squared minimization (projection) process, subject to the global conservation of both linear and angular momentum. With this in mind, two Lagrangian functions Π_I ($I = 0, 1$) are defined in terms of the (yet unknown) corrected values $\dot{\boldsymbol{p}}^c_{b,[0]}$ and $\dot{\boldsymbol{p}}^c_{b,[1]}$ for each of the two Runge–Kutta stages as follows:

(continued)

Example 7.7: *(cont.)*

$$\begin{aligned}
&\Pi_I\left(\dot{\boldsymbol{p}}^c_{1,[I]},\ldots,\dot{\boldsymbol{p}}^c_{N,[I]},\boldsymbol{\lambda}^I_{\text{lin}},\boldsymbol{\lambda}^I_{\text{ang}}\right)\\
&\quad=\frac{1}{2}\sum_{b=1}^{N} m^L_{bb}\left(\dot{\boldsymbol{p}}^c_{b,[I]}-\dot{\boldsymbol{p}}_{b,[I]}\right)\cdot\left(\dot{\boldsymbol{p}}^c_{b,[I]}-\dot{\boldsymbol{p}}_{b,[I]}\right)\\
&\qquad-\boldsymbol{\lambda}^I_{\text{lin}}\cdot\left(\sum_{a,b=1}^{N}\boldsymbol{M}_{ab}\,\dot{\boldsymbol{p}}^c_{b,[I]}\right)\\
&\qquad-\boldsymbol{\lambda}^I_{\text{ang}}\cdot\left(\sum_{a,b=1}^{N}\boldsymbol{M}_{ab}\left(\boldsymbol{x}_{a,n+I}\times\dot{\boldsymbol{p}}^c_{b,[I]}\right)\right),
\end{aligned}$$

where $\boldsymbol{\lambda}^I_{\text{lin}}$ and $\boldsymbol{\lambda}^I_{\text{ang}}$ are two Lagrange multiplier vectors used to enforce the global conservation of linear and angular momentum, respectively. The stationary conditions of the above Lagrangian with respect to nodal perturbations $\delta\dot{\boldsymbol{p}}^c_{a,[I]}$,

$$D\Pi_I\left(\dot{\boldsymbol{p}}^c_{1,[I]},\ldots,\dot{\boldsymbol{p}}^c_{N,[I]},\boldsymbol{\lambda}^I_{\text{lin}},\boldsymbol{\lambda}^I_{\text{ang}}\right)\left[\delta\dot{\boldsymbol{p}}^c_{a,[I]}\right]=0;\qquad a=1,\ldots,N,$$

result in an explicit expression for the nodal values $\dot{\boldsymbol{p}}^c_{a,[I]}$:

$$\dot{\boldsymbol{p}}^c_{a,[I]}=\dot{\boldsymbol{p}}_{a,[I]}+\boldsymbol{\lambda}^I_{\text{lin}}+\boldsymbol{\lambda}^I_{\text{ang}}\times\boldsymbol{x}_{a,n+I};\qquad a=1,\ldots,N.$$

Similarly, the stationary conditions of the above Lagrangian with respect to perturbations of the Lagrange multipliers $\delta\boldsymbol{\lambda}^I_{\text{lin}}$ and $\delta\boldsymbol{\lambda}^I_{\text{ang}}$,

$$\begin{aligned}
&D\Pi_I\left(\dot{\boldsymbol{p}}^c_{1,[I]},\ldots,\dot{\boldsymbol{p}}^c_{N,[I]},\boldsymbol{\lambda}^I_{\text{lin}},\boldsymbol{\lambda}^I_{\text{ang}}\right)\left[\delta\boldsymbol{\lambda}^I_{\text{lin}}\right]=0,\\
&D\Pi_I\left(\dot{\boldsymbol{p}}^c_{1,[I]},\ldots,\dot{\boldsymbol{p}}^c_{N,[I]},\boldsymbol{\lambda}^I_{\text{lin}},\boldsymbol{\lambda}^I_{\text{ang}}\right)\left[\delta\boldsymbol{\lambda}^I_{\text{ang}}\right]=0,
\end{aligned}$$

lead to the constraint equations representing the conservation of linear and angular momentum, respectively. Substitution of the above expression for the discrete nodal values $\dot{\boldsymbol{p}}^c_{a,[I]}$ into these constraint equations renders a system of equations in terms of the unknowns $\boldsymbol{\lambda}^I_{\text{lin}}$ and $\boldsymbol{\lambda}^I_{\text{ang}}$ as follows:

$$\begin{aligned}
&\sum_{a,b=1}^{N}\boldsymbol{M}_{ab}\left(\dot{\boldsymbol{p}}_{b,[I]}+\boldsymbol{\lambda}^I_{\text{lin}}+\boldsymbol{\lambda}^I_{\text{ang}}\times\boldsymbol{x}_{b,n+I}\right)=\boldsymbol{0},\\
&\sum_{a,b=1}^{N}\boldsymbol{M}_{ab}\left(\boldsymbol{x}_{a,n+I}\times\left(\dot{\boldsymbol{p}}_{b,[I]}+\boldsymbol{\lambda}^I_{\text{lin}}+\boldsymbol{\lambda}^I_{\text{ang}}\times\boldsymbol{x}_{b,n+I}\right)\right)=\boldsymbol{0}.
\end{aligned}$$

Once the above 6×6 system of equations is solved and values for $\boldsymbol{\lambda}^I_{\text{lin}}$ and $\boldsymbol{\lambda}^I_{\text{ang}}$ are obtained, these can be immediately used to update the discrete nodal values $\dot{\boldsymbol{p}}^a_{N,[I]}$.

7.4 A GENERAL THERMOELASTIC SYSTEM OF CONSERVATION LAWS

This section extends the methodology presented in the previous sections to the thermoelastic case. Following the content of Chapter 6 on thermodynamics, consider, in addition to the solution of the conservation laws for the linear momentum and the deformation gradient tensor given in Equations (7.48), the use of a new conservation law for the computation of the entropy η of the system (see Equation (6.15)), which is presented here again for convenience:

$$\frac{\partial \eta}{\partial t} = \frac{r_R}{\theta} - \frac{\mathrm{DIV}\boldsymbol{Q}}{\theta}. \tag{7.90}$$

In order to complete this extended system of conservation laws in terms of the conservation variables $\{\boldsymbol{p}_R, \boldsymbol{F}, \eta\}$, constitutive relationships are required for the definition of the vector $\boldsymbol{Q}$, the stress tensor $\boldsymbol{P}$, and the temperature θ in terms of the conservation variables. As far the heat flux vector $\boldsymbol{Q}$ is concerned, consider the typical Fourier law to hold, which can be defined in a total Lagrangian setting (see Boxes 6.1 and 6.3), as

$$\boldsymbol{Q} = -\boldsymbol{K}\boldsymbol{\nabla}_0\theta; \qquad \boldsymbol{K} = kJ\boldsymbol{C}^{-1}. \tag{7.91a,b}$$

Equations (7.91a,b) are written in terms of the so-called thermal conductivity coefficient k, the Jacobian J, and the right Cauchy–Green strain tensor $\boldsymbol{C} = \boldsymbol{F}^T\boldsymbol{F}$. At this juncture, it is important to emphasize that the (conductive) nature of the heat flux vector $\boldsymbol{Q}$ is very different to the (convective) nature of the fluxes featuring in the linear momentum and geometric conservation laws. Convective fluxes are functions of the conservation variables, whereas conductive fluxes depend on the gradients of these variables; see Example 5.2 in Chapter 5.

Regarding the definition of $\boldsymbol{P}$ and θ, we consider a distortional-volumetric energy decomposition, such as that introduced in Equation (6.72), as

$$\mathcal{E}(\boldsymbol{F}, \eta) = \hat{\mathcal{E}}(\boldsymbol{F}) + U(J, \eta), \tag{7.92}$$

so that we can deduce expressions for the temperature θ and the stress tensor $\boldsymbol{P}$ as follows:

$$\theta(J, \eta) = \frac{\partial U}{\partial \eta}; \qquad \boldsymbol{P}(\boldsymbol{F}, \eta) = \frac{\partial \mathcal{E}}{\partial \boldsymbol{F}} = \boldsymbol{P}'(\boldsymbol{F}) + p(J, \eta)J\boldsymbol{F}^{-T}, \tag{7.93a,b}$$

with the deviatoric component of the Piola–Kirchhoff stress tensor and the pressure given as

$$\boldsymbol{P}'(\boldsymbol{F}) = \frac{\partial \hat{\mathcal{E}}}{\partial \boldsymbol{F}}; \qquad p(J, \eta) = \frac{\partial U}{\partial J}. \tag{7.94a,b}$$

As a possible example, for the distortional component of the internal energy $\hat{\mathcal{E}}(\boldsymbol{F})$, adopt the deviatoric strain energy of an incompressible neo-Hookean constitutive model as

$$\hat{\mathcal{E}}(\boldsymbol{F}) = \frac{\mu}{2}\left(II_{\hat{\boldsymbol{F}}} - 3\right); \qquad \hat{\boldsymbol{F}} = J^{-1/3}\boldsymbol{F}, \tag{7.95a,b}$$

where μ is a positive material parameter which coincides with the shear modulus of the material. As for the volumetric internal energy contribution $U(J,\eta)$, we consider an additive decomposition into purely mechanical and coupled thermomechanical contributions (see Equation (6.74)) as

$$U(J,\eta) = U_0(J) + c_v\theta_0(J)\left[e^{\eta/c_v} - 1\right], \tag{7.96}$$

where the following mechanical contribution is considered:

$$U_0(J) = \frac{\kappa}{2}(J-1)^2. \tag{7.97}$$

For the particular case of a Mie–Grüneisen model the following relationship is obtained, after combination of Equations (6.37a,b)$_b$ and (6.90a,b)$_b$:

$$\theta_0(J) = \theta_R e^{\Gamma_0\left(\frac{1-J^q}{q}\right)}; \qquad q \in (0,1]. \tag{7.98}$$

In the above equations, θ_R denotes the reference temperature, c_v is the specific heat coefficient at constant volume, and Γ_0 and q are the so-called Mie–Grüneisen material parameters that describe the thermomechanical coupling. The above thermoelastic model leads to the following expressions for the deviatoric contribution to the Piola–Kirchhoff stress tensor $\boldsymbol{P}'$ and the pressure p:

$$\boldsymbol{P}'(\boldsymbol{F}) = \mu J^{-2/3}\left[\boldsymbol{F} - \frac{1}{3}II_{\boldsymbol{F}}\boldsymbol{F}^{-T}\right] \tag{7.99}$$

and

$$\begin{aligned} p(J,\eta) &= \frac{dU_0}{dJ} + c_v\left[e^{\eta/c_v} - 1\right]\frac{d\theta_0}{dJ} \\ &= \kappa(J-1) - c_v\theta_R\Gamma_0 J^{q-1}\left[e^{\eta/c_v} - 1\right]e^{\Gamma_0\left(\frac{1-J^q}{q}\right)}. \end{aligned} \tag{7.100}$$

Analogously, the temperature θ can be obtained as

$$\begin{aligned} \theta(J,\eta) &= \theta_0(J)e^{\eta/c_v} \\ &= \theta_R e^{\Gamma_0\left(\frac{1-J^q}{q}\right)} e^{\eta/c_v}. \end{aligned} \tag{7.101}$$

For isentropic deformations, i.e. when $\eta = 0$, substitution into Equation (7.100) implies that Equation (7.93a,b)$_b$ for the first Piola–Kirchhoff stress $\boldsymbol{P}$ reduces to the nearly incompressible neo-Hookean model presented in NL-Statics.**

** This can be obtained by transforming the second Piola–Kirchhoff stress tensor featuring in NL-Statics, Equation (6.54).

7.4.1 Weak Form Equations and Stabilization

In order to establish the system of weak forms necessary for space and time discretization, start by extending the system of residual Equations (7.54a,b) to include the entropy equation (7.90) as

$$\boldsymbol{\mathcal{R}} = 0; \qquad \boldsymbol{\mathcal{R}} = \begin{bmatrix} \boldsymbol{\mathcal{R}}_{\boldsymbol{p}} \\ \boldsymbol{\mathcal{R}}_{\boldsymbol{F}} \\ \mathcal{R}_{\eta} \end{bmatrix}, \tag{7.102a,b}$$

where the additional residual equation for the entropy is written as

$$\mathcal{R}_{\eta} = \frac{\partial \eta}{\partial t} + \frac{\text{DIV}\boldsymbol{Q}}{\theta} - \frac{r_R}{\theta}. \tag{7.103}$$

The *weak statement* can be defined by introducing compatible (with the boundary conditions) work conjugate virtual fields, which now include the virtual temperature as the energy conjugate variable to the entropy η as

$$\delta\boldsymbol{\mathcal{V}} = \begin{bmatrix} \delta\boldsymbol{v} \\ \delta\boldsymbol{P} \\ \delta\theta \end{bmatrix}, \tag{7.104}$$

in terms of virtual velocities $\delta\boldsymbol{v}$, stresses $\delta\boldsymbol{P}$, and temperature $\delta\theta$. This leads to the extended mixed *principle of virtual power* formulated as

$$\delta W(\delta\boldsymbol{\mathcal{V}}, \boldsymbol{\mathcal{U}}) = \int_V (\delta\boldsymbol{v} \cdot \boldsymbol{\mathcal{R}}_{\boldsymbol{p}} + \delta\boldsymbol{P} : \boldsymbol{\mathcal{R}}_{\boldsymbol{F}} + \delta\theta\, \mathcal{R}_{\eta})\, dV = 0. \tag{7.105}$$

It is interesting to note that the field $\delta\theta$ is work conjugate with respect to the entropy η of the system, and hence, when multiplied by $\mathcal{R}_{\eta}$, that it renders the expected units of power. Individual weak form statements can be obtained by considering individual virtual variations $\delta\boldsymbol{v}$, $\delta\boldsymbol{P}$, and $\delta\theta$. Proceeding analogously as in Section 7.3.1, identical expressions to those in Equations (7.58a) and (7.58b) (or Equation (7.59)) are obtained when considering the independent virtual fields $\delta\boldsymbol{v}$ and $\delta\boldsymbol{P}$. However, in order to ensure the stability of the formulation, we replace the standard evaluation of the first Piola–Kirchhoff stress tensor $\boldsymbol{P}$ in Equation (7.58a) by a stabilized counterpart $\boldsymbol{P}^{st}$, which is defined in a variational multi-scale (VMS) manner as

$$\boldsymbol{P}^{st} = \boldsymbol{P}(\boldsymbol{F}^{st}, \eta^{st}); \qquad \boldsymbol{F}^{st} = \boldsymbol{F} + \boldsymbol{F}'; \qquad \eta^{st} = \eta + \eta', \tag{7.106a,b,c}$$

with $\boldsymbol{F}'$ and η' defined as

$$\boldsymbol{F}' = -\tau_F \boldsymbol{\mathcal{R}}_{\boldsymbol{F}} - \alpha(\boldsymbol{F} - \boldsymbol{\nabla}_0 \boldsymbol{x}); \qquad \eta' = -\tau_{\eta} \mathcal{R}_{\eta}, \tag{7.107a,b}$$

where τ_{η} represents a new non-negative stabilization parameter. As can be easily observed, the above VMS stabilization Equations (7.106a,b,c) generalize the isothermal case given by Equations (7.64a,b,c) to incorporate the possibility of thermomechanical coupling. As for the weak form expression corresponding to the individual virtual field $\delta\theta$, it yields

$$\delta W(\delta\theta,\boldsymbol{\mathcal{U}}) = \int_V \delta\theta \frac{\partial \eta}{\partial t}\, dV + \int_V \delta\theta \frac{\text{DIV}\boldsymbol{Q}}{\theta}\, dV - \int_V \delta\theta \frac{r_R}{\theta}\, dV = 0. \quad (7.108)$$

In order to incorporate *naturally* possible heat flux boundary conditions, it is convenient to apply the Gauss divergence theorem to the heat flux term in Equation (7.108) so that

$$\int_V \delta\theta \frac{\partial \eta}{\partial t}\, dV = \int_V \boldsymbol{\nabla}_0 \delta\theta \cdot \frac{\boldsymbol{Q}}{\theta}\, dV - \int_V \delta\theta \left(\boldsymbol{\nabla}_0 \theta \cdot \frac{\boldsymbol{Q}}{\theta^2}\right) dV$$
$$- \int_{\partial V} \delta\theta \frac{Q_{N,B}}{\theta}\, dA + \int_V \delta\theta \frac{r_R}{\theta}\, dV, \quad (7.109)$$

where $Q_{N,B}$ represents the possible heat flux boundary conditions applied on part of the boundary ∂V. The rest of the boundary conditions are applied *essentially* either by the consideration of applied entropy or, more typically, applied temperature boundary conditions. In the latter case, inversion of Equation (7.101) (along with consideration of Equation (7.98)) is necessary, so that the entropy variable is expressed in terms of the temperature, which results in

$$\eta = c_v \ln \frac{\theta}{\theta_0(J)} = c_v \left[\ln \frac{\theta}{\theta_R} - \Gamma_0 \left(\frac{1 - J^q}{q} \right) \right]. \quad (7.110)$$

7.4.2 Three-Dimensional Spatial Discretization: Entropy

For the finite element discretization, semi-discrete expressions for the linear momentum and the geometric conservation laws were presented in Section 7.3.2. In this section, we present the additional semi-discrete expression for the time evolution equation of the entropy of the system. Considering $\delta\theta = \delta\theta_a N_a$, the a nodal equation ($a = 1, \ldots, N$) gives

$$\int_V N_a \frac{\partial \eta}{\partial t}\, dV = \int_V \boldsymbol{\nabla}_0 N_a \cdot \frac{\boldsymbol{Q}}{\theta}\, dV - \int_V N_a \left(\boldsymbol{\nabla}_0 \theta \cdot \frac{\boldsymbol{Q}}{\theta^2}\right) dV$$
$$- \int_{\partial V} N_a \frac{Q_{N,B}}{\theta}\, dA + \int_V N_a \frac{r_R}{\theta}\, dV, \quad (7.111)$$

After expanding the entropy and its time rate as

$$\eta = \sum_{b=1}^{N} N_b \eta_b(t); \quad \frac{\partial \eta}{\partial t} = \sum_{b=1}^{N} N_b \dot{\eta}_b(t); \quad \dot{\eta}_b(t) = \frac{d\eta_b}{dt}, \quad (7.112\text{a,b,c})$$

we obtain

$$\sum_{b=1}^{N} m_{ab}\dot{\eta}_b = F_a^\eta - T_a^\eta; \qquad a = 1, \ldots, N, \tag{7.113}$$

where the term F_a^η gathers the last two terms on the right-hand side of Equation (7.111) and T_a^η gathers the first two terms as follows:

$$F_a^\eta = -\int_{\partial V} N_a \frac{Q_{N,B}}{\theta}\, dA + \int_V N_a \frac{r_R}{\theta}\, dV, \tag{7.114a}$$

$$T_a^\eta = -\int_V \boldsymbol{\nabla}_0 N_a \cdot \frac{\boldsymbol{Q}}{\theta}\, dV + \int_V N_a \left(\boldsymbol{\nabla}_0 \theta \cdot \frac{\boldsymbol{Q}}{\theta^2}\right) dV. \tag{7.114b}$$

Finally, grouping Equation (7.113) into vector format results in

$$\mathbf{M}\dot{\mathbf{U}}^\eta = \mathbf{F}^\eta - \mathbf{T}^\eta, \tag{7.115}$$

where $\mathbf{U}^\eta$ is the vector of nodal entropy unknowns. For time integration, the same methodology as that shown in Section 7.3.3 is used, straightforwardly modified by incorporating the explicit advancement in time of Equation (7.115).

BOX 7.3: One-dimensional simple MATLAB program for conservation laws

```
clc; clear all; close all

L = 10;             % Domain
Nelem = 500;        % Number of elements
TauF  = 0.5;        % Stabilization coefficient
Rho   = 1;          % Density
K     = 1;          % Material constant
CFL   = 0.5;        % CFL number
Nstep = 12000;      % Number of time step
Traction = 1e-3;  % Specified traction force

% Preliminary and initiation
Npoin  = Nelem + 1;         % Number of nodes
cp     = sqrt(K/Rho);       % Pressure wave speed
p      = zeros(Npoin,1);    % Linear momentum
F      = zeros(Npoin,1);    % Deformation (or stretch)
Rp     = zeros(Npoin,1);    % Right hand side of linear momentum
RF     = zeros(Npoin,1);    % Right hand side of deformation
Le     = L/Nelem;           % Elemental length
Coor   = 0: Le: L;          % Coordinates
Dt     = CFL * Le / cp;     % Time increment
Lpoin  = [Le/2;  Le.*ones(Nelem-1,1); Le/2]; % Nodal length

% Initial and boundary conditions
F(:) = 1; p(:) = 0; Traction = Traction * sin(0.1*(0: Dt : Nstep * Dt));
```

(continued)

Box 7.3: *(cont.)*

```
% Plotting
Time_history = 0: Dt : Nstep * Dt;
Vel_history  = zeros(Nstep,1);
P_history    = zeros(Nstep,1);

% Main code
for istep = 1:Nstep
pold = p; Fold = F;
for i = 1:2
RFe       = 1/Rho * diff(p)/Le;
RF        = [RFe(1); RFe(1:end-1)+RFe(2:end); RFe(end)]/2*Le;
Fdot      = RF./ Lpoin;
Fe        = 0.5*(F(1:end-1)+F(2:end)) + TauF*Le/cp * (RFe - ...
0.5*(Fdot(1:end-1)+Fdot(2:end))) ;
Pe        = K * (Fe - 1);
Rp        = [0; -Pe(1:end-1)+Pe(2:end); -Pe(end)+ Traction(istep)];
pdot      = Rp ./ Lpoin;
F         = F + Dt * Fdot; p         = p + Dt * pdot;
end

% Next time step
p         = 0.5*(pold + p); F          = 0.5*(Fold + F);

% Store history (for plotting purposes)
Vel_history(istep+1) = p(Nelem/2+1) / Rho;
P_history(istep+1)   = K * (F(Nelem/2+1) -1);

end
figure(1); plot(Time_history,Vel_history); grid on; ylabel("Velocity")
figure(2); plot(Time_history,P_history); grid on; ylabel("Stress")
```

Exercises

1. Considering the deviatoric strain energy functional given by Equation (7.95a,b), compute the fourth-order constitutive tensor defined by $\hat{\boldsymbol{\mathcal{C}}} = \frac{\partial^2 \hat{\mathcal{E}}(\boldsymbol{F})}{\partial \boldsymbol{F} \partial \boldsymbol{F}}$ and show that it can be written as

$$\hat{\boldsymbol{\mathcal{C}}} = \gamma_1 \boldsymbol{\mathcal{I}} + \gamma_2 \boldsymbol{F}^{-T} \otimes \boldsymbol{F}^{-T} + \frac{3}{2}\gamma_2 \boldsymbol{\mathcal{H}} - \frac{2}{3}\gamma_1 \left(\boldsymbol{F} \otimes \boldsymbol{F}^{-T} + \boldsymbol{F}^{-T} \otimes \boldsymbol{F} \right),$$

in terms of the strain-dependent scalar functions

$$\gamma_1 = \mu J^{-2/3}; \qquad \gamma_2 = \frac{2}{9} \mu J^{-2/3} II_{\boldsymbol{F}}.$$

2. Considering the volumetric thermoelastic energy functional given by Equations (7.96), (7.97), and (7.98), compute the fourth-order constitutive tensor defined by $\boldsymbol{\mathcal{C}}_{vol} = \frac{\partial (pJ\boldsymbol{F}^{-T})}{\partial \boldsymbol{F}}$ and show that it can be written as

$$\boldsymbol{\mathcal{C}}_{vol} = \left(J^2 \frac{\partial p}{\partial J} + pJ \right) \boldsymbol{F}^{-T} \otimes \boldsymbol{F}^{-T} - pJ\boldsymbol{\mathcal{H}},$$

where

$$\frac{\partial p}{\partial J} = \frac{d^2 U_0}{dJ^2} + c_v \left[e^{\eta/c_v} - 1 \right] \frac{d^2 \theta_0}{dJ^2}.$$

3. Considering the results in Exercises 1 and 2, demonstrate that the acoustic tensor $\boldsymbol{\mathcal{C}}_{\nu\nu}$ is obtained as

$$\boldsymbol{\mathcal{C}}_{\nu\nu} = \gamma_1 \boldsymbol{I} + \left(J^2 \frac{\partial p}{\partial J} + \frac{5}{2}\gamma_2 \right) \boldsymbol{F}^{-T}\boldsymbol{\nu} \otimes \boldsymbol{F}^{-T}\boldsymbol{\nu}$$

$$- \frac{2}{3}\gamma_1 \left(\boldsymbol{F\nu} \otimes \boldsymbol{F}^{-T}\boldsymbol{\nu} + \boldsymbol{F}^{-T}\boldsymbol{\nu} \otimes \boldsymbol{F\nu} \right).$$

4. For the particular case where $\boldsymbol{\nu}$ is a material principal direction of deformation, show that the wave speeds c_p and c_s are given by

$$c_p = \sqrt{\frac{\gamma_3}{\rho_R}}; \qquad c_s = \sqrt{\frac{\gamma_1}{\rho_R}},$$

where

$$\gamma_3 = \left(J^2 \frac{\partial p}{\partial J} + \frac{5}{2}\gamma_2 \right) \Lambda_\nu^2 - \frac{1}{3}\gamma_1.$$

5. Based on the values for the c_p and c_s wave speeds obtained in Exercise 4, obtain expressions in terms of the material parameters to ensure the hyperbolicity of the thermoelastic constitutive model.

CHAPTER EIGHT

COMPUTER IMPLEMENTATION FOR DISPLACEMENT-BASED DYNAMICS

8.1 INTRODUCTION

This chapter introduces a computer program that can be used for the dynamic analysis of deformable solids undergoing large deformations. The program is developed using the computer language MATLAB and serves to crystallize the concepts described in the previous chapters of this book (specifically, those related to Chapter 4).

The computer program is built on the basis of the teaching program FLagSHyP (**F**inite Element **Lag**rangian **S**train **Hy**perelastoplastic **P**rogram), which is explained in the first volume (NL-Statics) of this series. Consequently, a basic understanding of that program is expected in order to fully appreciate what follows. Specifically, a new dynamic capability has been implemented where by it is possible for the user to simulate highly deformable (isothermal) hyperelastoplastic solids which are dynamically deformed. The unknowns of the problem are the nodal coordinates of the system $\mathbf{x}$, solved at a series of time instants t_n, with $n = 0, 1, \ldots$. The first addition in the new program (with respect to its original "Statics" version) is the computation of the mass (consistent or lumped) matrix of the system. Subsequently, the *static* algorithm of the original program is now complemented by a *dynamic* counterpart, where by the user can choose between four different time integrators, as described in Chapter 4, namely the leap-frog method, the Newmark method, the alpha method and the mid-point method. As in the original program, a variety of constitutive models and finite elements (i.e. truss-type elements, two-dimensional elements, and three-dimensional elements) can be employed.

This chapter will focus on the new capabilities introduced in the FLagSHyP program. Special attention will be paid to the description of the new variables and program functions necessary to carry out dynamic analysis. For completeness, the chapter will include user updated instructions, an updated dictionary of variables and MATLAB functions, and sample input and output for a few typical examples.

The program can be obtained, together with sample data, as a download from the website `www.flagshyp.com`.* Alternatively, it can be obtained by email request to the authors `a.j.gil@swansea.ac.uk` or `j.bonet@greenwich.ac.uk`.

The master m-file `FLagSHyP.m`, which controls the overall organization of the program, is divided into three sections. The first section includes a series of statements designed to add the necessary directories to the path of the program. The second section includes the function `input_data_and_initialisation.m`, which is devoted to the reading of input/control data and the initialization of critical variables, including equivalent nodal forces and the initial tangent stiffness matrix. The third section, depending on the value selected by the user for a question presented at the outset, chooses between two possible types of analysis: *statics* or *dynamics*. If the first (*statics*) option is chosen, the program proceeds as presented in NL-Statics and, thus, this will not be explained any further. Alternatively, if the second (*dynamics*) option is chosen, the function `Time_Integrator_algorithm.m` is called, where, first, the consistent mass matrix is computed by means of the function `consistent_mass_matrix_assembly.m`, and, second, a choice of different time integrators `OPTIONS` is made available to the user, namely:

- `OPTION 1` The function `LeapFrog.m` describes the leap-frog algorithm.
- `OPTION 2` The function `Newmark.m` describes the Newmark time integrator.
- `OPTION 3` The function `AlphaMethod.m` describes the alpha method time integrator.
- `OPTION 4` The function `MidPoint.m` describes the mid-point time integrator.

A feature of the FLagSHyP program is the arrangement of variables (along with the grouping of those which share a similar nature) into data structures. Note that while the use of data structures can greatly facilitate the understanding of all varibles in the program, their use can introduce some computational overheads, noticeable in a large-scale problem. In this enhanced version of the teaching program, the main new novelty from the data structure standpoint is the introduction of a new data structure `CON.DYNAMICS.`, where all the necessary variables for the activation of dynamic simulations are stored. We recall below the fundamental data structures (in capital letters) and their "children" (in lower case letters).

As an example, Table 8.1 details the new control variables that have been incorporated with respect to the original FLagSHyP program in order to run dynamics simulations. For completeness, a dictionary of all the variables of the program is included at the end of this chapter.

* Or `www.cambridge.org/9781107115620`

Name	Comments
`PRO.`	Program run mode, input/output files, problem title
`GEOM.`	Nodes, initial and spatial coordinates, elemental volumes
`FEM.`	Finite element information (shape functions) and connectivity
`QUADRATURE.`	Gauss point information: location and weights
`KINEMATICS.`	Kinematics quantities: $\boldsymbol{F}$, $\boldsymbol{b}$, J, ...
`MAT.`	Material properties: ρ, λ, μ, κ, ...
`BC.`	Boundary conditions: fixed and free
`LOAD.`	Loads: point loads, gravity, pressure load
`GLOBAL.`	Global vectors and matrices: residual and stiffness
`PLAST.`	Plasticity variables
`CON.`	Program control parameters: dynamics data ...

Parent	Child	Comments
`CON.DYNAMICS`	`dynamics`	Dynamics (yes or no)
`CON.DYNAMICS`	`time_integrator`	Type of time integrator
`CON.DYNAMICS`	`tmax`	Max. time of simulation
`CON.DYNAMICS`	`dt`	Time step
`CON.DYNAMICS`	`gamma`	γ coefficient for Newmark method
`CON.DYNAMICS`	`beta`	β coefficient for Newmark method
`CON.DYNAMICS`	`alpha`	α coefficient for α-method
`CON.DYNAMICS`	`load_type`	Type of time-varying load
`CON.DYNAMICS`	`load_parameter1`	Dynamic loading parameter 1
`CON.DYNAMICS`	`load_parameter2`	Dynamic loading parameter 2

8.2 USER INSTRUCTIONS

By executing the master m-file `FLagSHyP.m`, the directories (and subdirectories) `code` and `job_folder` are added to the directory path of the program. For simplicity, all the necessary MATLAB functions comprising the FLagSHyP program have been placed within subdirectories included in the directory `code`, whereas input/output data files have been located within subdirectories included in the directory `job_folder`. Moreover, the program expects the input file (e.g. `FLagSHyP_input_file.dat`) to be located in the directory `job_folder/FLagSHyP_input_file`. Note that although the name (e.g. `FLagSHyP_input_file`) or extension (e.g. `.dat`) of the input file are arbitrary,

TABLE 8.1 User input for FLagSHyP

Item 1: Title	**Number of lines:** 1
PRO.title	Problem title
Item 2: Type of element	**Number of lines:** 1
FEM.mesh.element_type	Element type
Item 3: Number of nodes	**Number of lines:** 1
GEOM.npoin	Number of mesh nodes
Item 4: Nodal information	**Number of lines:** GEOM.npoin
ip,	Node number
BC.icode(ip),	Boundary code
GEOM.x(i,ip)	x, y, z coordinates
[i=1:GEOM.ndime]	[Number of dimensions]
Item 5: Number of elements	**Number of lines:** 1
FEM.mesh.nelem	Number of elements
Item 6: Element information	**Number of lines:** FEM.mesh.nelem
ie,	Element number
MAT.matno(ie),	Material number
FEM.mesh.connectivity(ie,i)	Connectivities
[i=1:FEM.mesh.n_nodes_elem]	[Number of nodes per element]
Item 7: Number of materials	**Number of lines:** 1
MAT.nmats	Number of different materials
Item 8: Material properties	**Number of lines:** MAT.nmats
im,	Material number
MAT.matyp(im),	Constitutive equation type
MAT.props(ipr,im)	Properties
[ipr=1:npr(dependent upon material)]	[Number of properties]
Item 9: Load information	**Number of lines:** 1
n_point_loads,	Number of loaded nodes
BC.n_prescribed_displacements,	Number of non-zero prescribed displacements
LOAD.n_pressure_loads,	Number of surface (line) elements with pressure
LOAD.gravt(i)	Gravity vector
[i=1:GEOM.ndime]	[Number of dimensions]
Item 10: Nodal loads	**Number of lines:** n_point_loads
ip,	Node number
force_value(i),	Force vector
[i=1:GEOM.ndime]	[Number of dimensions]
Item 11: Prescribed displacements	**Number of lines:** BC.n_prescribed_displacements
ip,	Node number
id,	Spatial direction
prescribed_value	Nominal prescribed displacement
Item 12: Pressure loads	**Number of lines:** LOAD.n_pressure_loads
ie,	Surface element number
FEM.mesh.connectivity_faces(ie,in),	Force vector
LOAD.pressure(ie)	Nominal pressure
[in=1:FEM.mesh.connectivity_faces]	[Number of nodes per element]

Item 13: Control information	**Number of lines:** 1
CON.nincrm,	Number of load/displacement increments
CON.xlmax,	Maximum value of load scaling parameter
CON.dlamb,	Load parameter increment
CON.miter,	Maximum number of iterations per increment
CON.cnorm,	Convergence tolerance
CON.searc,	Line search parameter (if 0.0 not in use)
CON.ARCLEN.arcln,	Arc length parameter (if 0.0 not in use)
CON.OUTPUT.incout,	Output counter (5 for every 5 incr/time steps)
CON.ARCLEN.itarget,	Target iterations per increment (see note below)
CON.OUTPUT.nwant,	Single output node (0 if not used)
CON.OUTPUT.iwant	Output degree of freedom (0 if not used)
Item 14: Dynamic information	**Number of lines:** 1
CON.DYNAMICS.time_integrator,	Time integrator (see Note 1)
CON.DYNAMICS.tmax,	Maximum time of simulation
CON.DYNAMICS.dt,	Time increment
CON.DYNAMICS.gamma,	Newmark γ parameter
CON.DYNAMICS.beta,	Newmark β parameter
CON.DYNAMICS.alpha	α-method parameter
Item 15: Dynamic loading	**Number of lines:** 1
CON.DYNAMICS.load_type,	Dynamic analysis (if 0 not in use)
CON.DYNAMICS.load_parameter1,	Dynamic analysis (if 0.0 not in use)
CON.DYNAMICS.load_parameter2	Dynamic analysis (if 0.0 not in use)
Item 16: Velocity initial conditions	**Number of lines:** GEOM.npoin
ip,	Node number (see Note 2)
GEOM.velocity(i,ip)	v_x, v_y, v_z initial velocities
[i=1:GEOM.ndime]	[Number of dimensions]

the name of the input file must coincide precisely with the name given to the subdirectory which contains it (e.g. job_folder/FLagSHyP_input_file), otherwise an error will occur and the program will not progress. The input file required by FLagSHyP is described below. The file is free-formatted, so items within a line are simply separated by commas or spaces. Modifications with respect to the *statics* version of FLagSHyP have been added at the end of the input file. By doing this, old input files can still be used and run in FLagSHyP.

Note 1: When considering a dynamic analysis, a nonlinear structural solution is obtained in a number of time steps which permit advancing the solution over time. This is carried out until the current time CON.DYNAMICS.time exceeds the maximum allowable simulation time CON.DYNAMICS.tmax, at which instant the simulation stops.

The user has the choice of four different time integrators depending on the value given to the control parameter CON.DYNAMICS.time_ integrator, which varies from 1 to 4. If CON.DYNAMICS.time_integrator==1, the leap-frog time integrator is used; if CON.DYNAMICS.time_integrator==2, the Newmark time integrator is used; if CON.DYNAMICS.time_integrator==3, the alpha method time integrator is used; if CON.DYNAMICS.time_integrator==4, the mid-point time integrator is used.

When using the Newmark time integrator, the user is given the choice of its γ and β parameters (CON.DYNAMICS.gamma and CON.DYNAMICS.beta, respectively). For instance, $\gamma = 1/2$ and $\beta = 1/4$ lead to the so-called average acceleration method, and $\gamma = 1/2$ and $\beta = 1/6$ lead to the so-called linear acceleration method.

When using the α-method, the only necessary parameter is α (CON.DYNAMICS.alpha), which takes values in the interval $[-1/3, 0]$. If $\alpha = 0$, the method degenerates to the so-called average acceleration Newmark method.

Similar to the original FLagSHyP code, the applied *loads* can be point forces, pressure forces, or even prescribed displacements. Any input value of these items is nominal and the time variation is defined after multiplication by a relevant time-varying loading factor which is obtained based on the value given to the control parameters CON.DYNAMICS.load_type, CON.DYNAMICS.load_parameter1, and CON.DYNAMICS.load_parameter2.

If CON.DYNAMICS.load_type==1, then the loading factor grows linearly from zero (in a ramp type manner) until it reaches a maximum value defined by CON.DYNAMICS.load_parameter2, after which the loading factor remains constant. The time required to attain this maximum value is defined by CON.DYNAMICS.load_parameter1 according to the parameter.

$$\lambda(t) = \begin{cases} \frac{A}{T}t \text{ if } t \leq T \\ A \text{ if } t > T. \end{cases} \tag{8.1}$$

If CON.DYNAMICS.load_type==2, then the loading factor follows an endless sinusoidal function which starts at zero and is defined by two parameters, namely the amplitude CON.DYNAMICS.load_parameter2 and the period of oscillation CON.DYNAMICS.load_parameter1, as

$$\lambda(t) = A \sin\left(\frac{2\pi t}{T}\right). \tag{8.2}$$

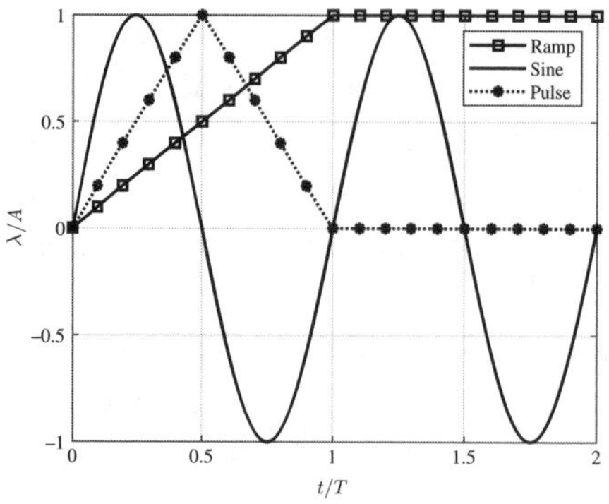

FIGURE 8.1 Time-varying loading factor.

If `CON.DYNAMICS.load_type==3`, then the loading factor is defined by a hat type (triangular step) function which starts at zero, grows linearly until reaching a maximum value `CON.DYNAMICS.load_parameter2`, and then decreases linearly until attaining a value of zero. The time required to return to zero is defined by `CON.DYNAMICS.load_parameter1`. Figure 8.1 shows a diagram depicting the three different loading options available:

$$\lambda(t) = \begin{cases} 2A\frac{t}{T} & \text{if} \quad t \leq T/2 \\ 2A\left(1 - \frac{t}{T}\right) & \text{if} \quad T/2 < t \leq T \\ 0 & \text{if} \quad t > T. \end{cases} \tag{8.3}$$

At each time step, and provided the selected time integrator is not the explicit leap-frog algorithm, the Newton–Raphson iteration attempts to achieve dynamic equilibrium within a maximum allowed number of iterations input as `CON.miter`. Equilibrium is achieved when the residual force $||\mathbf{R}|| <$ `CON.cnorm`, where `CON.cnorm` is the convergence tolerance, usually about `1.0E-06`.

To facilitate easy plotting of load displacement graphs, output from a single node and single degree of freedom (at that node) can be specified. Parameter `CON.OUTPUT.nwant` specifies the node and `CON.OUTPUT.iwant` specifies the degree of freedom. The output is in a file always called `flag.out`, which contains the time step number, the coordinate relating to the degree of freedom, the force relating to the degree of freedom, the current value of `CON.xlamb`, and the total energy of the system `CON.DYNAMICS.Energy` at every time step.

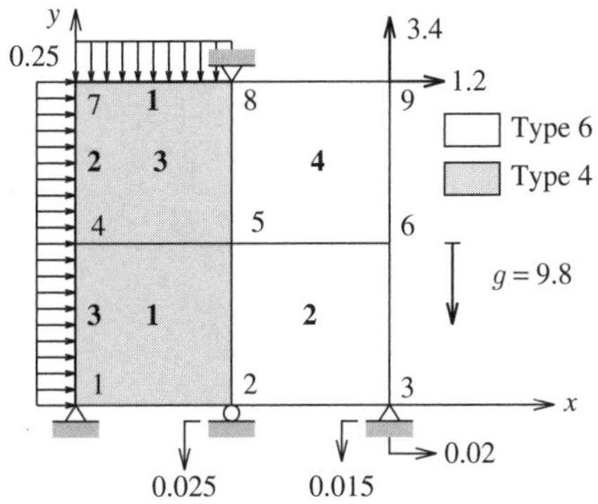

FIGURE 8.2 Simple two-dimensional example.

Note 2: Possible nonzero initial nodal velocities can be listed. By default, the program considers that these initial velocities are zero. However, if this is not the case, these can be included at the bottom of the input file.

The somewhat contrived example shown in Figure 8.2, already chosen in NL-Statics, helps illustrate as many diverse features of these input instructions as possible. The required input file is listed in Box 8.1, where the additional data required to carry out dynamic analysis is included at the bottom of the input file in bold font. Note that point loads, gravity loads, pressure loads, and prescribed displacements are all subject to the same time-varying loading parameter in the solution procedure.

BOX 8.1: Input file for example in Figure 8.2

```
2-D Example quad4
9
1 3 0.0 0.0
2 2 1.0 0.0
3 3 2.0 0.0
4 0 0.0 1.0
5 0 1.0 1.0
6 0 2.0 1.0
7 0 0.0 2.0
8 3 1.0 2.0
9 0 2.0 2.0
4
1 1 1 2 5 4
2 2 6 5 2 3
```

(continued)

Box 8.1: *(cont.)*

```
3 1 5 8 7 4
4 2 5 6 9 8
2
1 4
1.0 100. 100. 0.1
2 6
1.0 100. 0.1
1 3 3 0.0 −9.8
9 1.2 3.4
3 1 0.02
2 2 −0.025
3 2 −0.015
1 8 7 0.25
2 7 4 0.25
3 1 4 −0.25
2 10.0 5.0 25 1.e−10 0.0 0.0 1 5 7 1
3 1.0 0.01 0.25 0.5 0.0
1 0.06 1.0
1 0.0 0.0
2 0.0 0.0
3 0.0 0.0
4 0.0 0.0
5 0.0 0.0
6 0.0 0.0
7 0.0 0.0
8 0.0 0.0
9 0.0 0.0
```

As for output information, the program FLagSHyP produces a largely unannotated output file that is intended to be an input file for a postprocessor to be supplied by the user. The output file is only produced every `CON.OUTPUT.incout` time increments. The contents and structure of this file do not differ from those of the original FLagSHyP and hence are not discussed any further, except for the fact that the so-called load increments in the original *statics* version are now replaced with the concept of time increments.

8.3 SOLVER DETAILS

In this new dynamics version of FLagSHyP, the solver of the program builds upon that described in NL-Statics. In general terms, the different time integrators available in FLagSHyP can be classified into two groups: *implicit* (e.g. Newmark, α-method, and mid-point) and *explicit* (e.g. leap-frog).

Regarding the group of implicit time integrators, for every time step of the simulation a residual force vector is computed and is driven to the value of the convergence tolerance through a Newton–Raphson solver. For the solution of the resulting system of linear algebraic equations $\mathbf{K}\Delta\mathbf{a} = -\mathbf{R}$, FLagSHyP uses the MATLAB command $\Delta\mathbf{a} = -\mathbf{K}\backslash\mathbf{R}$, where $\mathbf{K}$ (`GLOBAL.K`)† denotes the assembled global tangent matrix, whose specific expression (i.e. linear combination of the mass and stiffness matrices) depends on the chosen time integrator, and $\mathbf{R}$ is the assembled global residual vector `GLOBAL.RESIDUAL`. MATLAB will aim to take advantage of possible symmetries in the problem, redirecting to the most appropriate linear algebra solver, with the final objective of minimizing the computational time while maintaining the accuracy of the final solution. Specifically, tailor-made solvers can be utilized for dealing with a sparse matrix $\mathbf{K}$, as is the present case.

In order to take advantage of the sparse solver capabilities, the final assembled tangent matrix is stored in a temporary one-dimensional vector array `global_stiffness`, along with two other temporary vector arrays (of the same length as `global_stiffness`), namely `indexi` and `indexj`, where the rows `indexi` and columns `indexj` corresponding to the entries of the tangent matrix are saved. By making use of the MATLAB intrinsic function `sparse.m`, these vectors are assembled into `GLOBAL.K`. Finally, it is important to remark that the linear algebra solver is not restricted to the typical case of dealing with a symmetric stiffness matrix. In other words, it allows for the most general case of considering pressure loads, which can lead, to an unsymmetric stiffness matrix.

NL-Statics described in detail the assembly process of the various contributions of the stiffness matrix, namely the constitutive matrix component $\mathbf{K}_c$ (refer to the function `constitutive_matrix.m`), the geometric or initial stress matrix component $\mathbf{K}_\sigma$ (refer to the function `geometric_matrix.m`), the volumetric mean dilatation stiffness component $\mathbf{K}_\kappa$ (refer to the function `volumetric_mean_dilatation_matrix.m`), and the external force pressure stiffness component $\mathbf{K}_p$. A new contribution is needed in the dynamics case corresponding to the so-called *consistent mass matrix* $\mathbf{M}$ (refer to the function `mass_matrix.m`). Algorithm 8.1 depicts in a simplified diagram the way in which the global mass matrix is formed as a series of loops, beginning with element and Gauss quadrature and cascading down to nodes and dimension loops. Upon completion of these calculations, the sparse assembly process is carried out.

† $\mathbf{K}$ is either $\mathbf{K}_a$ (4.61), $\mathbf{K}_\alpha$ (4.72) or $\mathbf{K}_m$ (4.99b) depending upon the time integration scheme employed.

Algorithm 8.1: Calculation and storage of assembled mass matrix entries

```
Input  : counter=0; indexi(:)=0; indexj(:)=0; global_mass(:)=0
Output: GLOBAL.Mass

for ielement=1:FEM.mesh.nelem do
    for igauss=1:QUADRATURE.element.ngauss do
        for anode=1:FEM.mesh.n_nodes_elem do
            for bnode=1:FEM.mesh.n_nodes_elem do
                Evaluate mass coefficient m;
                for i=1:GEOM.ndime do
                    Evaluate indexi(counter);
                    Evaluate indexj(counter);
                    Assign global_mass(counter) ← m;
                    counter ← counter + 1;
                end
            end
        end
    end
end

GLOBAL.M=sparse(indexi,indexj,global_mass)
```

8.4 PROGRAM STRUCTURE

In order to give an overview of the structure of FLagSHyP, Box 8.2 lists some of the most important functions comprising the program. In this list, emphasis has been placed on those functions necessary to carry out dynamics analysis. The remainder are either functions common to standard linear elasticity finite element codes or subsidiary functions that are not crucial to an understanding of the flow of the program and can be examined via a program download (see `www.flagshyp.com`). The functions in italic typeface are described in detail in the following sections.

BOX 8.2: FLagSHyP structure

```
FLagSHyP ............................................ m-file
└── code ............................................ Main code directory
    ├── solution_equations .......................... Solution algorithms directory
    │   ├── Time_Integrator_algorithm ............... Time integrators master function
    │   ├── LeapFrog ................................ Leap-frog function
    │   ├── Newmark ................................. Newmark method function
    │   ├── AlphaMethod ............................. α-method function
    │   ├── MidPoint ................................ Mid-point function
    │   ├── Newton_Raphson_algorithm ................ Newton–Raphson master function
    │   └── linear_solver ........................... Solver of linear system of equations
    ├── input_reading ............................... Reading of input data directory
    │   ├── boundary_codes .......................... Reading of boundary conditions
    │   ├── elinfo .................................. Preallocate element information
    │   ├── incontr ................................. Reading input control parameters
    │   ├── inelems ................................. Reading element information
    │   ├── inloads ................................. Reading loading information
    │   ├── innodes ................................. Reading nodal information
    │   ├── input_data_and_initialisation ........... Initialization of some variables
    │   ├── matprop ................................. Reading material properties
    │   ├── reading_input_file ...................... Reading input function
    │   ├── welcome ................................. Read input file name
    │   ├── initialvelocities ....................... Read input initial velocities
    │   └── TimeLoadIncrement ....................... Compute time load increment
    ├── initialisation .............................. Initialization functions directory
    ├── numerical_integration ....................... Numerical integration functions directory
    ├── FEM_shape_functions ......................... Shape functions directory
    ├── kinematics .................................. Kinematics calculations directory
    │   ├── gradients ............................... Computation of fundamental kinematics
    │   ├── kinematics_gauss_point .................. Extraction of quantities at Gauss point
    │   ├── normal_vector_boundary .................. Normal vector at surface
    │   └── thickness_plane_stress .................. Plane stress thickness computation
    ├── constitutive laws ........................... Constitutive behavior directory
    │   ├── Cauchy_type_selection ................... Cauchy stress tensor (material selection)
    │   ├── elasticity_modulus_selection ............ Elasticity tensor (material selection)
    │   ├── Energy_type_selection ................... Strain energy density (material selection)
    │   ├── muab_choice ............................. Checking function for materials in principal directions
    │   ├── elasticity_tensor ....................... Elastiticy tensor directory
    │   ├── stress .................................. Stress tensor directory
    │   └── Energy .................................. Strain energy directory
    │       ├── Energy1 ............................. Strain energy material 1
    │       ├── Energy17 ............................ Strain energy material 17
    │       ├── Energy3 ............................. Strain energy material 3
    │       ├── Energy4 ............................. Strain energy material 4
    │       ├── Energy5 ............................. Strain energy material 5
    │       ├── Energy6 ............................. Strain energy material 6
    │       ├── Energy7 ............................. Strain energy material 7
    │       └── Energy8 ............................. Strain energy material 8
    ├── plasticity .................................. Plasticity functions directory
    └── element_calculations ........................ Element calculations directory
        ├── element_energy .......................... Element internal energy
        ├── element_internal_energy_truss ........... Element truss internal energy
        ├── explicit_element_force .................. For leap-frog
        ├── explicit_element_force_truss ............ For leap-frog
        └── element_mass_matrix ..................... Element consistent mass matrix
```

(continued)

Box 8.2: *(cont.)*

```
│   ├── mass_matrix ........................................ Element mass matrix
│   ├── constitutive_matrix .......................... Element constitutive matrix
│   ├── geometric_matrix ........................... Element initial stress matrix
│   ├── mean_dilatation_pressure ............ Mean dilatation pressure computation
│   ├── mean_dilatation_pressure_addition ........... Adds pressure to stress tensor
│   ├── explicit_mean_dilatation_pressure_addition ................. For leap-frog
│   ├── mean_dilatation_volumetric_matrix .................... Mean dilatation matrix
│   ├── pressure_element_load_and_stiffness ............ Pressure load and stiffness
│   ├── pressure_load_matrix ......................... Element pressure load matrix
│   └── explicit_pressure_element_load .............................. For leap-frog
├── global_assembly ................................ Assembly functions directory
│   ├── consistent_mass_matrix_assembly .......... Assembly consistent mass matrix
│   ├── lump_mass_matrix_assembly ................... Assembly lumped mass matrix
│   ├── explicit_pressure_load_assembly ............................. For leap-frog
│   ├── explicit_residual_assembly .................................. For leap-frog
│   └── TotalEnergyComputation ........................... Total energy computation
├── solution_update ................................................ Update geometry
├── convergence_check ........................ Newton–Raphson convergence check
├── solution_write ................................................. Output solution
└── support ...................................................... Auxiliary functions
```

The first section sets up the directory path, which has been previously explained in the user instructions. The second section is the function `input_data_and_initialisation.m`, for reading the input/control data and initialization of variables including the initial assembled tangent stiffness and equivalent nodal forces. Depending on the value chosen by the user for some of the problem-dependent control variables, this redirects the program to either a *dynamic* or a *static* simulation. The static option is fully described in NL-Statics and will not be discussed any further. If the *dynamics* option is selected, the program calls the function `Time_Integrator_algorithm.m`, where the main actions take place.

BOX 8.3: FLagSHyP: Master m-file

```
clear all;close all;clc

% SECTION 1
if isunix()
   dirsep =  '/';
else
   dirsep =  '\';
end
basedir_fem = mfilename('fullpath');
basedir_fem = fileparts(basedir_fem);
basedir_fem = strrep(basedir_fem,['code' dirsep 'support'],'');
addpath(fullfile(basedir_fem));
addpath(genpath(fullfile(basedir_fem,'code')));
addpath((fullfile(basedir_fem,'job_folder')));
% SECTION 2
[PRO,FEM,GEOM,QUADRATURE,BC,MAT,LOAD,CON,CONSTANT,GLOBAL,...
```

(continued)

Box 8.3: *(cont.)*

```
 PLAST,KINEMATICS] = input_data_and_initialisation (basedir_fem);
% SECTION 3
if CON.DYNAMICS.dynamics
   Time_Integrator_algorithm(PRO,FEM,GEOM,QUADRATURE,BC,MAT,LOAD,
   CON,...CONSTANT,GLOBAL,PLAST,KINEMATICS);
else
   if (abs(CON.ARCLEN.arcln)==0)
      if ~CON.searc
        Newton_Raphson_algorithm(PRO,FEM,GEOM,QUADRATURE,BC,MAT,...
        LOAD,CON,CONSTANT,GLOBAL,PLAST,KINEMATICS);
      else
        Line_Search_Newton_Raphson_algorithm(PRO,FEM,GEOM,...
        QUADRATURE,BC,MAT,LOAD,CON,CONSTANT,GLOBAL,PLAST,KINEMATICS);
       end
   else
      Arc_Length_Newton_Raphson_algorithm(PRO,FEM,GEOM,QUADRATURE,...
     BC,MAT,LOAD,CON,CONSTANT,GLOBAL,PLAST,KINEMATICS);
   end
end
```

The function `Time_Integrator_algorithm.m` starts by computing the consistent mass matrix by calling function `consistent_mass_matrix_assembly.m`. This mass matrix is saved and used for the remainder of the simulation (i.e. time steps). Subsequently, the function permits the user to select among the four different time integrators available, depending on the selected function: `LeapFrog.m`, `Newmark.m`, `AlphaMethod.m`, or `MidPoint.m`.

BOX 8.4: FLagSHyP: Time integrator algorithm

```
function Time_Integrator_algorithm(PRO,FEM,GEOM,QUADRATURE,BC,MAT,
LOAD,...CON,CONSTANT,GLOBAL,PLAST,KINEMATICS)

GLOBAL = consistent_mass_matrix_assembly(GEOM,MAT,FEM,...
GLOBAL,QUADRATURE.element,KINEMATICS);

if (CON.DYNAMICS.time_integrator==1)
    LeapFrog(PRO,FEM,GEOM,QUADRATURE,BC,MAT,LOAD,...
    CON,CONSTANT,GLOBAL,PLAST,KINEMATICS);
elseif (CON.DYNAMICS.time_integrator==2)
    Newmark(PRO,FEM,GEOM,QUADRATURE,BC,MAT,LOAD,...
    CON,CONSTANT,GLOBAL,PLAST,KINEMATICS);
elseif (CON.DYNAMICS.time_integrator==3)
    AlphaMethod(PRO,FEM,GEOM,QUADRATURE,BC,MAT,LOAD,...
    CON,CONSTANT,GLOBAL,PLAST,KINEMATICS);
elseif (CON.DYNAMICS.time_integrator==4)
    MidPoint(PRO,FEM,GEOM,QUADRATURE,BC,MAT,LOAD,...
    CON,CONSTANT,GLOBAL,PLAST,KINEMATICS);
end
```

As an example of the different time integrators, function `AlphaMethod.m` is listed and explained below, where its flow has been split into the *predictor* and *corrector* steps. Regarding the predictor step, and provided a `while` loop condition is satisfied, the physical time and its corresponding time increment are advanced. Clearly, once the physical time step exceeds a maximum allowable time, the simulation exits this `while` loop and it finishes. Then, the loading increment `CON.xlamb` is computed in `TimeLoadIncrement.m`, taking one of three different choices: ramp, sinusoidal, or step. In addition, possible displacement boundary conditions are applied. In this case, the function `update_prescribed_displacements.m` will reset the current geometry for those nodes with prescribed displacements based on their initial coordinates, the nominal value of the prescribed displacement, and the current load-scaling parameter `CON.xlamb`.

The geometry and the velocity fields are then predicted in preparation for the Newton–Raphson iterative corrector algorithm. The imposition of an increment of point or gravity loads immediately creates a residual force. This is added to any small residual carried over from the previous increment by the function `external_force_update.m`. Then, the function `residual_and_stiffness_assembly.m` recomputes the residual vector and the global tangent stiffness matrix due to the new updated configuration. Subsequently, function `pressure_load_and_stiffness_assembly.m` is now called with the last argument set to `CON.xlamb`, to add the total value of the nodal forces due to surface pressure and obtain the corresponding initial surface pressure tangent matrix component.

BOX 8.5: FLagSHyP: Alpha method, predictor step

```
while (CON.DYNAMICS.tmax-CON.DYNAMICS.time) > 1.0e-10

time_old              =  CON.DYNAMICS.time;
CON.DYNAMICS.time     =  CON.DYNAMICS.time + CON.DYNAMICS.dt;
CON.incrm             =  CON.incrm + 1;

timealpha =  (1 + alpha)*CON.DYNAMICS.time - alpha*time_old;
[CON.xlamb,CON.dlamb]  =  TimeLoadIncrement(timealpha,...
CON.xlamb,CON.DYNAMICS);

GEOM.x = update_prescribed_displacements(BC.dofprescribed,...
GEOM.x0,GEOM.x,CON.xlamb,BC.presc_displacement);
xn          =  GEOM.x;
vn          =  GEOM.velocity;

GEOM.x        =  xn + (1 + alpha)*dt*(vn + dt/2*GEOM.acceleration);
GEOM.velocity    =  vn + (1 + alpha)*dt*GEOM.acceleration;
[GLOBAL.Residual,GLOBAL.external_load] = ...
```

(continued)

Box 8.5: *(cont.)*

```
external_force_update(GLOBAL.nominal_external_load,...
 GLOBAL.Residual,GLOBAL.external_load,CON.dlamb);

[GLOBAL,updated_PLAST] = residual_and_stiffness_assembly(CON.xlamb,...
 GEOM,MAT,FEM,GLOBAL,CONSTANT,QUADRATURE.element,PLAST,KINEMATICS);

if LOAD.n_pressure_loads
   GLOBAL = pressure_load_and_stiffness_assembly(GEOM,MAT,FEM,...
   GLOBAL,LOAD,QUADRATURE.boundary,CON.xlamb);
end
```

For the corrector step, a `while` loop controls the Newton–Raphson iteration process. This continues until the residual norm `rnorm` is smaller than the tolerance `CON.cnorm` and, of course, while the iteration number `CON.niter` is smaller than the maximum allowed `CON.miter`. Note that `rnorm` is initialized (`rnorm = 2*CON.cnorm`) in such a way that this loop is completed at least once. The function `linear_solver.m` solves the linear system of equations to obtain the incremental accelerations $\Delta\mathbf{a}$, from which the corrected accelerations, velocities, and displacements are obtained. Given the new geometry, the two functions `residual_and_stiffness_assembly.m` and `pressure_load_and_stiffness_assembly.m` serve the same purpose as described above. Finally, `checkr_residual_norm.m` computes the residual norm `rnorm` given the residual force vector `GLOBAL.Residual` and the total external force vector (including pressure loading). Finally, if convergence is achieved, the geometry and the velocity are recovered at the correct time step.

BOX 8.6: FLagSHyP: Alpha method, corrector step

```
CON.niter = 0;
rnorm = 2*CON.cnorm;

GLOBAL.K  =  GLOBAL.Mass + (1 + alpha)*gamma*dt*0 + ...
             (1 + alpha)*beta*dt^2*GLOBAL.K;
GLOBAL.Residual  =  GLOBAL.Mass*GEOM.acceleration(:) ...
                 +  GLOBAL.Residual;
while((rnorm > CON.cnorm) && (CON.niter < CON.miter))
  CON.niter = CON.niter + 1;

  da       = linear_solver(GLOBAL.K,-GLOBAL.Residual,BC.fixdof);

  GEOM.acceleration(BC.freedof) = GEOM.acceleration(BC.freedof) + da;
  GEOM.velocity(BC.freedof) = GEOM.velocity(BC.freedof) ...
                            + (1 + alpha)*gamma*dt*da;
  GEOM.x(BC.freedof) = GEOM.x(BC.freedof) ...
                     + (1 + alpha)*beta*dt^2*da;
```

(continued)

Box 8.6: *(cont.)*

```
    [GLOBAL,updated_PLAST] = residual_and_stiffness_assembly...
    (CON.xlamb,GEOM,MAT,FEM,GLOBAL,CONSTANT,QUADRATURE.element,PLAST,...
    KINEMATICS);
    if LOAD.n_pressure_loads
       GLOBAL = pressure_load_and_stiffness_assembly(GEOM,MAT,FEM,...
       GLOBAL,LOAD,QUADRATURE.boundary,CON.xlamb);
    end

    GLOBAL.K          =  GLOBAL.Mass + (1 + alpha)*beta*dt^2*GLOBAL.K;
    GLOBAL.Residual   =  GLOBAL.Mass*GEOM.acceleration(:) ...
                      +  GLOBAL.Residual;

    [rnorm,GLOBAL] = check_residual_norm(CON,BC,GLOBAL,BC.freedof);

    if (abs(rnorm)>1e7 || isnan(rnorm))
        CON.niter=CON.miter;
    break;
    end
  end

  if (CON.niter >=CON.miter)
  else
      GEOM.x             =  1/(1 + alpha)*(GEOM.x + alpha*xn);
      GEOM.velocity      =  1/(1 + alpha)*(GEOM.velocity + alpha*vn);

      PLAST = save_output(updated_PLAST,PRO,FEM,GEOM,QUADRATURE,BC,...
              MAT,LOAD,CON,CONSTANT,GLOBAL,PLAST,KINEMATICS);
  end
```

For every time step, the load increment is computed within the function `TimeLoadIncrement.m` which is listed below. The load increment can be varied among three different options, depending on the value of the variable `DYNAMICS.load_type`, namely, 1, 2 or 3. Otherwise, no load increment is applied.

BOX 8.7: FLagSHyP: Time load increment

```
function [xlamb,dlamb]  =  TimeLoadIncrement(time,xlamb,DYNAMICS)

old_xlamb  =  xlamb;
T=DYNAMICS.load_parameter1;
A=DYNAMICS.load_parameter2;

if DYNAMICS.load_type==1
   if DYNAMICS.time<T
      xlamb   =  time/T*A;
   else
      xlamb   =  A;
   end

elseif DYNAMICS.load_type==2
   xlamb     =  A*sin(2*pi*time/T);
```

(continued)

Box 8.7: *(cont.)*

```
elseif DYNAMICS.load_type==3
  if DYNAMICS.time<T/2
     xlamb    =  time/(T/2)*A;
  elseif (DYNAMICS.time>=T/2 && DYNAMICS.time<T)
     xlamb    =  A*(1- (time-T/2)/(T/2));
  else
     xlamb    =  0;
  end
end
dlamb  =  xlamb - old_xlamb;
```

The computation of the consistent mass matrix is carried out element by element through the assembly process detailed above. The function element_mass_matrix.m listed below computes the entries to the elemental mass matrix. This starts by calling the function gradients.m, where some of the fields in the variable KINEMATICS are updated. By looping over the quadrature points QUADRATURE.ngauss, the entries to the element mass matrix are computed by calling the function mass_matrix.m, which is also subsequently included.

BOX 8.8: FLagSHyP: Element mass matrix

```
function [indexi,indexj,global_mass,counter] = ...
element_mass_matrix(FEM,xlocal,x0local,element_connectivity,...

QUADRATURE,properties,dim,matyp,counter,KINEMATICS,indexi,...
indexj,global_mass)

KINEMATICS = gradients(xlocal,x0local,...
FEM.interpolation.element.DN_chi,QUADRATURE,KINEMATICS);
for igauss=1:QUADRATURE.ngauss
    kinematics_gauss = kinematics_gauss_point(KINEMATICS,igauss);
    JW = kinematics_gauss.Jx_chi*QUADRATURE.W(igauss)*...
    thickness_plane_stress(properties,kinematics_gauss.J,matyp);
    [indexi,indexj,global_mass,counter] = ...
     mass_matrix(FEM,dim,element_connectivity,...
     FEM.interpolation.element.N(:,igauss),JW,properties(1),...
     counter,indexi,indexj,global_mass);
end
end
```

BOX 8.9: FLagSHyP: Mass matrix

```
function [element_indexi,element_indexj,element_stiffness,...
counter] = mass_matrix(FEM,dim,element_connectivity,...
Nshape,JW,density,counter,element_indexi,element_indexj,...
element_stiffness)
```

(continued)

Box 8.9: *(cont.)*

```
for bnode=1:FEM.mesh.n_nodes_elem
    for anode=1:FEM.mesh.n_nodes_elem
        mass = density*Nshape(anode)*Nshape(bnode)*JW;
        element_indexi(counter:counter+dim-1)=...
        FEM.mesh.dof_nodes(:,element_connectivity(anode));
        element_indexj(counter:counter+dim-1)=...
        FEM.mesh.dof_nodes(:,element_connectivity(bnode));
        element_stiffness(counter:counter+dim-1) = mass;
        counter = counter + dim;
    end
end
```

FLagSHyP incorporates the facility to compute the total energy of the system as a function of time. Although this calculation is not strictly necessary for the flow of the program, it is useful in order to study the conservation of total energy of the system, which will be dependent upon the selected time integrator. The total energy of the system is the result of the sum of three components, namely the internal strain energy, the kinetic energy, and the external work. The computation of the elemental internal strain energy, is carried out in two different functions, depending on whether the element is of `truss2` type or not. For the latter case, the function `element_energy.m` is used and listed subsequently.

BOX 8.10: FLagSHyP: Total energy computation

```
function [Energy,K,U,Wext] = TotalEnergyComputation(GEOM,MAT,FEM,...
GLOBAL,QUADRATURE,KINEMATICS,CON)
U  =  0;
for ielement=1:FEM.mesh.nelem
    global_nodes    = FEM.mesh.connectivity(:,ielement);
    material_number = MAT.matno(ielement);
    matyp           = MAT.matyp(material_number);
    properties      = MAT.props(:,material_number);
    xlocal          = GEOM.x(:,global_nodes);
    x0local         = GEOM.x0(:,global_nodes);
    Ve              = GEOM.Ve(ielement);

    switch FEM.mesh.element_type
      case 'truss2'
      U = U + element_internal_energy_truss(properties,xlocal,x0local);
      otherwise
      U = U + element_energy(FEM,xlocal,x0local,Ve,QUADRATURE,...
                              properties,GEOM.ndime,matyp,KINEMATICS);
    end
end
if CON.DYNAMICS.time_integrator==1
   K   =  0.5*GEOM.velocity(:)'*(GLOBAL.Mass.*GEOM.velocity(:));
```

(continued)

Box 8.10: *(cont.)*

```
else
   K     =  0.5*GEOM.velocity(:)'*(GLOBAL.Mass*GEOM.velocity(:));
end
Wext     =  CON.xlamb*GLOBAL.nominal_external_load'*...
            (GEOM.x(:) - GEOM.x0(:));
Energy      =  U + K - Wext;
end
```

The computation of the internal strain energy for a given element of type other than truss2 is carried out within the function element_energy.m. A similar function called element_internal_energy_truss.m performs the same task for the truss2 type. This is done by first calling the function gradients.m to update the fields within the variable KINEMATICS, then looping over the quadrature points, and, finally, selecting the appropriate strain energy density within the function Energy_type_selection.m

BOX 8.11: FLagSHyP: Element energy

```
function Energy = element_energy(FEM,xlocal,x0local,Ve,QUADRATURE,...
properties,dim,matyp,KINEMATICS)

KINEMATICS = gradients(xlocal,x0local,...
FEM.interpolation.element.DN_chi,QUADRATURE,KINEMATICS);
ve = 0;
for igauss=1:QUADRATURE.ngauss
    JW = KINEMATICS.Jx_chi(igauss)*QUADRATURE.W(igauss);
    ve = ve + JW;
end
Energy = 0;
for igauss=1:QUADRATURE.ngauss
    kinematics_gauss = kinematics_gauss_point(KINEMATICS,igauss);
    W = Energy_type_selection(kinematics_gauss,properties,dim,matyp);
    JW = kinematics_gauss.Jx_chi*QUADRATURE.W(igauss)*...
    thickness_plane_stress(properties,kinematics_gauss.J,matyp);
    Energy = Energy + W*JW;
end
Energy = mean_dilatation_energy_addition(Energy,matyp,properties,...
ve,Ve);
end
```

The function Energy_type_selection.m selects the strain energy density per quadrature point depending on the material type model.

BOX 8.12: FLagSHyP: Energy type selection

```
function Energy = Energy_type_selection(kinematics,properties,dim,...
matyp) switch matyp
case 1
Energy = Energy1(kinematics,properties);
case 3
Energy = Energy3(kinematics,properties,dim);
case 4
Energy = Energy4(kinematics,properties);
case 5
Energy = Energy5(kinematics,properties,dim);
case 6
Energy = Energy6(kinematics,properties);
case 7
Energy = Energy7(kinematics,properties,dim);
case 8
Energy = Energy8(kinematics,properties,dim);
case 17
Energy = Energy17(kinematics,properties,dim);
end
```

As an example, function `Energy5.m` included below computes the strain energy density for the material type 5. Similar functions have been implemented for the other material types.

BOX 8.13: FLagSHyP: Energy5

```
function Energy = Energy5(kinematics,properties,dimension)
mu     = properties(2);
J      = kinematics.J;
b      = kinematics.b;
Energy =  mu/2*(J^(-2/3)*trace(b) - dimension);
end
```

8.5 EXAMPLES

A limited number of examples are described in this section in order to show the capabilities of FLagSHyP and illustrate some of the difficulties that may arise in the dynamic analysis of highly nonlinear problems. However, the reader is encouraged to explore via further examples the potential of the program FLagSHyP .

8.5.1 Two Dimensional Free Oscillating Pulsating Square

We consider a square two-dimensional plate of unit side length defined as $\Omega_0 = [0,1] \times [0,1]$ under plane strain conditions (i.e. $u_3(X_1, X_2) = 0$). The left and bottom boundaries are allowed to move only tangentially (i.e. $u_1(0, X_2) = 0$ and $u_2(X_1, 0) = 0$), whereas the top and right boundaries are restricted to move normally (i.e. $u_1(X_1, 1) = 0$ and $u_2(1, X_2) = 0$), as shown in Figure 8.3(a). Under the assumption of small strains, the problem has an analytical solution (displacement) of the form

$$\boldsymbol{x} = \boldsymbol{X} + \boldsymbol{u}; \qquad \boldsymbol{u}(X_1, X_2, t) = \lambda(t)\boldsymbol{s}(X_1, X_2), \tag{8.4}$$

with the time-dependent scalar function $\lambda(t)$ defined as

$$\lambda(t) = U_0 \sin\left(\frac{c_s \pi t}{\sqrt{2}}\right); \quad c_s = \sqrt{\frac{\mu}{\rho_0}}, \tag{8.5}$$

and the spatially dependent vector field $\boldsymbol{s}(X_1, X_2)$ defined as

$$\boldsymbol{s}(X_1, X_2) = \begin{bmatrix} \sin\left(\frac{\pi X_1}{2}\right)\cos\left(\frac{\pi X_2}{2}\right) \\ -\cos\left(\frac{\pi X_1}{2}\right)\sin\left(\frac{\pi X_2}{2}\right) \end{bmatrix}. \tag{8.6}$$

The deformation process is initiated with an initial velocity field which is obtained after computing the time derivative of the above displacement field at time $t = 0$, namely

$$\boldsymbol{v}(X_1, X_2)|_{t=0} = \left.\frac{d\lambda(t)}{dt}\right|_{t=0} \boldsymbol{s}(X_1, X_2). \tag{8.7}$$

For small values of U_0 (around or below 0.001), the solution can be considered to be linear and can be used to assess the convergence of the computer program. In this case, a value $U_0 = 0.001$ is used. The domain is discretized into a series of successively refined meshes of quadrilateral four-node elements (i.e. 2×2, 4×4, 8×8, 16×16). Figure 8.3(b) shows a representation of the finest 16×16 mesh. The problem is modeled with a compressible neo-Hookean model with generic material parameters $\lambda = 1$, $\mu = 1$, and $\rho_0 = 1$, and the problem is solved in time using the leap-frog method with successively reduced time steps for the different employed discretizations (i.e. $\Delta t = 0.05$, $\Delta t = 0.025$, $\Delta t = 0.0125$, $\Delta t = 0.00625$).

Figure 8.4 shows a series of snapshots (for the finest discretization) in which the deformation of the domain can be observed. Note that, for visualization purposes, a scaling factor of $\times 100$ has been used to represent the displaced shape. Figure 8.5(a) shows the time history of the vertical component of the coordinate of the point $[0.5, 0.5]$ for the different meshes used, where a clear convergence pattern is observed as the discretization is refined. Figure 8.5(b) shows the error measured as the difference between the vertical displacement of point $[0.5, 0.5]$ at time $t = 5$ for the different meshes and the analytical solution. As can clearly be seen, the error decreases with order 2 (in the asymptotic region), as is expected for the four-node quadrilateral mesh discretization.

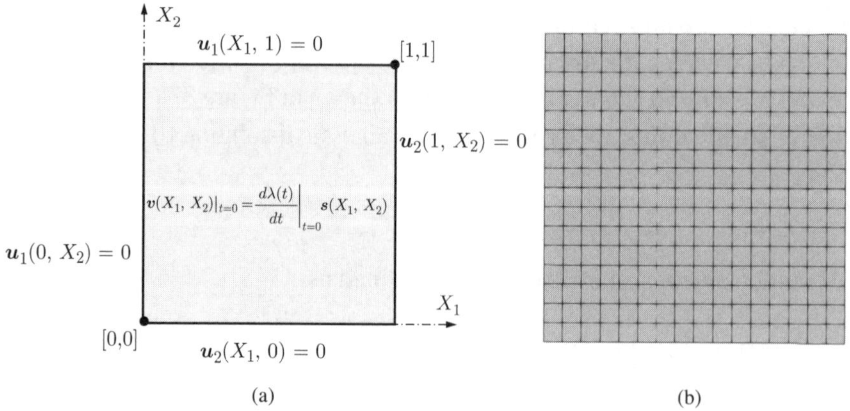

FIGURE 8.3 Swinging square plate: (a) Diagrammatic representation of the problem; (b) Computational mesh.

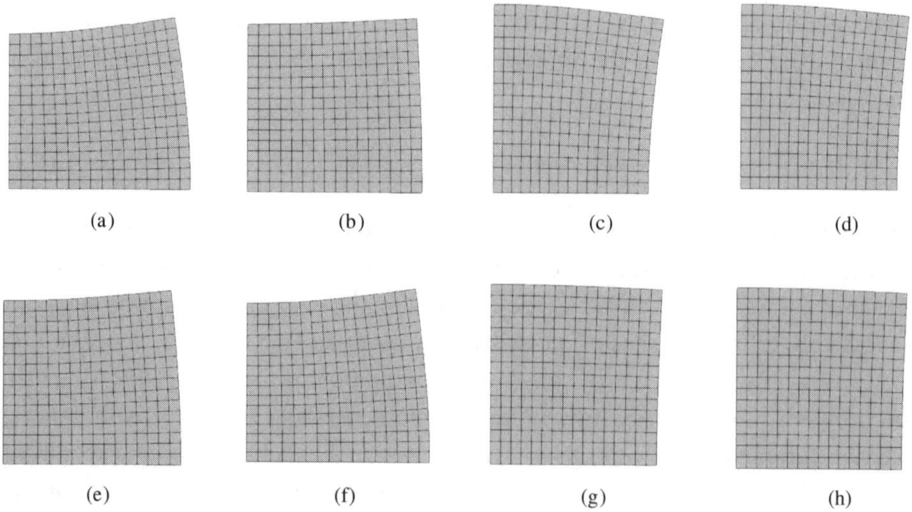

FIGURE 8.4 Swinging square plate: deformation history ($\Delta t = 0.00625$ s). (a) $t = 1000\Delta t$, (b) $t = 200\Delta t$, (c) $t = 300\Delta t$, (d) $t = 400\Delta t$, (e) $t = 500\Delta t$, (f) $t = 600\Delta t$, (g) $t = 700\Delta t$, (h) $t = 800\Delta t$.

8.5.2 Two Dimensional Free Oscillating Column

In this example, we model the dynamic deformation of a hyperelastic column clamped at its bottom end and left freely oscillating as a result of the application of initial velocity boundary conditions. This problem can be used in order to assess both the bending capability and the conservation properties of the algorithm. Specifically, we consider a two-dimensional rectangular domain with material

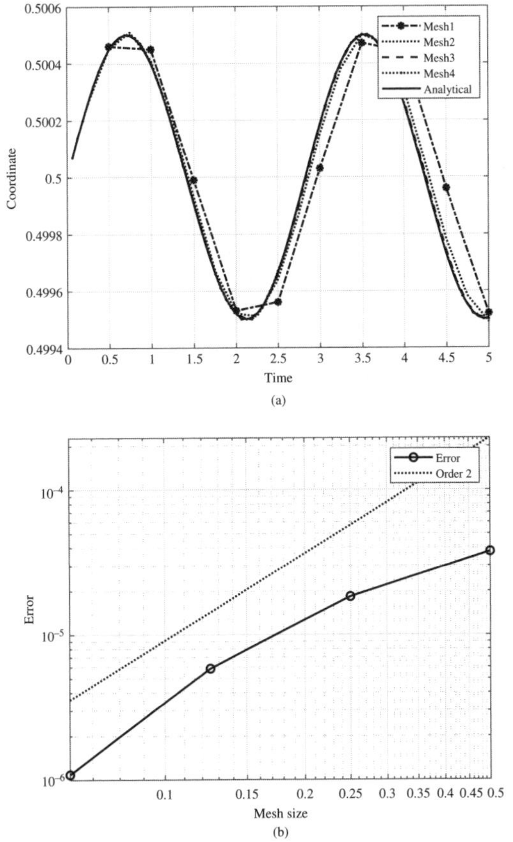

FIGURE 8.5 Swinging square plate: (a) x_2 coordinate evolution of point $[0.5, 0.5]$; (b) Error convergence.

configuration defined by $\Omega_0 = [-t/2, t/2] \times [0, h]$, with $t = 1\,\text{m}$ and $h = 6\,\text{m}$. The domain is made of a nearly incompressible neo-Hookean material and it is simulated under plane strain conditions, with fictitious density $\rho_0 = 1\,\text{kg/m}^3$ and material properties defined by $\mu = 0.4225\,\text{N/m}^2$ and $\kappa = 10\,\text{N/m}^2$. The domain is clamped at its bottom end and left free on the remaining boundaries. The domain is not subjected to any time-varying loads, but its dynamic deformation is triggered by initial conditions defined by a horizontal velocity field described by a linearly varying profile defined by $\boldsymbol{v}(X_1, X_2)|_{t=0} = [v_0 \frac{X_2}{h}, 0]\,\text{m/s}$ and $v_0 = 0.2\,\text{m/s}$ (see Figure 8.6(a)). The domain is discretized with a coarse 4×24 mesh of four-node quadrilateral elements (see Figure 8.6(b)). The problem is solved in time by means of the mid-point time integrator along the time interval $[0, t_{max}]$, with $t_{max} = 200\,\text{s}$ and $\Delta t = 0.125\,\text{s}$.

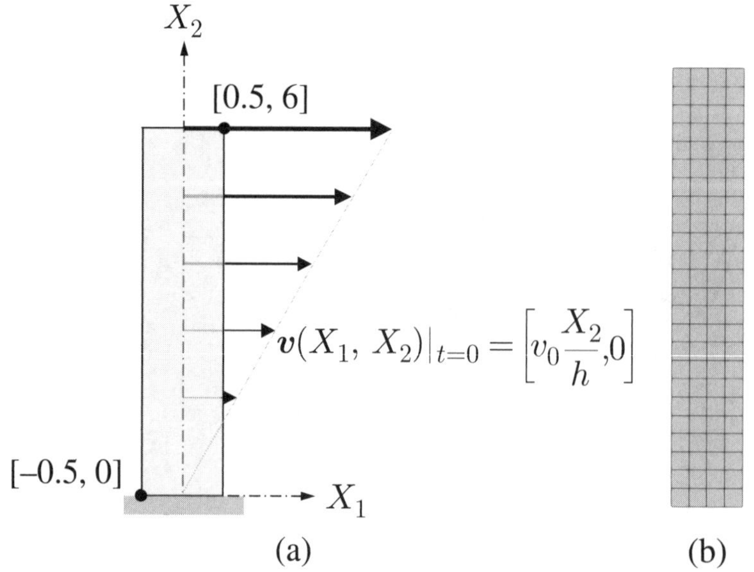

FIGURE 8.6 Bending column: (a) Diagrammatic representation of the problem; (b) Computational mesh.

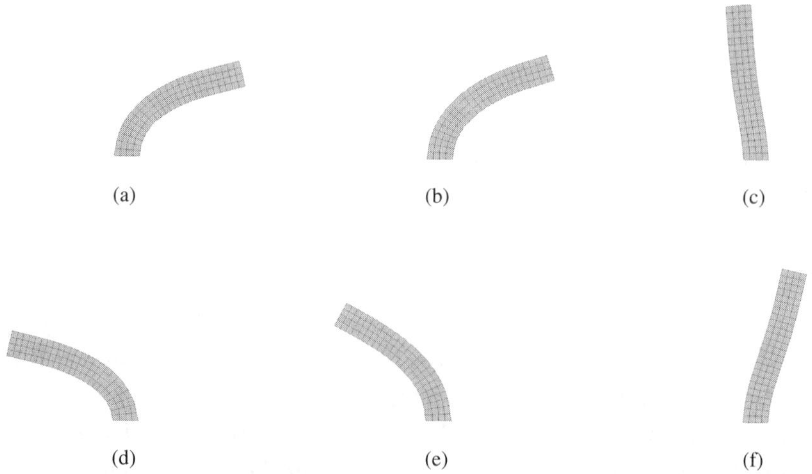

FIGURE 8.7 Bending column: deformation history ($\Delta t = 0.125\,\text{s}$). (a) $t = 250\Delta t$, (b) $t = 500\Delta t$, (c) $t = 750\Delta t$, (d) $t = 1000\Delta t$, (e) $t = 1250\Delta t$, (f) $t = 1500\Delta t$.

Figure 8.7 displays a series of time snapshots representing the time deformation history of the problem. In addition, Figure 8.8 shows the evolution of the total, external, internal, and kinetic energy contributions of the system. As can be observed, the total energy of the system is relatively well conserved despite using a non-energy conserving time integrator and a relatively coarse spatial discretization.

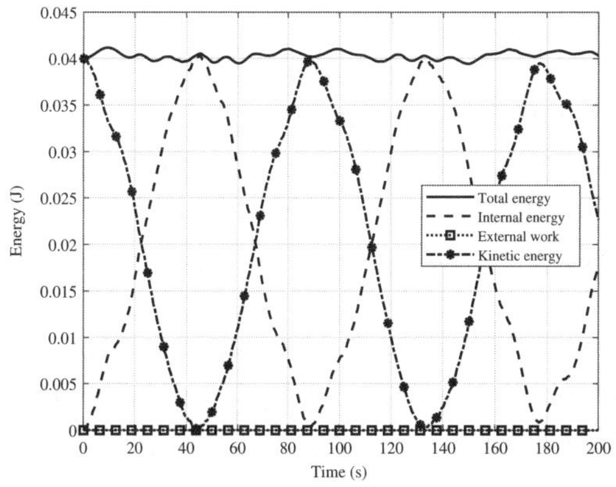

FIGURE 8.8 Bending column: evolution of the different energy contributions.

8.5.3 Three-Dimensional Free Oscillating Twisting Column

In this example, we study the performance of the FLagSHyP program when modeling a three-dimensional example subjected to a complex deformation pattern. Similar to the previous example, the column comprising a hyperelastic material and initially subjected to a velocity profile is left to oscillate freely. Specifically, we consider a three-dimensional domain with material configuration defined by $\Omega_0 = [-t/2, t/2] \times [-t/2, t/2] \times [0, h]$ with $t = 1\,\text{m}$ and $h = 6\,\text{m}$. The domain is

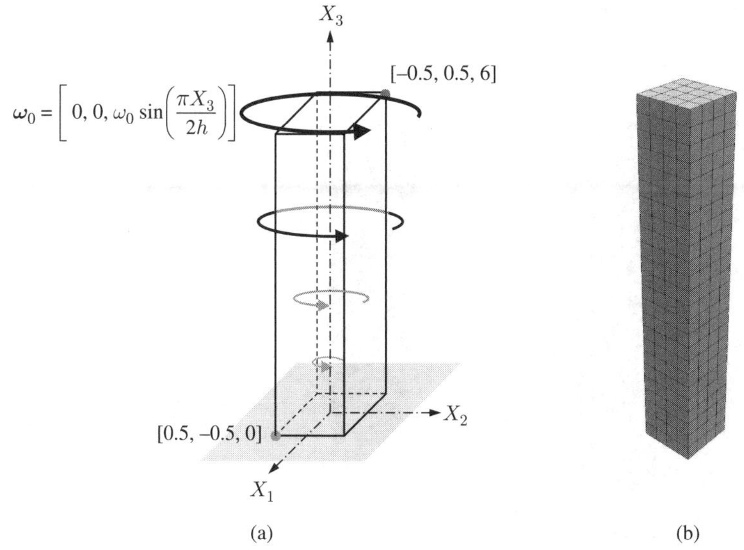

FIGURE 8.9 Twisting column: (a) Diagrammatic representation of the problem; (b) Computational mesh.

made of a compressible neo-Hookean material, with density $\rho_0 = 1100\,\mathrm{kg/m^3}$ and material properties defined by $\mu = 6.5385{\cdot}10^6\ \mathrm{N/m^2}$ and $\lambda = 9.87077{\cdot}10^6\,\mathrm{N/m^2}$. The domain is clamped at its bottom end and left free on the remaining boundaries. The domain is not subjected to any time-varying loads, but its dynamic deformation is triggered by the initial conditions defined by the velocity profile $\boldsymbol{v}(\boldsymbol{X})|_{t=0} = \boldsymbol{\omega}_0(\boldsymbol{X}) \times \boldsymbol{X}$, with $\boldsymbol{\omega}_0 = [0, 0, w_0 \sin\left(\frac{\pi X_3}{2h}\right)]\,\mathrm{rad/s}$ and $w_0 = 105\,\mathrm{rad/s}$ (see Figure 8.9(a)). The domain is discretized with a $4 \times 4 \times 24$ mesh of eight-node hexahedral elements (see Figure 8.9(b)) and is solved in time by means of the generalized-α time integrator method ($\alpha = -0.25$) along the time interval $[0, t_{max}]$, with $t_{max} = 25\,\mathrm{s}$ and $\Delta t = 0.00125\,\mathrm{s}$.

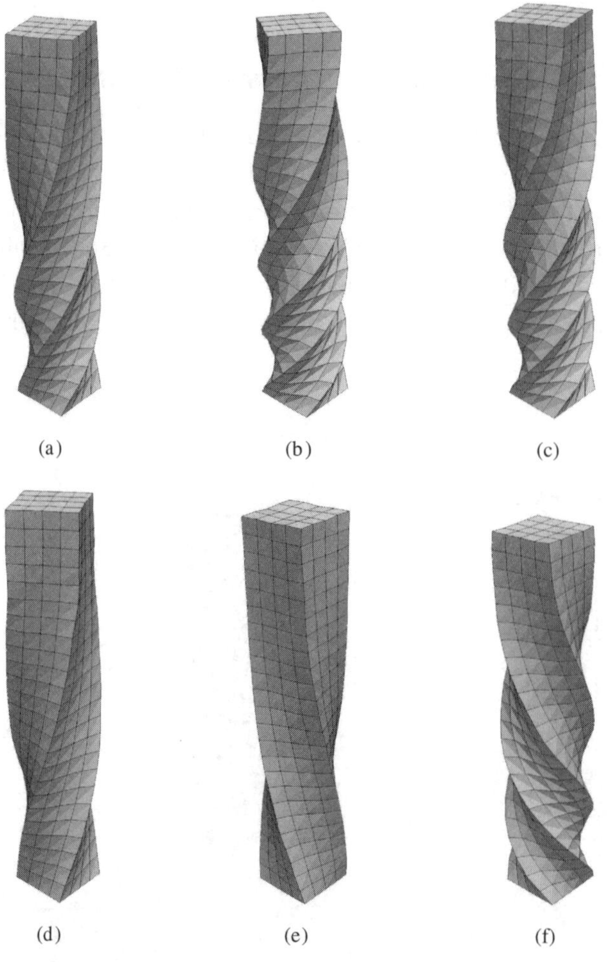

FIGURE 8.10 Twisting column: deformation history ($\Delta t = 0.00125\,\mathrm{s}$). (a) $t = 50\Delta t$, (b) $t = 100\Delta t$, (c) $t = 150\Delta t$, (d) $t = 200\Delta t$, (e) $t = 250\Delta t$, (f) $t = 300\Delta t$.

Figure 8.10 displays a series of time snapshots representing the time deformation history of the problem, and Figure 8.11 shows the evolution of the total, external, internal, and kinetic energy contributions of the system. Once again, the total energy of the system is relatively well conserved, despite using a very coarse spatial discretization.

8.5.4 Three-Dimensional L-Shape Block

This problem was originally proposed by Simo and Tarnow‡ in order to assess the conservation properties of a time integration algorithm. It depicts the free vibration of a hyperelastic three-dimensional domain in the form of an L-shaped block, as presented in Figure 8.12(a). The deformation of the solid is triggered by the application, for a short initial period of time, of globally canceling uniform surface traction forces acting on the two boundary faces defined by $X_1 = 6\,\text{m}$ and $X_2 = 10\,\text{m}$. In this case, and for simplicity, the load is defined by means

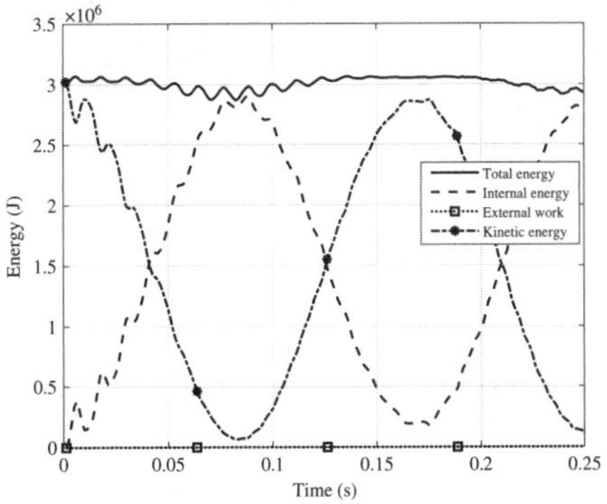

FIGURE 8.11 Twisting column: evolution of the different energy contributions.

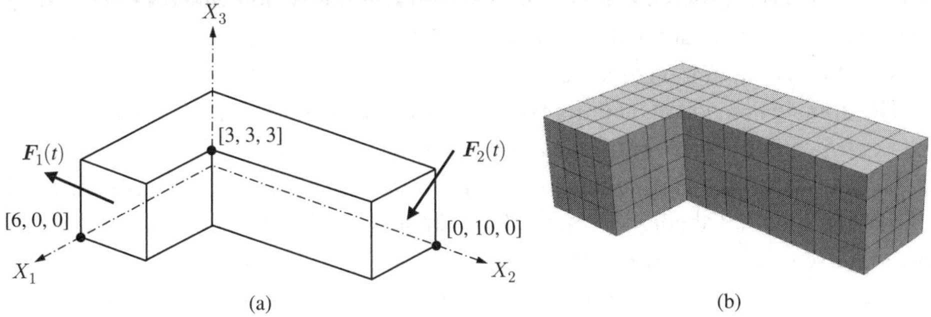

FIGURE 8.12 L-shaped block: Diagrammatic representation of the problem; (b) Computational mesh.

‡ Simo and Tarnow (1994), pp. 2527–2549.

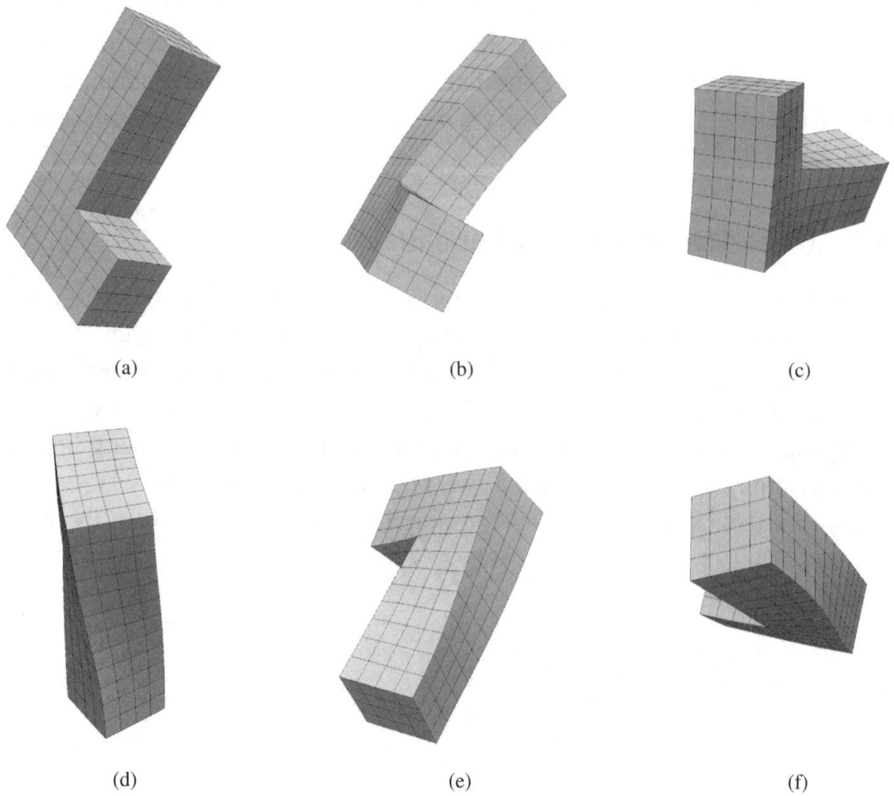

FIGURE 8.13 L-shaped block: deformation history ($\Delta t = 0.01\,\mathrm{s}$). (a) $t = 0$, (b) $t = 500\Delta t$, (c) $t = 1000\Delta t$, (d) $t = 1500\Delta t$, (e) $t = 2000\Delta t$, (f) $t = 2500\Delta t$.

of point forces acting on all the nodes of the two boundary faces. For the face defined by $X_1 = 6\,\mathrm{m}$, the surface traction takes the value $\boldsymbol{F}_1(t) = \boldsymbol{F}_0\lambda(t)$, with $\boldsymbol{F}_0 = [150, 300, 450]\,\mathrm{N/m^2}$ and $\lambda(t)$ is a time-varying scalar function given by a hat type function as defined in Equation (8.3) with $A = 2.5$ and $T = 5\,\mathrm{s}$. The surface traction acting on face $\boldsymbol{X}_2 = 10\,\mathrm{m}$ is given by $\boldsymbol{F}_2(t) = -\boldsymbol{F}_1(t)$.

The domain is made of a compressible neo-Hookean material, with density $\rho_0 = 10^3\,\mathrm{kg/m^3}$ and material properties defined by $\lambda = 2.91354 \cdot 10^4\,\mathrm{N/m^2}$ and $\mu = 1.9423 \cdot 10^4\,\mathrm{N/m^2}$. The domain is discretized with a relatively coarse mesh of 256 eight-node hexahedral elements (see Figure 8.12(b) and is solved in time by means of the leap-frog time integrator method along the time interval $[0, t_{max}]$, with $t_{max} = 25\,\mathrm{s}$ and $\Delta t = 0.01\,\mathrm{s}$. Figure 8.13 displays a series of time snapshots representing the time deformation history of the problem. As can be observed, the domain freely oscillates in space, displaying a very smooth deformation pattern. Figure 8.14 shows the evolution of the energy contributions of the system. As can be seen, for the first $5\,\mathrm{s}$, during which time the surface traction forces act upon

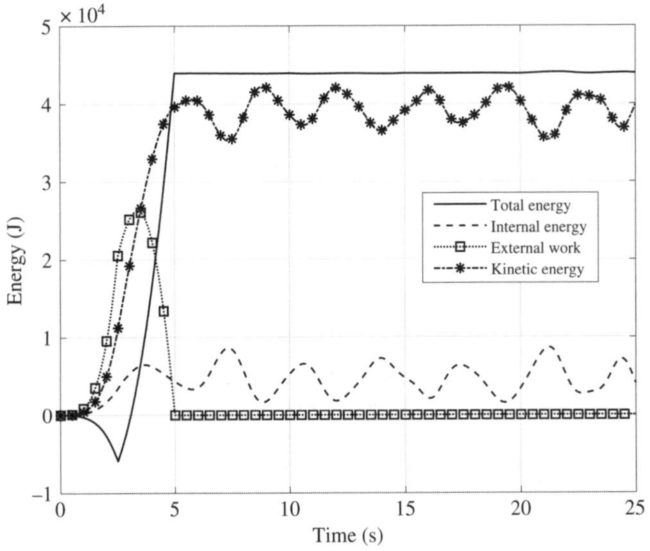

FIGURE 8.14 L-shaped block: evolution of the different energy contributions.

two opposing faces, the total energy of the system varies, reaching a maximum which is then maintained during the free oscillation phase. The total energy of the deformable solid is very well preserved despite the leap-frog method not being an exact energy conserving time integrator.

8.6 DICTIONARY OF MAIN VARIABLES

The main variables of FLagSHyP are listed below; they are grouped based on their parent data structure. Some other important variables are also included.

`PRO.rest`	-	Restart file indicator: `.true.` if problem is restarted, `.false.` if problem started from scratch
`PRO.title`	-	Title of the problem
`PRO.inputfile_name`	-	Name of input file
`PRO.outputfile_name`	-	Name of output file
`PRO.outputfile_name_flagout`	-	Name of single degree of freedom output file
`GEOM.ndime`	-	Number of dimensions
`GEOM.npoin`	-	Number of nodes in the mesh
`GEOM.x`	-	Current coordinates
`GEOM.x0`	-	Initial coordinates
`GEOM.Ve`	-	Initial elemental volume
`GEOM.V_total`	-	Total initial volume
`GEOM.acceleration`	-	Current acceleration
`GEOM.velocity`	-	Current velocity
`GEOM.Energy`	-	Total energy

GEOM.Kinetic	-	Kinetic energy
GEOM.IntEnergy	-	Internal energy
GEOM.ExtEnergy	-	External energy
FEM.mesh.element_type	-	Type of element
FEM.mesh.n_nodes_elem	-	Number of nodes per element
FEM.mesh.n_face_nodes_elem	-	Number of nodes per surface (line) element subjected to pressure loads
FEM.mesh.nelem	-	Number of elements
FEM.mesh.connectivity	-	Element connectivity
FEM.mesh.connectivity_faces	-	Connectivity of surface (line) elements subjected to surface loads
FEM.interpolation.element.N	-	Element shape functions
FEM.interpolation.element.DN_chi	-	Element shape function derivatives
FEM.interpolation.boundary.N	-	Boundary element shape functions
FEM.interpolation.boundary.DN_chi	-	Shape function derivatives of surface (line) elements subjected to pressure loads
QUADRATURE.element.Chi	-	Gauss point location
QUADRATURE.element.W	-	Gauss point weights
QUADRATURE.element.ngauss	-	Number of Gauss points per element
QUADRATURE.boundary.Chi	-	Gauss point location (boundary element)
QUADRATURE.boundary.W	-	Gauss point weights (boundary element)
QUADRATURE.boundary.ngauss	-	Number of Gauss points per boundary element
KINEMATICS.DN_x	-	Current Cartesian derivatives of shape functions
KINEMATICS.Jx_chi	-	Volume ratio (Jacobian) between current and nondimensional isoparametric domain
KINEMATICS.F	-	Deformation gradient tensor
KINEMATICS.J	-	Volume ratio (Jacobian) between current and initial domain ($J = \det\boldsymbol{F}$)
KINEMATICS.b	-	Left Cauchy–Green deformation tensor
KINEMATICS.Ib	-	First invariant of the left Cauchy–Green deformation tensor
KINEMATICS.lambda	-	Principal stretches
KINEMATICS.n	-	Principal directions
MAT.matno	-	Element material number identifier
MAT.nmats	-	Number of materials
MAT.props	-	Material properties: the first is always the initial density, the rest depend on the material type
MAT.matyp	-	Material number identifier
MAT.n_nearly_incompressible	-	Number of nearly incompressible materials
BC.icode	-	Nodal boundary condition identifier
BC.freedof	-	Nodal free degrees of freedom
BC.fixdof	-	Nodal fixed degrees of freedom
BC.n_prescribed_displacements	-	Number of prescribed displacements
BC.presc_displacement	-	Prescribed displacement vector
BC.dofprescribed	-	Prescribed degree of freedom
LOAD.n_pressure_loads	-	Number of pressure load surface elements
LOAD.gravt	-	Gravity vector
LOAD.pressure	-	Value of pressure in surface element
LOAD.pressure_element	-	Element subjected to pressure
GLOBAL.nominal_external_load	-	Nominal external load (without pressure)
GLOBAL.Residual	-	Global residual vector
GLOBAL.external_load	-	Current load vector
GLOBAL.nominal_pressure	-	Equivalent nodal force due to pressure
GLOBAL.T_int	-	Equivalent internal force vector

GLOBAL.K	-	Global stiffness matrix
GLOBAL.R_pressure	-	Equivalent internal force vector due to pressure
GLOBAL.K_pressure	-	Global stiffness matrix due to pressure
GLOBAL.Mass	-	Global mass matrix
PLAST.yield.f	-	Yield surface value
PLAST.yield.Dgamma	-	Incremental plastic multiplier
PLAST.yield.nu_a	-	Direction vector
PLAST.trial.lambdae	-	Trial elastic stretches
PLAST.trial.tau	-	Trial Kirchhoff stress tensor
PLAST.trial.n	-	Trial eigenvectors
PLAST.UPDATED.invCp	-	Updated plastic right Cauchy–Green tensor
PLAST.UPDATED.epbar	-	Updated equivalent plastic strain
PLAST.stress.Cauchy	-	Cauchy stress tensor
PLAST.stress.Cauchyaa	-	Cauchy principal stresses
CON.nincr	-	Number of load increments
CON.incrm	-	Current load increment
CON.xlmax	-	Maximum load parameter
CON.dlamb	-	Incremental load value
CON.miter	-	Maximum number of iterations per increment
CON.niter	-	Current iteration per increment
CON.cnorm	-	Convergence criterion for Newton–Raphson
CON.searc	-	Line search parameter
CON.msearch	-	Maximum number of line search iterations
CON.incrm	-	Current load increment
CON.xlamb	-	Current load parameter
CON.ARCLEN.farcl	-	Logical fixed arc length indicator
CON.ARCLEN.arcln	-	Arc length parameter (0.0 means no arc length)
CON.ARCLEN.itarget	-	Target iteration/increment for variable arc length option
CON.ARCLEN.iterold	-	Number of iterations in previous increment for arc length
CON.ARCLEN.xincr	-	Total displacement over the load increment
CON.ARCLEN.afail	-	Logical arc length failure indicator
CON.OUTPUT.incout	-	Output counter
CON.OUTPUT.nwant	-	Single output node
CON.OUTPUT.iwant	-	Output degree of freedom
CON.DYNAMICS.time_integrator	-	Time integrator
CON.DYNAMICS.tmax	-	Maximum time of simulation
CON.DYNAMICS.dt	-	Time increment
CON.DYNAMICS.gamma	-	Newmark γ parameter
CON.DYNAMICS.beta	-	Newmark β parameter
CON.DYNAMICS.alpha	-	α-method parameter
CON.DYNAMICS.load_type	-	Dynamic analysis
CON.DYNAMICS.load_parameter1	-	Dynamic analysis
CON.DYNAMICS.load_parameter2	-	Dynamic analysis
da	-	Newton–Raphson iterative acceleration
eta	-	Scaling factor or displacement factor
eta0	-	Previous value of the parameter eta
rtu	-	Current dot product of $\mathbf{R}$ by $\mathbf{u}$
rtu0	-	Initial dot product of $\mathbf{R}$ by $\mathbf{u}$
displ	-	Newton–Raphson displacement vector $\mathbf{u}$
dispf	-	Load component of the displacement vector $\mathbf{u}_F$
rnorm	-	Current residual norm
c	-	Fourth-order tensor
indexi	-	Vector storing the row entry of the stiffness matrix coefficient
indexj	-	Vector storing the column entry of the stiffness matrix coefficient

`counter`	-	Counter to travel through the vectors storing the stiffness matrix coefficients
`global_stiffness`	-	Vector storing each coefficient of the stiffness matrix
`mu`	-	μ coefficient (material parameter)
`lambda`	-	λ coefficient (material parameter)
`kappa`	-	κ coefficient (material parameter)
`H`	-	Strain hardening (material parameter)

8.7 CONSTITUTIVE STRAIN ENERGY EQUATION SUMMARY

This section briefly includes the strain energy functions used for the various material type models when computing the total energy of the system as an output variable. Although this calculation is not strictly necessary for FLagSHyP to function, it is necessary when displaying the evolution of the energy of the system as a function of time as an output. In the following boxes, Material 8.14, for example, refers to constitutive equation 1 as listed in Note 3 on p. 266 of NL-Statics. Likewise, Material 8.17 refers to constitutive equation 17. Further details of the constitutive equations are given in Section 10.14 of NL-Statics.

Material 8.1: Energy 1: three-dimensional or plane strain compressible neo-Hookean

$$\Psi = \frac{\mu}{2}(I_b - 3) - \mu \ln J + \frac{\lambda}{2}(\ln J)^2; \qquad \text{for plane strain, } \boldsymbol{b} = \begin{bmatrix} b_{11} & b_{12} & 0 \\ b_{21} & b_{22} & 0 \\ 0 & 0 & 1 \end{bmatrix}$$

Material 8.2: One-dimensional stretch-based hyperelastic plastic (truss2 only)

$$\Psi = \frac{E}{2}(\ln \lambda_e)^2; \qquad \ln \lambda_e = \ln \lambda - \ln \lambda_p; \qquad \lambda = \frac{l}{L}$$

Material 8.3: Energy 3: plane strain or three-dimensional hyperelastic in principal directions

$$\Psi = \mu \sum_{\alpha=1}^{3} (\ln \lambda_\alpha)^2 + \frac{\lambda}{2}(\ln J)^2; \qquad \text{for plane strain, } \lambda_3 = 1$$

Material 8.4: Energy 4: plane stress hyperelastic in principal directions

$$\Psi = \mu \sum_{\alpha=1}^{2} (\ln \lambda_\alpha)^2 + \frac{\bar{\lambda}}{2}(\ln j)^2; \qquad \gamma = \frac{2\mu}{\lambda + 2\mu}; \qquad \bar{\lambda} = \gamma\lambda$$

Material 8.5: Energy 5: plane strain or three-dimensional nearly incompressible neo-Hookean

$$\Psi = \frac{\mu}{2}(J^{-2/3} I_{\boldsymbol{b}} - 3); \qquad U(\bar{J}) = \frac{\kappa}{2}(\bar{J} - 1)^2; \qquad \bar{J} = \frac{v^{(e)}}{V^{(e)}}$$

Material 8.6: Energy 6: plane stress incompressible neo-Hookean

$$\Psi = \frac{\mu}{2}\left(I_{\boldsymbol{b}} + \frac{1}{j^2} - 3\right); \qquad j^2 = \det_{2\times 2} \boldsymbol{b}; \qquad \boldsymbol{b} = \begin{bmatrix} b_{11} & b_{12} \\ b_{21} & b_{22} \end{bmatrix}$$

Material 8.7: Energy 7: plane strain or three-dimensional nearly incompressible hyperelastic in principal directions

$$\Psi = \mu \sum_{\alpha=1}^{3} (\ln \lambda_\alpha)^2 - \frac{\mu}{3}(\ln J)^2; \qquad U(\bar{J}) = \frac{\kappa}{2}(\ln \bar{J})^2; \qquad \bar{J} = \frac{v^{(e)}}{V^{(e)}}$$

Material 8.8: Energy 8: plane stress incompressible hyperelasticity in principal directions

$$\Psi = \mu \left((\ln \lambda_1)^2 + (\ln \lambda_2)^2 + \left(\ln \left(\frac{1}{\lambda_1 \lambda_2} \right) \right)^2 \right)$$

Material 8.17: Energy 17: plane strain or three-dimensional nearly incompressible hyperelastic plastic in principal directions

$$\Psi = \mu \sum_{\alpha=1}^{3} \ln(\lambda_{e,\alpha})^2 - \frac{\mu}{3}(\ln J)^2; \qquad U(\bar{J}) = \frac{\kappa}{2}(\ln \bar{J})^2; \qquad \bar{J} = \frac{v^{(e)}}{V^{(e)}}$$

CHAPTER NINE

COMPUTATIONAL IMPLEMENTATION FOR CONSERVATION-LAW-BASED EXPLICIT FAST DYNAMICS

9.1 INTRODUCTION

This chapter introduces a new computer program, entitled PG_DYNA_LAWS (**P**etrov–**G**alerkin explicit **DYNA**mics for conservation **LAWS**), that can be used for the fast dynamic analysis of three-dimensional hyperelastic deformable solids undergoing large deformations in both isothermal and non-isothermal scenarios.

PG_DYNA_LAWS is designed based on the principles presented in Chapter 7, where the deformation of a solid is described via a set of first-order conservation laws, written in terms of the linear momentum, the deformation gradient, and the entropy density per unit undeformed volume $\{\boldsymbol{p}_R, \boldsymbol{F}, \eta\}$. For simplicity and ease of understanding, the focus will be on solids described by the constitutive models used in Chapter 7, namely the isothermal compressible neo-Hookean model and the thermoelastic Mie–Grüneisen model. As for the spatial discretization, this is restricted to the use of three-dimensional linear tetrahedral finite elements in conjunction with a Streamline Upwinding Petrov–Galerkin (Variational Multi-Scale) stabilization based method. Regarding the time integration, a two-stage Runge–Kutta method is used in order to explicitly evolve the equations in time.

The basic structure and the most important functions of this program will be presented in order to illustrate the main differences with respect to alternative dynamics implementations, presented in the FLagSHyP software in Chapter 8. The program description includes user instructions, a dictionary of variables and functions, and sample input and output for a few simple examples. The computer program, in conjunction with sample data, can be downloaded from the website `www.flagshyp.com`.* Alternatively, it can be obtained by email request to the authors `a.j.gil@swansea.ac.uk` or `j.bonet@greenwich.ac.uk`.

The master m-file `PG_DYNA_LAWS.m`, which controls the overall organization of the program, is divided into three sections. The first section includes a series of

* Or `www.cambridge.org/9781107115620`.

Name	Comments
`PRO.`	Program run mode, input/output files, problem title
`GEO.`	Nodes, initial coordinates, number of elements, element connectivity, and boundary connectivity
`DAT.`	Conservation variables $\{\boldsymbol{p}, \boldsymbol{F}, \eta\}$, SUPG parameters $\{\tau_{\boldsymbol{F}}, \alpha\}$, material properties $\{\rho_R, E, \nu\} \ldots$
`COMP.`	Wave speeds $\{c_p, c_s\}$, material parameters μ, λ, κ, lumped mass, time control parameters
`FUNC.`	Various sub-functions
`CON.`	Program control parameters: dynamics data...
`OUT.`	Output results: conservation variables, spatial coordinates...

statements designed to add the necessary directories to the path of the program. The second section includes the function `input_data_and_initialisation.m`, which is devoted to the reading of input and control data and the initialization of variables including equivalent nodal forces and the lumped mass matrix. The third section encompasses the main explicit stabilized finite element algorithm.

A feature of the program PG_DYNA_LAWS is the arrangement of variables (along with the grouping of those which share a similar nature) into data structures. As already noted when describing the alternative dynamics code FLagSHyP, although the use of data structures can greatly facilitate the understanding of all variables in the program, their use can introduce some computational overheads, noticeable in a large-scale problem. For completeness, some of the fundamental data structures used in the program PG_DYNA_LAWS are listed below.

As an example, the table below shows the user input variables that must be provided in order to run explicit dynamics simulations. Note that only two stabilization coefficients, $\tau_{\boldsymbol{F}}$ and α, are needed ($\tau_\eta = 0$). For completeness, a dictionary of all the variables of the program is included at the end of this chapter.

9.2 USER INSTRUCTIONS

By executing the master m-file `PG_DYNA_LAWS.m`, the directories (and subdirectories) `code` and `job_folder` are added to the directory path of the program. For simplicity, all the necessary MATLAB functions comprising the PG_DYNA_LAWS program have been placed within subdirectories included in the directory `code`, whereas input/output data files have been located within subdirectories included in the directory `job_folder`. Moreover, the program expects two input files: the first is the user-defined input control file `PG_DYNA_LAWS_input_file.txt` and the second is the mesh information file suitably formatted (generated via any appropriate mesh generator), both to be located within the directory `job_folder/PG_DYNA_LAWS`. Note that while the name (e.g. `PG_DYNA_LAWS_input_file`) or extension (e.g. `.txt` or

Parent	Child	Comments
CON	Restart_run	New run (yes or no)
DAT	Model	Type of constitutive model
DAT	CF	Specific heat
DAT	MG_q	q coefficient for Mie–Grüneisen model
DAT	Conduct	Thermal conductivity coefficient h
DAT	E	Young's modulus
DAT	Poisson	Poisson's ratio
DAT	Den	Initial density
CON	Nstep	Number of time steps
CON	Type_dt	Type of time interval (fixed CFL or Δt)
CON	CFL	CFL stability number (7.89)
CON	dt	Fixed time increment
DAT	SUPG_tauF	SUPG coefficient τ_F
DAT	SUPG_xiF	SUPG coefficient α
CON	OutInc	Output increment

.mat) of the input file are arbitrary, the name of the input file must coincide precisely with the name given to the subdirectory which contains it (e.g. job_folder/PG_DYNA_LAWS, otherwise, an error will occur and the program will not progress. Examples of the structure of the two required input files are included below. In addition, Box 9.1 includes the control input file used for a typical numerical simulation.

BOX 9.1: Example of input file

```
Starting option (1 = New; 2 = Restart)
1
Type of Constitutive Model
1
Thermal parameters
1
2.223 × 10^-4
10
1
Material properties
50500
0.3
1000
Time parameters
1000
1
0.3
5 × 10^-4
```

(continued)

Box 9.1: *(cont.)*

```
SUPG coefficients
0.5
0.0
Output increments
10
```

Input I: User input parameters PG_DYNA_LAWS_input_file.txt

Item 1: New run (yes or no)	**Number of lines:** 1
`CON.Restart_Run`	Starting options
Item 2: Type of constitutive model	**Number of lines:** 1
`DAT.Model`	Constitutive model
Item 3: Thermal parameters	**Number of lines:** 4
`DAT.CF`	Specific heat coefficient
`DAT.MG_Thermal_expansion`	Thermal expansion coefficient
`DAT.Conduct`	Thermal conductivity coefficient
`DAT.MG_q`	Mie–Grüneisen parameter q
Item 4: Material properties	**Number of lines:** 3
`DAT.E`	Young's modulus
`DAT.Poisson`	Poisson's ratio
`DAT.Den`	Density
Item 5: Time parameters	**Number of lines:** 4
`CON.Nstep`	Number of time steps
`CON.Type_dt`	Fixed CFL or fixed time increment Δt
`CON.CFL`	CFL stability number
`CON.dt`	Time increment Δt
Item 6: SUPG coefficients	**Number of lines:** 2
`DAT.SUPG_tauF`	SUPG coefficient τ_F
`DAT.SUPG_xiF`	Time integrated SUPG coefficient α
Item 7: Output increment	**Number of lines:** 1
`CON.OutInc`	Output time increment

Input II: Geometry, mesh, and initial conditions PG_DYNA_LAWS.mat

Item 1: Geometry `GEO.X`	Initial geometry
Item 2: Number of elements `GEO.Nelem`	Number of elements
Item 3: Number of nodes `GEO.Nnode`	Number of nodes
Item 4: Element connectivity `GEO.T`	Element connectivity
Item 5: Number of boundary surfaces `GEO.Nface`	Number of boundary faces
Item 6: Boundary connectivity `GEO.T_B`	Surface boundary connectivity
Item 7: Boundary conditions `GEO.Bcode`	Type of boundary conditions
Item 8: Initial conditions for velocity `DAT.v`	Velocity vector
Item 9: Initial conditions for fibre map `DAT.F`	Deformation gradient tensor
Item 10: Initial conditions for entropy `DAT.Eta`	Entropy scalar field

Algorithm 9.1: Main algorithm

Input : Initial geometry $\mathbf{X}$; Material properties; Solution parameters;
Initial conditions $\{\mathbf{U}_0^p, \mathbf{U}_0^F, \mathbf{U}_0^\eta\}$, Mass matrix

Output: Current value of conservation variables $\{\mathbf{U}_N^p, \mathbf{U}_N^F, \mathbf{U}_N^\eta\}$ and current geometry $\mathbf{x}_N$

for itime=1:CON.Nstep **do**

- Compute maxinum wave speed `COMP.upmax`;
- Evaluate time increment `COMP.dt`;
- **for** `iRK=1:2` **do**
  - Incorporate boundary contributions of `pdot` $(\dot{\mathbf{U}}^p)$ and `Etadot` $(\dot{\mathbf{U}}^\eta)$ (refer to Algorithm 9.2);
  - Evaluate right-hand side contributions: RF $(\mathbf{R}^F = \mathbf{F}^F - \mathbf{T}^F)$ and `REta` $(\mathbf{R}^\eta = \mathbf{F}^\eta - \mathbf{T}^\eta)$ (refer to Algorithm 9.3);
  - Compute $\{\dot{\mathbf{U}}^F, \dot{\mathbf{U}}^\eta\}$ `Fdot, Etadot`;
  - Evaluate right-hand side contribution: Rp $(\mathbf{R}^p = \mathbf{F}^p - \mathbf{T}^p)$ (refer to Algorithm 9.4);
  - Angular momentum projection algorithm Rp;
  - Compute time update `pdot` $(\dot{\mathbf{U}}^p)$;
  - Update mesh coordinates `UNK.X` $(\mathbf{x})$;
  - Update conservation variables `UNK.p, UNK.F, UNK.Eta`;
  - Apply strong boundary conditions on linear momentum and entropy (only when imposing temperature) `UNK.p, UNK.Eta`;
- **end**
- Update conservation variables, mesh coordinates, and time `UNK.p, UNK.F, UNK.Eta, UNK.X, COMP.time`;
- Apply strong boundary conditions on linear momentum and entropy (only when imposing temperature) `UNK.p, UNK.Eta`;
- Output results `OUT`;

end

Algorithm 9.2: Evaluate boundary contributions

Input : Finite Element boundary information and connectivity;
Conservation variables $\{\mathbf{U}^p, \mathbf{U}^F, \mathbf{U}^\eta\}$; Material properties
Output: Update Right-hand side of $\mathbf{R}^p$ and $\mathbf{R}^\eta$ (include boundary contributions)

for `iface=1:GEO.Nface` **do**
    Extract area and its unit normal vector corresponding to the centroid of boundary face f, namely `GEO.A_B, GEO.N_B`;
    for `in=1:3` **do**
        Interpolate variables at the centroid of boundary face f $\{\boldsymbol{p}_f, \boldsymbol{F}_f, \eta_f\}$;
    end
    Compute first Piola–Kirchhoff stress tensor $\boldsymbol{P}_f$, dependent upon $\{\boldsymbol{F}_f, \eta_f\}$;
    Imposition of traction and heat flux boundary contributions;
    for `in=1:3` **do**
        Incorporate appropriate boundary contributions to the right-hand side of $\mathbf{R}^p$ and $\mathbf{R}^\eta$
    end
end

9.3 SOLVER DETAILS

The main algorithm is listed above (refer to Algorithm 9.1), where an incremental time outer loop is used to advance the solution from a time step t_n to time t_{n+1}. Once the size of the time step Δt is carefully evaluated (that is, making use of the Courant–Friedrichs–Lewy stability condition (α_{CFL}) in order to prevent instability of the algorithm, the two-stage Runge–Kutta (RK) time integrator inner loop is initiated. Within this inner Runge–Kutta loop, the unknown conservation variables $\{\mathbf{U}^p, \mathbf{U}^F, \mathbf{U}^\eta\}$ are updated from one Runge–Kutta stage to the next by computation of their time rates $\{\dot{\mathbf{U}}^p, \dot{\mathbf{U}}^F, \dot{\mathbf{U}}^\eta\}$ and application of appropriate boundary conditions.

The time rates require first the evaluation of the right-hand sides of the relevant conservation equations, $\{\mathbf{R}^p, \mathbf{R}^F, \mathbf{R}^\eta\}$, consisting of internal $\{\mathbf{T}^p, \mathbf{T}^F, \mathbf{T}^\eta\}$ and external $\{\mathbf{F}^p, \mathbf{F}^F, \mathbf{F}^\eta\}$ contributions. The evaluation of these right-hand sides is summarized in Algorithms 9.2–9.4. For problems involving long term rotations, an angular momentum projection procedure is also included as part of the solution algorithm (see Examples 7.6 and 7.7).

Algorithm 9.3: Right-hand side evaluation for $\{\boldsymbol{F}, \eta\}$ equations

Input : Finite Element information and connectivity; Conservation variables $\{\mathbf{U}^p, \mathbf{U}^F, \mathbf{U}^\eta\}$; Material properties
Output: Right-hand side $\mathbf{R}^F$ and $\mathbf{R}^\eta$ (without accounting for boundary contributions)

for `ielem=1:GEO.Nelem` **do**
    Extract Jacobian and weight at the centroid of tetrahedral element (e) `DAT.ISO_JOgp`;
    for `in=1:4` **do**
        Interpolate variables at the centroid of tetrahedral element (e) $\{\boldsymbol{F}_e, \boldsymbol{\nabla}_0\boldsymbol{v}|_e, \theta_e, \boldsymbol{\nabla}_0\theta|_e\}$;
    end
    Compute the elemental area map and volume map $\{\boldsymbol{H}_e, J_e\}$;
    Evaluate the elemental heat flux vector $\boldsymbol{Q}_e := -kJ_e^{-1}\boldsymbol{H}_e^T\boldsymbol{H}_e\boldsymbol{\nabla}_0\theta|_e$;
    for `in=1:4` **do**
        Update the right-hand side of $\mathbf{R}^F$ and $\mathbf{R}^\eta$ (without accounting for boundary contributions)
    end
end

9.4 PROGRAM STRUCTURE

This section gives an overview of the structure of the program PG_DYNA_LAWS, with an emphasis on some of the most important functions. The remainder are functions not strictly necessary for the understanding of the flow of the program and can be examined via a program download (see `www.flagshyp.com`). The master m-file `PG_DYNA_LAWS.m` is divided into three segments, presented below.

BOX 9.2: PG_DYNA_LAWS segment 1 – Master m-file

```
clear all;close all;clc;

%%SECTION 1

if isunix()
   dirsep =  '/';
```

(continued)

Box 9.2: *(cont.)*

```
else
   dirsep =  '\';
end
basedir_fem = mfilename('fullpath');
basedir_fem = fileparts(basedir_fem);
basedir_fem = strrep(basedir_fem,['code' dirsep 'support'],'');
addpath(fullfile(basedir_fem));
addpath(genpath(fullfile(basedir_fem,'code')));
addpath((fullfile(basedir_fem,'job_folder')));

PRO.inputfile_name = ...
input([' Enter the data file name    : ' ' \n'],'s');
PRO.job_folder = ...
fullfile(basedir_fem,['job_folder\PRO.inputfile_name]);

if ~exist(PRO.job_folder,'dir')
   error(['The folder ' PRO.job_folder ' has not been ' ...
   'created. Please define the folder and the input file (not '...
   'including the extension) identical'])
end
cd(PRO.job_folder)

%%SECTION 2

[CON,DAT,GEO,FUNC,UNK,COMP,PRO] = input_data_and_initialization(PRO);

%%SECTION 3

Run_Jobs_Batch_Background(CON,DAT,GEO,FUNC,UNK,COMP,PRO)
```

The first segment sets up the directory path, which has been previously explained in the user instructions. The second segment is the function `input_data_and_initialisation.m`, for the reading of input/control data and the initialization of variables. The third segment calls the function `Run_Jobs_Batch_Background.m`, which encompasses the main solver algorithm. The second segment, is presented below.

BOX 9.3: PG_DYNA_LAWS segment 2 – Input data and initialization

```
function [CON,DAT,GEO,FUNC,UNK,COMP,PRO] = ...
          input_data_and_initialisation(PRO)

[CON,FUNC,DAT]                 = Read_Control_File(PRO);
[GEO,DAT,FUNC]                 = Read_Geometry_File(FUNC,CON,DAT,PRO);
[FUNC,CON,DAT,GEO,UNK,COMP]    = Initialisation(FUNC,CON,DAT,GEO);

end
```

Algorithm 9.4: Right-hand side evaluation for linear momentum equation

Input : Finite Element information and connectivity; Conservation variables $\{\mathbf{U}^p, \mathbf{U}^F, \mathbf{U}^\eta\}$ and time rate $\dot{\mathbf{U}}^F$; Material properties
Output: Right-hand side of $\mathbf{R}^p$ (without boundary contributions)

for `ielem=1:GEO.Nelem` **do**
  Extract Jacobian and weight at the centroid of tetrahedral element (e)
  `DAT.ISO_J0gp`;
  for `in=1:4` **do**
    Interpolate variables at the centroid of tetrahedral element (e)
    $\{\boldsymbol{F}_e, \eta_e, \boldsymbol{\nabla}_0\boldsymbol{x}|_e, \dot{\boldsymbol{F}}_e, \boldsymbol{\nabla}_0\boldsymbol{v}|_e\}$;
  end
  Evaluate stabilized deformation gradient $\boldsymbol{F}_e^{st}$;
  Compute first Piola–Kirchhoff stress tensor $\boldsymbol{P}_e$, dependent upon the fields $\{\boldsymbol{F}_e^{st}, \eta_e\}$;
  for `in=1:4` **do**
    Update the right-hand side of $\mathbf{R}^p$ (without boundary contributions)
  end
end

The program can start either from the input data files or, when convenient, using data from a restart file written during a previous incomplete analysis, based on value of the variable `CON.Restart_Run`.

BOX 9.4: PG_DYNA_LAWS segment 3 – Run the simulation

```
if(CON.Restart_Run == 1)
   [FUNC,CON,DAT,GEO]        = Initialisation(FUNC,CON,DAT,GEO);
   save('Mesh.mat','CON','GEO','DAT','FUNC')

   Main_Algorithm(GEO,DAT,CON,FUNC,UNK,COMP);

else

  Nstep                      = CON.Nstep;
  OutInc                     = CON.OutInc;
  load('Mesh.mat')
  load('Result.mat')
```

(continued)

Box 9.4: *(cont.)*

```
    CON.Nstep                   = Nstep;
    CON.OutInc                  = OutInc;

    Main_Algorithm(GEO,DAT,CON,FUNC,UNK,COMP);

  end
```

The program can restart from a previously converged incremental step in order to progress to subsequent increments. This can be useful when wishing to continue the analysis to obtain further incremental solutions. In the following it is assumed that the data is read for the first time. The function `Main_Algorithm.m` includes the main time incremental time loop, which advances the simulation explicitly in time. Within every time step, the function `Main_Algorithm.m` carries out three important tasks. First, the size of the time step is continuously updated via the Courant–Friedrichs–Lewy stability condition `FUNC.dt_method.m`. Second, the function `TVD_RK.m`, which includes the main Runge–Kutta time integration algorithm, is called. Third, the function `Save_Output_Paraview.m` is called in order to output the data for visualization purposes.

BOX 9.5: PG_DYNA_LAWS segment 4 – Main algorithm

```
function Main_Algorithm(GEO,DAT,CON,FUNC,UNK,COMP)

COMP.itime0                         = COMP.itime;
COMP.t_start                        = tic;
for i = COMP.itime0 + 1 : COMP.itime0 + CON.Nstep

    COMP.itime                      = i;
    [stret]                         = Stretch(GEO.Nnode,UNK.F);
    COMP.upmax                      = COMP.up_temp / min(stret);
    [COMP]                          = FUNC.dt_method(GEO,CON,COMP);
    [UNK,COMP]                      = TVD_RK(UNK,COMP,GEO,FUNC,DAT);
    Save_Output_Paraview(UNK,COMP,CON,FUNC,GEO,DAT,i);
end
end
```

Within the function `TVD_RK.m`, the update of the conservation variables is carried out for each of the two stages of the Runge–Kutta time integrator algorithm. The fundamental actions within every Runge–Kutta stage are listed below. The update of the conservation variables requires the evaluation of the right-hand side of the various conservation equations, which is carried out as part of the function `RHS_all.m`. Once the conservation variables are updated, the function `Apply_strong_boundary_conditions.m` is called in order to apply the strong (essential) boundary conditions.

BOX 9.6: PG_DYNA_LAWS segment 5 – Total variation diminishing Runge–Kutta

```
UNK.pold   = UNK.p;
UNK.Fold   = UNK.F;
UNK.Etaold = UNK.Eta;
UNK.Xold   = UNK.X;
UNK.X_AM   = UNK.X;
[UNK,COMP] = RHS_all(UNK,COMP,GEO,DAT,FUNC);

UNK.X_AM    = UNK.Xold + 0.5 * COMP.dt * COMP.irho*(UNK.pold + UNK.p);
[UNK,COMP]  = RHS_all(UNK,COMP,GEO,DAT,FUNC);

UNK.p       = 0.5 * (UNK.pold + UNK.p);
UNK.F       = 0.5 * (UNK.Fold + UNK.F);
UNK.Eta     = 0.5 * (UNK.Etaold + UNK.Eta);
UNK.X       = 0.5 * (UNK.Xold + UNK.X);
COMP.time   = COMP.time + COMP.dt;
[UNK]       = Apply_strong_boundary_conditions(UNK,GEO);
```

The fundamental actions within the function `RHS_all.m` are listed below. These involve first the update of the variables $\boldsymbol{F}$ and η, followed by the update of the linear momentum variable $\boldsymbol{p}_R$. Note that updating the various conservation variables requires the use of the lumped mass matrix, which is stored in the variable `COMP.mass`. The boundary term contributions are evaluated within the function `Boundary_Contribution.m`. A function specifically designed to ensure the conservation of global angular momentum is included and named `Angular_Momentum_Projection.m`

BOX 9.7: PG_DYNA_LAWS segment 6 – Right hand-side of the algorithm

```
[RpBoundary, REta]  = Boundary_Contribution(FUNC,UNK,GEO,DAT,COMP);

[RF,REta] = RHS_F_Eta(UNK,GEO,DAT,COMP,FUNC,REta);
 Fdot     = Solve_System(RF,GEO,COMP);
 Etadot   = REta ./ COMP.mass;
[Rp]      = RHS_p(FUNC,UNK,GEO,DAT,COMP,Fdot,Etadot);
[Rp]      = Angular_Momentum_Projection(Rp,COMP.mass,UNK.X_AM,GEO.
            Nnode);
 Rp       = Rp + RpBoundary;
 pdot     = zeros(3,GEO.Nnode);
 pdot(1,:)= Rp(1,:) ./ COMP.mass';
 pdot(2,:)= Rp(2,:) ./ COMP.mass';
 pdot(3,:)= Rp(3,:) ./ COMP.mass';

 UNK.X    = UNK.X  + COMP.dt * COMP.irho * UNK.p;
 UNK.p    = UNK.p  + pdot   * COMP.dt;
```

(continued)

Box 9.7: *(cont.)*

```
UNK.F     = UNK.F  + Fdot   * COMP.dt;
UNK.Eta   = UNK.Eta+ Etadot * COMP.dt;

[UNK]     = Apply_strong_boundary_conditions(UNK,GEO);
```

The evaluation of the right-hand side (internal) contributions to the deformation gradient and entropy equations is carried out by the function RHS_F_Eta.m, given below.

BOX 9.8: PG_DYNA_LAWS segment 7 – Right-hand side of F and η equations

```
function [RF,REta] = RHS_F_Eta(UNK,GEO,DAT,COMP,FUNC,REta)

RF              = zeros(3,3,GEO.Nnode);

for ielem = 1:GEO.Nelem
    F_gp          = [0 0 0; 0 0 0; 0 0 0];
    H_gp          = [0 0 0; 0 0 0; 0 0 0];
    J_gp          = 0;
    Grad_v_gp     = [0 0 0; 0 0 0; 0 0 0];
    Theta_gp      = 0;
    Grad_Theta_gp = [0;0;0];
    J0gp          = DAT.ISO_J0gp(ielem);

    for in = 1:4
        Node_ID     = GEO.T(in,ielem);
        Grad0       = DAT.ISO_Grad0(in,:,ielem);
        p           = UNK.p(:,Node_ID);
        F           = UNK.F(:,:,Node_ID);
        Eta         = UNK.Eta(Node_ID);
        H           = 0.5 * tensor_cross_product(F,F);
        J           = det(F);
        Eta_R       = FUNC.Eta_R(DAT.MG_gamma, DAT.MG_q, F, H, J,...
        DAT.Den, DAT.CF, DAT.Temp_R(Node_ID), DAT, COMP);
        Theta0      = DAT.Temp_R(Node_ID) * exp(-Eta_R/DAT.Den/DAT.CF);
        Theta       = Theta0 * exp(Eta/DAT.Den/DAT.CF);
        F_gp        = F_gp          + F / 4;
        Theta_gp    = Theta_gp      + Theta / 4;
        Grad_v_gp = Grad_v_gp       + COMP_irho * p * Grad0;
        Grad_Theta_gp = Grad_Theta_gp  + Theta * Grad0';
    end

    H_gp  = 0.5 * tensor_cross_product(F_gp,F_gp);
    J_gp  = det(F_gp);
    Q_gp  = - DAT.Conduct * H_gp' * H_gp/J_gp * Grad_Theta_gp;
```

(continued)

Box 9.8: *(cont.)*

```
    for in = 1:4
        Node_ID          = GEO.T(in,ielem);
        Grad0            = DAT.ISO_Grad0(in,:,ielem);
        REta(Node_ID)    = REta(Node_ID)     + ...
        Grad0 * Q_gp / Theta_gp * J0gp - ...
        Grad_Theta_gp' * Q_gp / Theta_gp / Theta_gp / 4 * J0gp;
        RF(:,:,Node_ID)= RF(:,:,Node_ID)  + Grad_v_gp    / 4 * J0gp;
    end
end
end
```

Similarly, the (internal) right-hand side contributions for the update of the linear momentum equation are computed in the function `RHS_p.m`, which is given below.

BOX 9.9: PG_DYNA_LAWS segment 8 – Right-hand side of linear momentum equation

```
function [Rp] = RHS_p(FUNC,UNK,GEO,DAT,COMP,RF,REta)

Rp    = zeros(3,GEO_Nnode);
for ielem = 1:GEO.Nelem
    F_gp       = [0 0 0; 0 0 0; 0 0 0];
    Eta_gp     = 0;
    Fdot_gp    = [0 0 0; 0 0 0; 0 0 0];
    Fx_gp      = [0 0 0; 0 0 0; 0 0 0];
    Grad_v_gp = [0 0 0; 0 0 0; 0 0 0];
    ThetaR_gp = 0;
    J0gp       = DAT.ISO_J0gp(ielem);

    for in = 1:4
        Node_ID    = GEO.T(in,ielem);
        Grad0      = DAT.ISO_Grad0(in,:,ielem);
        p          = UNK.p(:,Node_ID);
        F          = UNK.F(:,:,Node_ID);
        Fdot       = RF(:,:,Node_ID);
        Eta        = UNK.Eta(Node_ID);
        X          = UNK.X(:,Node_ID);
        ThetaR     = DAT.Temp_R(Node_ID);
        F_gp       = F_gp         + F / 4;
        Eta_gp     = Eta_gp       + Eta / 4;
        Fdot_gp    = Fdot_gp      + Fdot / 4;
        ThetaR_gp = ThetaR_gp     + ThetaR / 4;
        Fx_gp      = Fx_gp        + X     * Grad0;
        Grad_v_gp = Grad_v_gp     + COMP_irho * p * Grad0;
    end

    C_gp     = F_gp' * F_gp;
    Stret    = eig(C_gp);
```

(continued)

Box 9.9: *(cont.)*

```
      Stret    = min(sqrt(Stret));
      cp_max   = COMP.up_temp / Stret;
      dt       = GEO.h(ielem) / cp_max;
      H_gp     = 0.5 * tensor_cross_product(F_gp,F_gp);
      J_gp     = det(F_gp);
      F_st     = F_gp + ...
        DAT.SUPG_tauF*dt*(Grad_v_gp-Fdot_gp)+DAT.SUPG_xiF*(Fx_gp-F_gp);
      H_st     = 0.5 * tensor_cross_product(F_st,F_st);
      J_st     = det(F_st);
     [P_gp]   = FUNC.Piola_kirchhoff(COMP,DAT,F_gp,H_gp,J_gp,F_st,...
            H_st,J_st,Eta_gp,ThetaR_gp);

     for in = 1:4
         Node_ID        = GEO.T(in,ielem);
         Grad0          = DAT.ISO_Grad0(in,:,ielem);
         Rp(:,Node_ID) = Rp(:,Node_ID) - (P_gp * Grad0') * J0gp;
     end
  end
  end
```

Finally, the evaluation of the external right-hand side contributions to the linear momentum and entropy equation is evaluated as part of the function `Boundary_Contribution.m`, which is shown below.

BOX 9.10: PG_DYNA_LAWS segment 9 – Boundary contribution

```
function [Rp,REta] = Boundary_Contribution(FUNC,UNK,GEO,DAT,COMP)

Rp              = zeros(3,GEO.Nnode);
REta            = zeros(GEO.Nnode,1);
for iface = 1:GEO.Nface
    Area        = GEO.A_B(iface);
    Normal      = GEO.N_B(:,iface);
    p_gp        = zeros(3,1);
    F_gp        = zeros(3,3);
    Eta_gp      = 0;
    ThetaR_gp = 0;
    Theta_gp    = 0;

    for innode = 1:3
        Node_ID = GEO.T_B(innode,iface);
        p       = UNK.p(:,Node_ID);
        F       = UNK.F(:,:,Node_ID);
        Eta     = UNK.Eta(Node_ID);
        ThetaR  = DAT.Temp_R(Node_ID);
        H       = 0.5 * tensor_cross_product(F,F);
        J       = det(F);
        Eta_R   = FUNC.Eta_R(DAT.MG_gamma, DAT.MG_q, F, H, J, ...
                  DAT.Den, DAT.CF, ThetaR, DAT, COMP);
        Theta0  = ThetaR * exp(-Eta_R/DAT.Den/DAT.CF);
        Theta   = Theta0 * exp(Eta/DAT.Den/DAT.CF);
        p_gp    = p_gp + p / 3;
```

(continued)

Box 9.10: *(cont.)*

```
            F_gp     = F_gp + F / 3;
            Eta_gp   = Eta_gp + Eta / 3;
            ThetaR_gp = ThetaR_gp + ThetaR / 3;
            Theta_gp  = Theta_gp  + Theta  / 3;
        end

        H_gp  = 0.5 * tensor_cross_product(F_gp,F_gp);
        J_gp  = det(F_gp);
        [P]    = FUNC.Piola_kirchhoff(COMP,DAT,F_gp,H_gp,J_gp,...
        F_gp,H_gp,J_gp,Eta_gp,ThetaR_gp);
        Traction = P * Normal;
        [Traction,Heat_Flux] = Type_of_boundary(Traction,Normal,F_gp,...
        GEO.Bcode(iface),DAT.ExtPres,DAT.TB,DAT.TB_Neg,COMP.time,DAT.Q);

        for in = 1:3
            Node_ID    = GEO.T_B(in,iface);
            Rp(:,Node_ID) = Rp(:,Node_ID) + Traction * Area / 3;
            REta(Node_ID) = REta(Node_ID)-Heat_Flux / Theta_gp * Area/ 3;
        end
    end
    end
```

9.5 EXAMPLES

Three test cases, designed to grow in complexity, are included. Three test cases, designed to grow in complexity, are included in this section to show the capabilities of the program PG_DYNA_LAWS. The examples help to illustrate some of the difficulties that may arise in the analysis of highly geometrically nonlinear scenarios, with and without thermal effects.

9.5.1 Three Dimensional Free Oscillating Bending Column

In this example, the performance of the program PG_DYNA_LAWS is assessed in a problem dominated by bending deformation. A 1 m by 1 m square cross-section column of height $h = 6\,\text{m}$ is clamped at the bottom and left free to oscillate on all other sides, as presented in Figure 9.1(a). The column undergoes bending deformation through the application of an initial linearly varying (in height) velocity profile given by $\boldsymbol{v}(\boldsymbol{X})|_{t=0} = [v_0 \frac{X_3}{h}, 0, 0]^T$ m/s, with $v_0 = 10\,\text{m/s}$. A neo-Hookean constitutive law is used to describe the response of the solid, for which the selected material parameters are: density $\rho_R = 1100\,\text{kg/m}^3$, Young's modulus $E = 17\,\text{MPa}$, and Poisson's ratio $\nu = 0.3$ (equivalent to Lamé parameters $\lambda = 9.87077 \cdot 10^6\,\text{N/m}^2$ and $\mu = 6.5385 \cdot 10^6\,\text{N/m}^2$). The domain is discretized with a coarse discretization comprising 288 linear tetrahedral elements (with 117 nodes) (see Figure 9.1(b) for a view of the discretization). The time step used in the Runge–Kutta time integrator is computed on the basis of a dimensionless stability coefficient $\alpha_{CFL} = 0.3$.

The time evolution of the deformation of the column is shown in Figure 9.2 through a sequence of snapshots. As can be observed, the deformation is very smooth and very well captured by the numerical simulation, despite the coarse discretization. The time history of the energy evolution is depicted in Figure 9.3(a), showing a reasonably good pattern. Figure 9.3(b) shows the energy evolution for the same problem when considering a finer discretization of 2304 tetrahedral elements. Mesh refinement clearly leads to a more accurate satisfaction of the total energy conservation.

9.5.2 Three-Dimensional Free Oscillating Twisting Column

In this example, we study the problem of a free oscillating twisting column; it was already investigated in Chapter 8 via the alternative program FLagSHyP. The geometry and boundary conditions are shown in Figure 9.4(a). A neo-Hookean constitutive law is used to describe the response of the solid, in which the selected material parameters are: density $\rho_R = 1100\,\text{kg/m}^3$, Young's modulus $E = 17\,\text{MPa}$, and Poisson's ratio $\nu = 0.3$ (equivalent to Lamé parameters $\lambda = 9.87077 \cdot 10^6\,\text{N/m}^2$ and $\mu = 6.5385 \cdot 10^6\,\text{N/m}^2$). The twisting deformation is triggered by the initial conditions defined via the velocity profile $\boldsymbol{v}(\boldsymbol{X})|_{t=0} =$

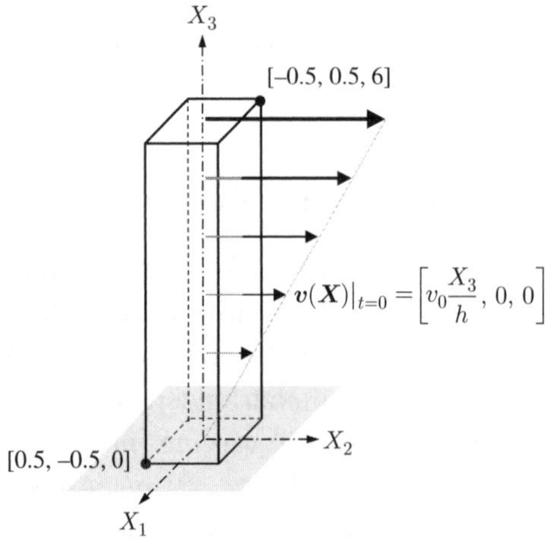

FIGURE 9.1 Bending column: (a) Diagrammatic representation of the problem; (b) Computational mesh.

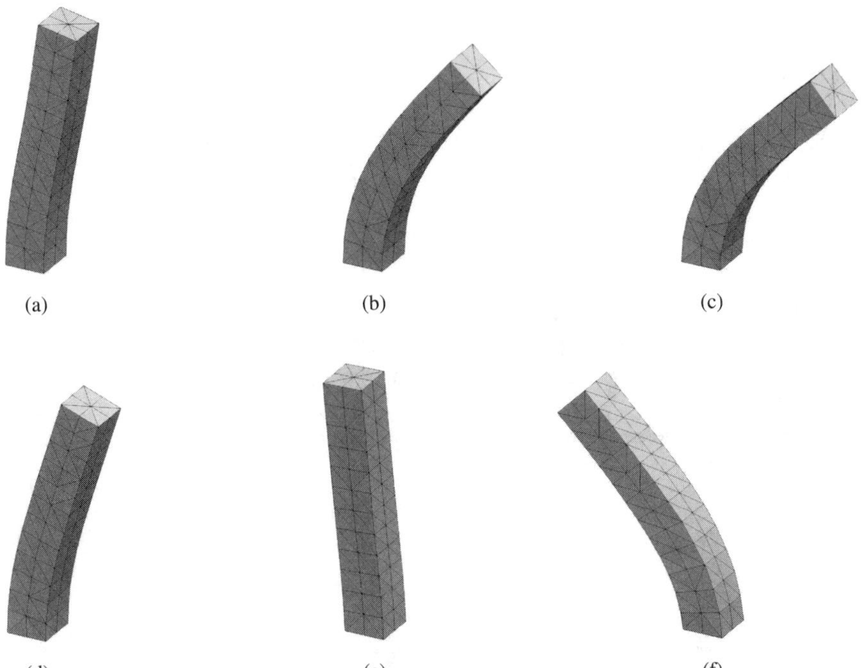

FIGURE 9.2 Bending column: deformation history. (a) $t = 0.1\,\mathrm{s}$; (b) $t = 0.3\,\mathrm{s}$; (c) $t = 0.6\,\mathrm{s}$; (d) $t = 0.9\,\mathrm{s}$; (e) $t = 1.1\,\mathrm{s}$; (f) $t = 1.5\,\mathrm{s}$.

$\boldsymbol{\omega}_0(\boldsymbol{X}) \times \boldsymbol{X}$, with $\boldsymbol{\omega}_0 = [0, 0, w_0 \sin\left(\frac{\pi X_3}{2h}\right)]\,\mathrm{rad/s}$ and $w_0 = 105\,\mathrm{rad/s}$ (see Figure 9.4(a)). The domain is discretized with two different linear tetrahedral meshes: a coarse mesh comprising 2304 elements (625 nodes) and a finer mesh of 147 456 elements (28 033 nodes). A representation of the coarse mesh is shown in Figure 9.4(b). The time step used in the Runge–Kutta time integrator is computed on the basis of a dimensionless stability coefficient $\alpha_{CFL} = 0.3$.

Figure 9.5 displays the deformation of the column, with three time snapshots for each level of discretization. As can be observed, the displayed results are very consistent regardless of the level of discretization employed. Moreover, the deformation also agrees extremely well with that displayed for the same example in Chapter 8 (see Figure 8.10). The time history of the various energy contributions is displayed in Figure 9.3 for the two discretizations. Mesh refinement clearly leads to a more accurate satisfaction of the total energy conservation.

9.5.3 Three-Dimensional L-Shaped Block

In this section, we investigate again the problem already explored in Chapter 8 using the FLagSHyP program. The problem involves the free vibration of a

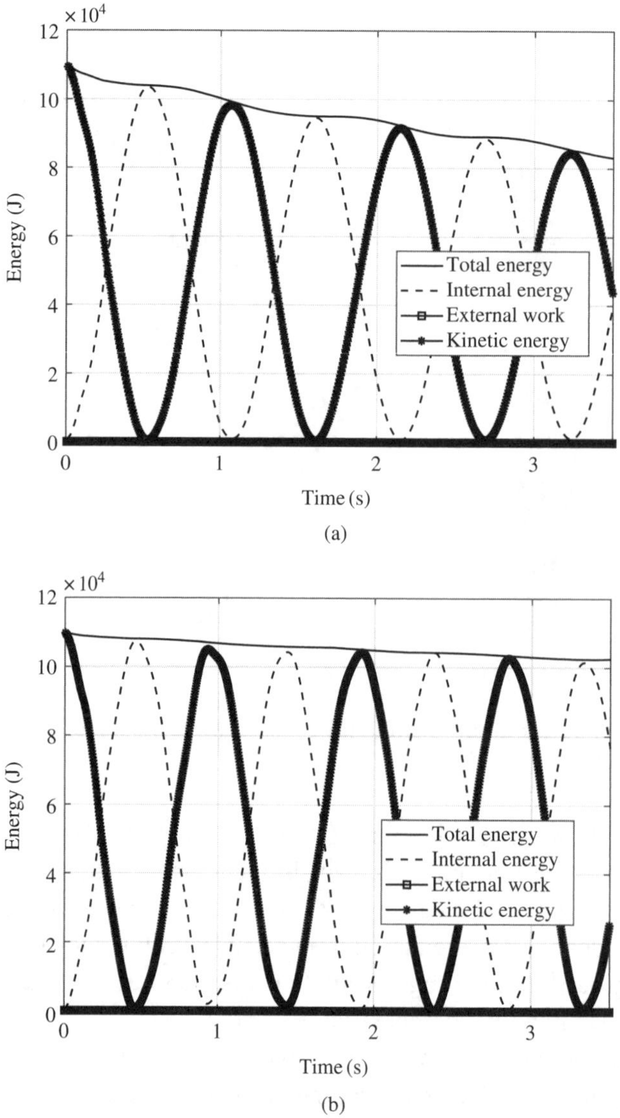

FIGURE 9.3 Bending column energy history: (a) Coarse discretization (288 elements); (b) Fine discretization (2304 elements).

hyperelastic three-dimensional deformable solid in the form of an L-shaped block. Geometry and boundary conditions are displayed in Figure 9.7(a), where globally canceling traction forces are applied on two boundary faces. For the face defined by $X_1 = 6\,\text{m}$, the surface traction takes the value $\boldsymbol{F}_1(t) = \boldsymbol{F}_0\lambda(t)$, with $\boldsymbol{F}_0 = [150, 300, 450]\,\text{N/m}^2$ and $\lambda(t)$ is a time-varying scalar function given by a hat type function, as defined in Equation (8.3), with $A = 2.5$ and

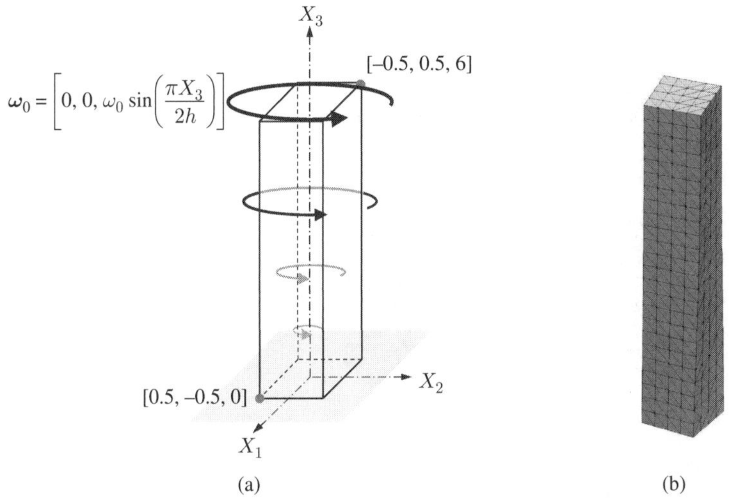

FIGURE 9.4 Twisting column: (a) Diagrammatic representation of the problem; (b) Computational mesh.

$T = 5\,\text{s}$. The surface traction acting on the face $X_2 = 10\,\text{m}$ is given by $\boldsymbol{F}_2(t) = -\boldsymbol{F}_1(t)$. The domain is discretized with 5616 linear tetrahedral elements (1323 nodes) (see Figure 9.7(b)). The time step used in the Runge–Kutta time integrator is computed on the basis of a dimensionless stability coefficient $\alpha_{CFL} = 0.3$.

For the numerical simulation, two different material models will be analyzed, namely isothermal and non-isothermal.

Isothermal Elasticity Model

In this particular case, a neo-Hookean constitutive law is used, where the material parameters are: density $\rho_R = 1100\,\text{kg/m}^3$, Young's modulus $E = 17\,\text{MPa}$, and Poisson's ratio $\nu = 0.3$ (equivalent to Lamé parameters $\lambda = 9.87077 \cdot 10^6\,\text{N/m}^2$ and $\mu = 6.5385 \cdot 10^6\,\text{N/m}^2$). Figure 9.8 displays a series of time snapshots representing the time deformation history of the problem. As can be observed, results agree very well with those obtained for the same problem in Chapter 8 (see Figure 8.13). The time evolution of the various energy contributions is displayed in Figure 9.9(a). As can be seen, for the first 5 s, during which time the surface traction forces act upon two opposing faces, the total energy of the system varies, reaching a maximum which is then maintained during the free oscillation phase.

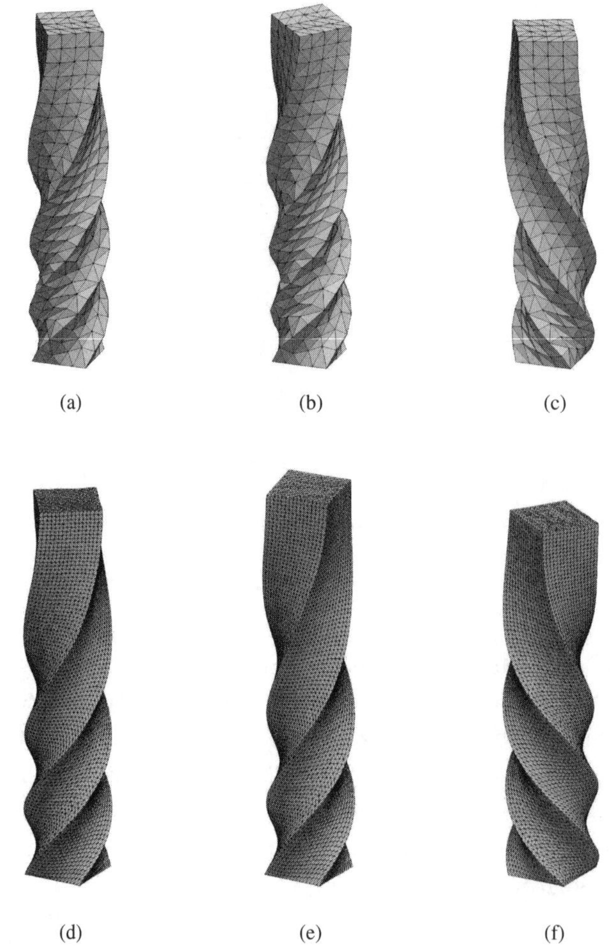

FIGURE 9.5 Twisting column: deformation history for coarse (top row) and fine (bottom row) discretizations. (a), (d) $t = 0.125\,\mathrm{s}$; (b), (e) $t = 0.1875\,\mathrm{s}$; (c), (f) $t = 0.375\,\mathrm{s}$.

Thermoelasticity Model

To examine the program in a thermoelastic scenario, a Mie–Grüneisen neo-Hookean material model is used with material parameters identical to those of the isothermal elasticity model described above, but now with the addition of extra thermal parameters, namely specific heat coefficient $c_v = 1\,\mathrm{J/(m^3\,K)}$, a thermal coefficient $\alpha = 2.223 \times 10^{-4}\,\mathrm{K}^{-1}$, a Mie–Grüneisen parameter $q = 1$, and a thermal conductivity coefficient $k = 10\,\mathrm{J/(ms\,K)}$. For this scenario, the activation of the problem through the presence of boundary traction forces is complemented by the consideration of thermal effects, namely (1) all boundary faces are treated

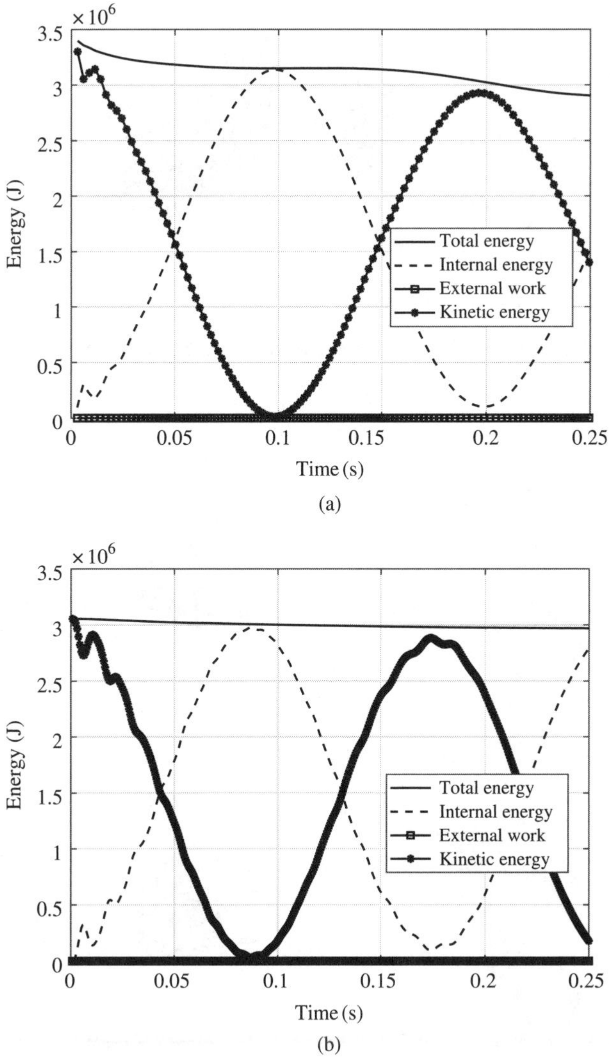

FIGURE 9.6 Twisting column energy history: (a) Coarse discretization (2304 elements); (b) Fine discretization (147 456 elements).

as thermal insulators, and (2) an initial temperature profile is considered across the domain as follows:

$$\theta(\boldsymbol{X})|_{t=0} = \begin{cases} 250\,\mathrm{K} & \text{if} \quad X_1 = 6\,\mathrm{m} \\ 300\,\mathrm{K} & \text{if} \quad X_2 = 10\,\mathrm{m} \\ 293.15\,\mathrm{K} & \quad\text{elsewhere.} \end{cases} \tag{9.1}$$

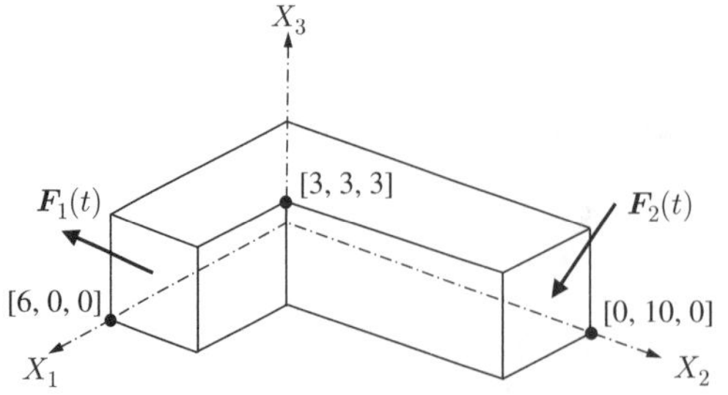

(a)

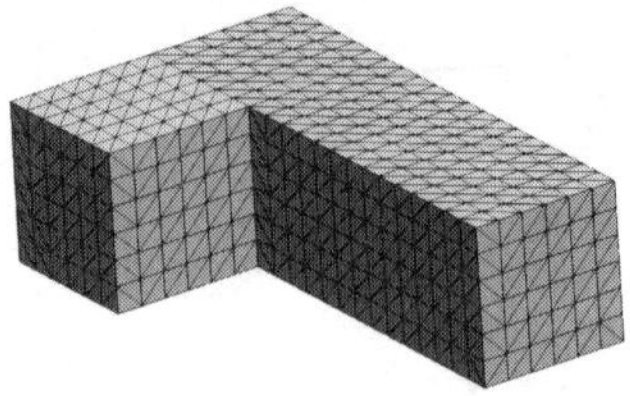

(b)

FIGURE 9.7 L-shaped block: (a) Diagrammatic representation of the problem; (b) Computational mesh.

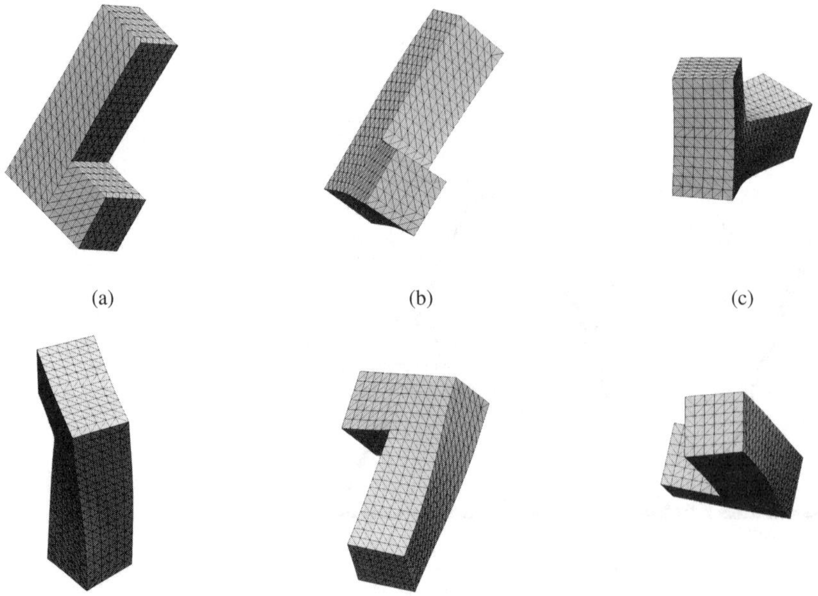

(a) (b) (c) (d) (e) (f)

FIGURE 9.8 L-shaped block: deformation history. (a) $t = 0\,\mathrm{s}$; (b) $t = 3\,\mathrm{s}$; (c) $t = 6\,\mathrm{s}$; (d) $t = 9\,\mathrm{s}$; (e) $t = 12\,\mathrm{s}$; (f) $t = 15\,\mathrm{s}$.

For the thermoelastic simulation, and in terms of deformation, results are extremely similar to those displayed in Figure 9.8, and hence are not displayed again. The time evolution of the various energy contributions is displayed in Figure 9.9(b), with very little differences to that of the isothermal case (see Figure 9.9(a)). Figure 9.10(a) displays the time evolution of the temperature in two different locations placed at the boundary faces $X_1 = 6\,\mathrm{m}$ and $X_2 =$

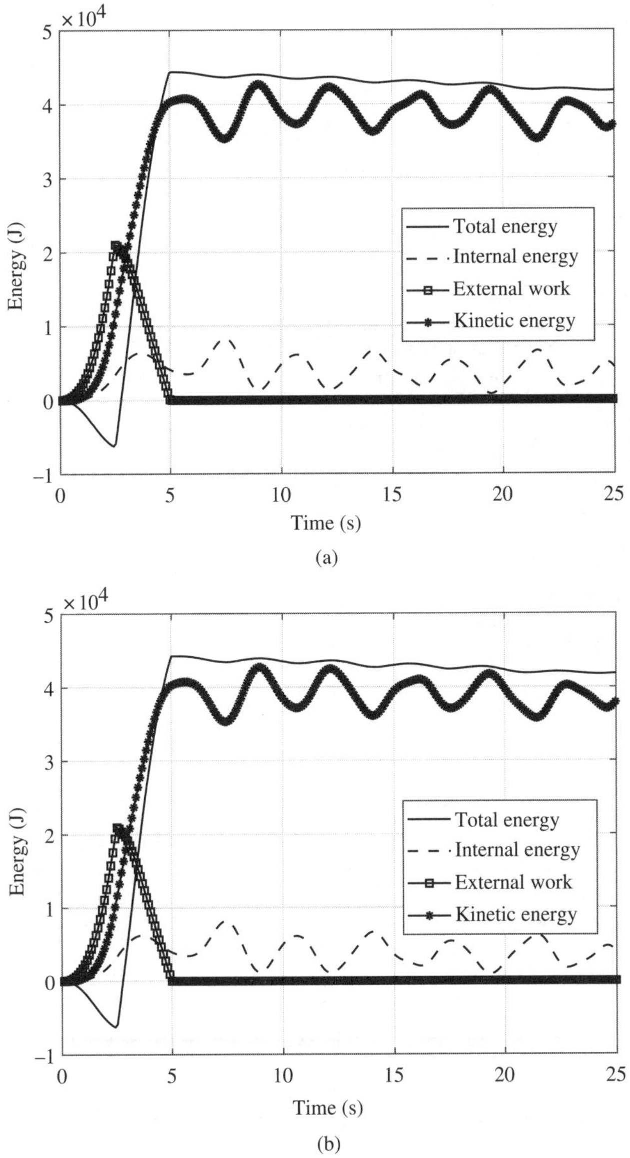

FIGURE 9.9 L-shaped block energy history: (a) Isothermal problem; (b) thermo-elastic problem.

10 m. As expected, the temperature field at these two points evolves (due to conduction) seeking a homogeneous temperature distribution. Figure 9.10(b) shows the evolution of the global entropy of the system, which grows over the time due to the conduction effect of the temperature across the domain.

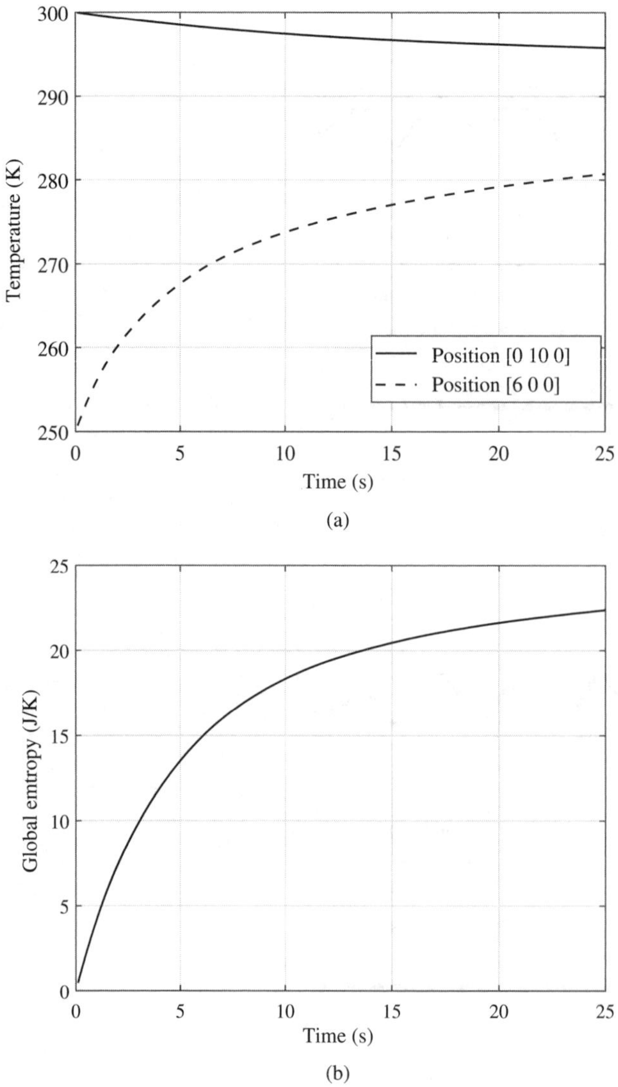

FIGURE 9.10 L-shaped block thermoelastic problem. (a) Temperature evolution; (b) Global entropy evolution.

9.6 DICTIONARY OF MAIN VARIABLES

The main variables of PG_DYNA_LAWS are listed below; they are grouped based on their parent data structure. Some other important variables are also included.

PRO.inputfile_name	-	Title of the problem
PRO.job_folder	-	Job folder
COMP.time	-	Time
COMP.dt	-	Time increment
COMP.mu	-	Shear modulus
COMP.lambda	-	Lamé parameter
COMP.kappa	-	Bulk modulus
COMP.mass	-	Lumped mass
COMP.up_temp	-	Pressure wave speed
COMP.us_temp	-	Shear wave speed
COMP.upmax	-	Maximum pressure wave speed
CON.Restart_Run	-	Restart file indicator
CON.Nstep	-	Number of time steps
CON.Type_dt	-	Type of time increment
CON.CFL	-	CFL stability number
CON.dt	-	User specified time increment
CON.OutInc	-	Output time increment
DAT.Model	-	Type of constitutive model
DAT.CF	-	Specific heat
DAT.MG_Thermal_expansion	-	Thermal expansion coefficient
DAT.Conduct	-	Heat conductivity coefficient
DAT.E	-	Young's modulus
DAT.Poisson	-	Poisson's ratio
DAT.Den	-	Material density
DAT.SUPG_tauF	-	SUPG coefficient
DAT.SUPG_xiF	-	SUPG coefficient
DAT.v	-	Initial condition for velocity
DAT.F	-	Initial condition for deformation gradient tensor
DAT.Eta	-	Initial condition for entropy
DAT.Temp_R	-	Initial condition for reference temperature
DAT.TB	-	Boundary traction vector
DAT.Q	-	Boundary normal component of heat flux
FUNC.dt_method	-	Subroutine for type of time increment used
FUNC.Piola_kirchhoff	-	Subroutine for the computational of first Piola–Kirchhoff
FUNC.Compute_Pressure	-	Subroutine for pressure computation
FUNC.Eta_R	-	Subroutine for the evaluation of entropy at reference temperature
GEO.X	-	Material coordinates
GEO.Nelem	-	Number of elements
GEO.Nnode	-	Number of nodes
GEO.T	-	Element connectivity
GEO.Nface	-	Number of boundary surfaces
GEO.T_B	-	Boundary connectivity
GEO.Bcode	-	Boundary conditions
GEO.A_B	-	Boundary area vector
GEO.N_B	-	Boundary outward unit normal vector
UNK.p	-	Linear momentum unknown variable
UNK.F	-	Deformation gradient unknown variable
UNK.Eta	-	Entropy unknown variable
UNK.X	-	Geometry

APPENDIX

SHOCKS

A.1 INTRODUCTION

Previous chapters in this text have proceeded under the assumption that the variables describing the state of the solid in motion, such as velocity $\boldsymbol{v}$, deformation gradient $\boldsymbol{F}$, or stresses $\boldsymbol{\sigma}$ or $\boldsymbol{P}$, are continuous functions throughout the solid. It is therefore possible to find their spatial derivatives as required by the divergence operators that appear in the conservation laws described in Chapter 5. This is, of course, usually the case, but situations can arise when these variables experience sudden jumps in value, that is, they become discontinuous across surfaces which move across the body. These jumps are know as *shocks* and are the result of sudden physical phenomena such as impacts or other forms of suddenly applied loads such as explosions. This appendix considers the equations that govern the transmission of these surfaces and the relationships between jumps of different variables across these surfaces which can be derived from the basic conservation laws. The material contained herein is somewhat advanced and not generally required when dealing with continuous finite element discretizations such as the Petrov–Galerkin technique presented in Chapter 7. However, there are other discretization techniques, such as discontinuous Galerkin or finite volume methods where jumps in the solution appear as a consequence of the nature of the discretization itself; hence these require the development of equations describing shocks. Moreover, even in the case of continuous approximations of variables, the correct representation of shocks requires the introduction of artificial viscosity to account for the shock effects smoothed by the absence of jumps in the discrete representation of the variables.

A.2 GENERIC SPATIAL JUMP CONDITION

Consider an arbitrary solid having a spatial volume v in which the variable $\mathcal{U}$ satisfies a general conservation law in terms of the fluxes $\boldsymbol{\mathcal{F}}$ and source vector $\mathcal{S}$, see Figure A.1, as

$$\frac{d}{dt}\int_v \mathcal{U}\, dv = \int_v \mathcal{S}\, dv - \int_{\partial v} \boldsymbol{\mathcal{F}} \cdot \boldsymbol{n}\, da. \tag{A.1}$$

Now envisage the possibility of a discontinuity in $\mathcal{U}$, $\mathcal{S}$, and $\boldsymbol{\mathcal{F}}$ across a moving surface Γ with normal $\boldsymbol{n}$ within the body and moving at a normal velocity U.* The surface Γ separates the body volume v into two parts, with v^+ ahead of the shock and v^- behind. Note that the normal $\boldsymbol{n}$ has been chosen in the direction of the motion of the shock, that is, pointing away from v^- into v^+. The jump in any variable, e.g. $\mathcal{U}$, across Γ is denoted by

$$[\![\mathcal{U}]\!] = \mathcal{U}^+ - \mathcal{U}^-. \tag{A.2}$$

If the variation of $\mathcal{U}$ and $\boldsymbol{\mathcal{F}}$ are governed by the conservation law given by Equation (A.1), it is possible to find a simple relationship between the jump in $\mathcal{U}$ and $\boldsymbol{\mathcal{F}}$ across Γ in terms of the propogation speed U and the normal $\boldsymbol{n}$. In order to show this, consider first the left-hand side of Equation (A.1) as

$$\frac{d}{dt}\int_v \mathcal{U}\, dv = \frac{d}{dt}\int_{v^+(t)} \mathcal{U}\, dv + \frac{d}{dt}\int_{v^-(t)} \mathcal{U}\, dv. \tag{A.3}$$

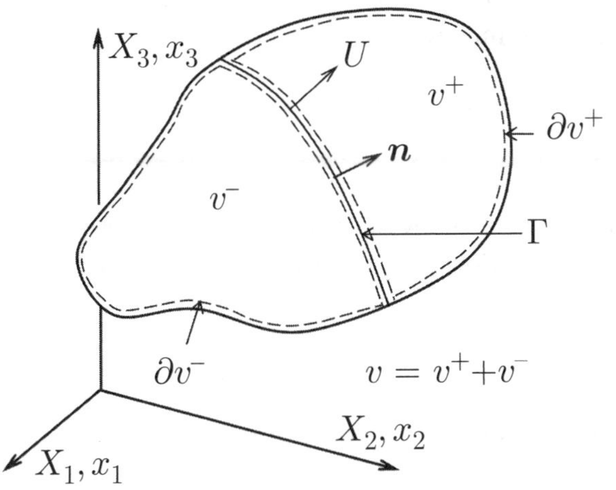

FIGURE A.1 Propagation of a shock across a solid.

* Throughout this appendix, the symbol U will always be used to represent the normal velocity of the discontinuity, not to be confused with the volumetric energy of a deformable solid, as appears in Chapters 6 and 7 or in NL-Statics.

Due to the motion of the interface Γ, the left and right volume components v^+ and v^- are not constant. In order to evaluate these time derivatives it is necessary to use the Reynolds transport theorem derived in Chapter 5, Equation (5.49), to give

$$\begin{aligned}
\frac{d}{dt}\int_v \mathcal{U}\,dv &= \frac{d}{dt}\int_{v^-(t)} \mathcal{U}\,dv + \frac{d}{dt}\int_{v^+(t)} \mathcal{U}\,dv \\
&= \int_{v^-} \frac{\partial \mathcal{U}}{\partial t}\,dv + \int_\Gamma U\mathcal{U}^-\,da + \int_{v^+} \frac{\partial \mathcal{U}}{\partial t}\,dv - \int_\Gamma U\mathcal{U}^+\,da \\
&= \int_v \frac{\partial \mathcal{U}}{\partial t}\,dv - \int_\Gamma U[\![\mathcal{U}]\!]\,da. \qquad \text{(A.4)}
\end{aligned}$$

Note that the negative sign before the last term in the middle line above is because the normal vector $\boldsymbol{n}$ points outwards for v^- but inwards for v^+ so that the outward normal vector for v^+ is $-\boldsymbol{n}$.

Consider now the last term in the conservation Equation (A.1) as

$$\begin{aligned}
\int_{\partial v} \mathcal{F}\cdot\boldsymbol{n}\,da &= \int_{\partial v^-} \mathcal{F}\cdot\boldsymbol{n}\,da - \int_\Gamma \mathcal{F}^-\cdot\boldsymbol{n}\,da + \int_{\partial v^+} \mathcal{F}\cdot\boldsymbol{n}\,da + \int_\Gamma \mathcal{F}^+\cdot\boldsymbol{n}\,da \\
&= \int_{v^-} \operatorname{div}\mathcal{F}\,dv + \int_{v^+} \operatorname{div}\mathcal{F}\,dv + \int_\Gamma [\![\mathcal{F}]\!]\cdot\boldsymbol{n}\,da \\
&= \int_v \operatorname{div}\mathcal{F}\,dv + \int_\Gamma [\![\mathcal{F}]\!]\cdot\boldsymbol{n}\,da. \qquad \text{(A.5)}
\end{aligned}$$

Substituting Equations (A.4) and (A.5) into the conservation law Eqution (A.1), after some simple algebra separating volume from surface terms, gives

$$\int_v \left(\frac{\partial \mathcal{U}}{\partial t} + \operatorname{div}\mathcal{F} - \mathcal{S}\right) dv = \int_\Gamma \left(U[\![\mathcal{U}]\!] - [\![\mathcal{F}]\!]\cdot\boldsymbol{n}\right) da. \qquad \text{(A.6)}$$

The fact that this equation can be established for arbitrary volume v^* inside v, which may or may not contain the jump component, implies that the regions where $\mathcal{U}$ and $\mathcal{F}$ are smooth and differentiable allow the standard pointwise conservation law to be recovered as

$$\frac{\partial \mathcal{U}}{\partial t} + \operatorname{div}\mathcal{F} = \mathcal{S}, \qquad \text{(A.7)}$$

together with, where applicable, the so-called jump condition

$$U[\![\mathcal{U}]\!] = [\![\mathcal{F}]\!]\cdot\boldsymbol{n}. \qquad \text{(A.8)}$$

A similar expression can be derived for the case where $\mathcal{U}$ is a vector $\boldsymbol{\mathcal{U}}$ and $\mathcal{F}$ is the tensor $\mathscr{F}$:

$$U[\![\boldsymbol{\mathcal{U}}]\!] = [\![\mathscr{F}]\!]\boldsymbol{n}. \qquad \text{(A.9)}$$

A.3 JUMP CONDITIONS IN TOTAL LAGRANGIAN SOLID DYNAMICS

The generic jump condition given by Equation (A.9) can be applied to the conservation laws in solid mechanics expressed in a total Lagrangian setting (see Equations (5.16), (5.26), and (5.22)), namely†

$$\frac{d}{dt}\int_V \boldsymbol{p}_R\, dV = \int_V \boldsymbol{f}_R\, dV + \int_{\partial V} \boldsymbol{P}\boldsymbol{N}\, dA; \quad \boldsymbol{p}_R = \rho_R \boldsymbol{v}, \tag{A.10a}$$

$$\frac{d}{dt}\int_V \boldsymbol{F}\, dV = \int_{\partial V} \boldsymbol{v} \otimes \boldsymbol{N}\, dA, \tag{A.10b}$$

$$\frac{d}{dt}\int_V E\, dV = \int_V (r_R + \boldsymbol{f}_R \cdot \boldsymbol{v})\, dV - \int_{\partial V} (\boldsymbol{Q} - \boldsymbol{P}^T\boldsymbol{v}) \cdot \boldsymbol{N}\, dA. \tag{A.10c}$$

The application of Equations (A.8) and (A.9) to the above set of conservation variables and corresponding fluxes gives

$$U[\![\boldsymbol{p}_R]\!] = -[\![\boldsymbol{P}]\!]\boldsymbol{N}, \tag{A.11a}$$

$$U[\![\boldsymbol{F}]\!] = -[\![\boldsymbol{v}]\!] \otimes \boldsymbol{N}, \tag{A.11b}$$

$$U[\![E]\!] = [\![\boldsymbol{Q}]\!] \cdot \boldsymbol{N} - [\![\boldsymbol{P}^T\boldsymbol{v}]\!] \cdot \boldsymbol{N}, \tag{A.11c}$$

where U represents the speed of the shock surface across the Lagrangian reference configuration. In the context of fluid mechanics, these jump conditions are sometimes referred to as *Rankine–Hugoniot* equations describing the behavior of a material across a shock. Equation (A.11c) is generally simplified to cases where either there is no heat flux or it is continuous, giving the jump in energy across the shock as

$$U[\![E]\!] = -[\![\boldsymbol{v} \cdot \boldsymbol{P}\boldsymbol{N}]\!]. \tag{A.12}$$

Note that the shock surface normal $\boldsymbol{N}$ is the same across Γ so it can be moved from inside to outside the jump operator $[\![\ ]\!]$ and vice versa. In addition to the jump in energy, the second law of thermodynamics expressed in volumetric form as the Clausius–Duhem inequality, and in the absence of jump in the heat flux, leads to an inequality constraint for entropy (refer to Example A.1 below):

$$[\![\eta]\!] \leq 0. \tag{A.13}$$

This equation implies that the entropy behind the shock must be greater than that ahead, that is, the shock creates entropy as it propagates. As such, this constraint

† Note that although the reference domain does not move, the shock can nevertheless still travel through it.

places restrictions on the nature of possible shocks. For instance, a shock can generally propagate only from a compressed state into an unstressed state, but it cannot propagate in the opposite direction; likewise, heat can flow only from a hot state toward a cold state. It is not possible for this to happen in reverse. The special case in Equation (A.13) where the entropy jump vanishes, that is, no entropy is generated and the discontinuity surface can travel reversibly in both directions, is strictly speaking known as a "contact wave" rather than a shock. Such cases arise almost exclusively in linear elasticity rather than hyperelasticity or thermoelasticity.

EXAMPLE A.1: Entropy jump condition

Consider the second law of thermodynamics as given by Equation (6.21):

$$\frac{d}{dt}\int_V \eta\, dV + \int_{\partial V} \frac{Q_N}{\theta}\, dA - \int_V \frac{r_R}{\theta}\, dV \geq 0.$$

Introducing a jump into the above equation, the first two terms can be rewritten as

$$\frac{d}{dt}\int_V \eta\, dV = \int_V \frac{\partial \eta}{\partial t} dV - \int_\Gamma U[\![\eta]\!]\, dA$$

and

$$\int_{\partial V} \frac{Q_N}{\theta}\, dA = \int_V \text{DIV}\left(\frac{\boldsymbol{Q}}{\theta}\right) dV + \int_\Gamma [\![\boldsymbol{Q}/\theta]\!] \cdot \boldsymbol{N}\, dA.$$

Substituting the above two equations into Equation (6.21) and re-arranging yields

$$\int_V \left(\frac{\partial \eta}{\partial t} + \text{DIV}\left(\frac{\boldsymbol{Q}}{\theta}\right) - \frac{r_R}{\theta}\right) dV - \int_\Gamma \left(U[\![\eta]\!] - [\![\boldsymbol{Q}/\theta]\!] \cdot \boldsymbol{N}\right) dA \geq 0.$$

The fact that the above inequality can be established for arbitrary volume (either the whole volume or any part of it), which may or may not contain the jump component, implies that, for those regions where η and $\boldsymbol{Q}/\theta$ are smooth and differentiable, the standard local entropy inequality holds,

$$\frac{\partial \eta}{\partial t} + \text{DIV}\left(\frac{\boldsymbol{Q}}{\theta}\right) - \frac{r_R}{\theta} \geq 0,$$

along with the jump condition

$$-\left(U[\![\eta]\!] - [\![\boldsymbol{Q}/\theta]\!] \cdot \boldsymbol{N}\right) \geq 0.$$

Finally, in the absence of jump in the heat flux component $\boldsymbol{Q}/\theta$, and recalling that the speed of the shock is considered non-negative ($U \geq 0$), it results in $[\![\eta]\!] \leq 0$.

A.4 SPEED OF PROPAGATION OF VOLUMETRIC AND SHEAR SHOCKS

Equations (A.11) provide relationships between unknown mechanical variables on either side of a shock front; in addition, they contain the speed of propagation U, which is also an unknown that needs to be determined from the equations. This can be achieved by first combining Equation (A.11a) multplied by $\boldsymbol{N}$ with Equation (A.11b):

$$\begin{aligned} U[\![\boldsymbol{F}]\!]\boldsymbol{N} &= -[\![\boldsymbol{v}]\!] \\ &= \frac{1}{\rho_R U}[\![\boldsymbol{P}]\!]\boldsymbol{N}, \end{aligned} \tag{A.14}$$

which leads to a relationship between the jump in traction vector $\boldsymbol{PN}$ and the jump of the component of the deformation gradient normal to the propagating surface $\boldsymbol{FN}$ in terms of the coefficient $\rho_R U^2$, that is,

$$\rho_R U^2[\![\boldsymbol{F}]\!]\boldsymbol{N} = [\![\boldsymbol{P}]\!]\boldsymbol{N}. \tag{A.15}$$

In order to proceed further it is necessary to examine individually the normal and tangential components of the above jump terms in relation to the Lagrangian propagation direction $\boldsymbol{N}$. For this purpose consider two arbitrary but orthogonal unit vectors $\boldsymbol{T}_1$ and $\boldsymbol{T}_2$ tangential to the propagation surface as represented in the reference configuration, and therefore orthogonal to $\boldsymbol{N}$ in the reference configuration, as shown in Figure A.2. The vectors $\boldsymbol{T}_1$ and $\boldsymbol{T}_2$ are mapped into spatial vectors

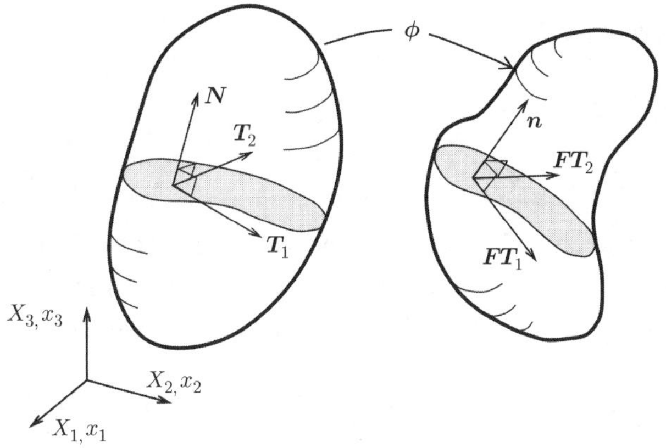

FIGURE A.2 Shock surface in reference and deformed configurations.

$\boldsymbol{FT}_1$ and $\boldsymbol{FT}_2$ in the deformed setting, thus defining the normal $\boldsymbol{n}$ to the deformed propagating surface by

$$\boldsymbol{n} = \frac{\boldsymbol{FT}_1 \times \boldsymbol{FT}_2}{\|\boldsymbol{FT}_1 \times \boldsymbol{FT}_2\|}. \tag{A.16}$$

By virtue of Equation (A.11b), $[\![\boldsymbol{FT}_{1,2}]\!] = 0$, so that this tangential vector is the same on both sides of the shock surface, and therefore $[\![\boldsymbol{n}]\!] = \boldsymbol{0}$, which is an obvious geometric imperative of the shock surface propagating in the deformed configuration. Note also that, on using Nanson's formula (see NL-Statics, Equation (4.68)) relating the elements of area in the reference and spatial configurations, $\boldsymbol{n}$ and $\boldsymbol{N}$ can also be related as follows:

$$da\,\boldsymbol{n} = J\boldsymbol{F}^{-T} dA\,\boldsymbol{N}\,; \quad \Lambda_A \boldsymbol{n} = J\boldsymbol{F}^{-T}\boldsymbol{N}\,; \quad \Lambda_A = \frac{da}{dA}. \tag{A.17a,b,c}$$

Equation (A.17a,b,c)$_b$ together with the fact that $[\![\boldsymbol{n}]\!] = \boldsymbol{0}$, implies that

$$[\![J\boldsymbol{F}^{-T}]\!]\boldsymbol{N} = \boldsymbol{0}. \tag{A.18}$$

With the help of the above geometric expressions it is possible to derive specific equations for the speed of propagation of different types of shocks, namely shear and pressure shocks. Shear shocks are discontinuities in the shear component of the traction and, correspondingly, the component of the velocity tangential to the propagating surface. In contrast, volumetric or pressure shocks represent jumps in the normal component of the traction and particle velocity. It will be shown that these two speeds are generally different, which implies that they separate into a pure pressure shock, which travels faster, followed by a pure shear shock that lags behind.

First, consider the case of a shear shock wave by multiplying the right-hand side of Equation (A.15) by a tangential vector $\boldsymbol{FT}_\alpha$, $\alpha = 1, 2$. The jump in the traction vector $[\![\boldsymbol{P}]\!]\boldsymbol{N}$ across the shock surface in the tangential direction becomes

$$\begin{aligned}
[\![\boldsymbol{P}]\!]\boldsymbol{N} \cdot \boldsymbol{FT}_\alpha &= [\![\boldsymbol{T}_\alpha \cdot \boldsymbol{F}^T\boldsymbol{P}]\!]\boldsymbol{N} \\
&= [\![\boldsymbol{F}^T\boldsymbol{P}]\!] : (\boldsymbol{T}_\alpha \otimes \boldsymbol{N}) \\
&= [\![\boldsymbol{CS}]\!] : (\boldsymbol{T}_\alpha \otimes \boldsymbol{N}),
\end{aligned} \tag{A.19}$$

where $\boldsymbol{C} = \boldsymbol{F}^T\boldsymbol{F}$ is the right Cauchy–Green tensor and $\boldsymbol{S}$ is the second Piola–Kirchhoff tensor related to $\boldsymbol{P}$ via $\boldsymbol{P} = \boldsymbol{FS}$ (from Equations (5.45a, b) in NL-Statics). Similarly, the leftnd side of Equation (A.15) projected on to the tangential direction gives

$$\begin{aligned}
\rho_R U_s^2 [\![\boldsymbol{F}]\!]\boldsymbol{N} \cdot \boldsymbol{FT}_\alpha &= \rho_R U_s^2 [\![\boldsymbol{F}^T\boldsymbol{F}]\!] : (\boldsymbol{T}_\alpha \otimes \boldsymbol{N}) \\
&= \rho_R U_s^2 [\![\boldsymbol{C}]\!] : (\boldsymbol{T}_\alpha \otimes \boldsymbol{N}),
\end{aligned} \tag{A.20}$$

where U has been replaced by U_s to denote that the speed being evaluated corresponds to a jump in the shear component of the traction vector. Combining Equations (A.19) and (A.20) gives the speed of the tangential shock as

$$\rho_R U_s^2 = \frac{[\![\boldsymbol{CS}]\!] : (\boldsymbol{T}_\alpha \otimes \boldsymbol{N})}{[\![\boldsymbol{C}]\!] : (\boldsymbol{T}_\alpha \otimes \boldsymbol{N})}. \tag{A.21}$$

As a simple example, consider the case of a neo-Hookean material for which the second Piola–Kirchhoff stress $\boldsymbol{S}$ is (see NL-Statics, Equation (6.28))

$$\boldsymbol{S} = \mu(\boldsymbol{I} - \boldsymbol{C}^{-1}) + \lambda(\ln J)\boldsymbol{C}^{-1}. \tag{A.22}$$

Substituting this expression into Equation (A.21) and noting that the tangential vector $\boldsymbol{T}_\alpha$ is orthogonal to the normal vector $\boldsymbol{N}$ (i.e. $\boldsymbol{T}_\alpha \cdot \boldsymbol{N} = 0$), yields

$$\begin{aligned} \rho_R U_s^2 &= \frac{(\mu[\![\boldsymbol{C}]\!] + \lambda[\![\ln J]\!]\boldsymbol{I}) : (\boldsymbol{T}_\alpha \otimes \boldsymbol{N})}{[\![\boldsymbol{C}]\!] : (\boldsymbol{T}_\alpha \otimes \boldsymbol{N})} \\ &= \frac{\mu[\![\boldsymbol{C}]\!] : (\boldsymbol{T}_\alpha \otimes \boldsymbol{N}) + \lambda[\![\ln J]\!](\boldsymbol{T}_\alpha \cdot \boldsymbol{N})}{[\![\boldsymbol{C}]\!] : (\boldsymbol{T}_\alpha \otimes \boldsymbol{N})} \\ &= \mu. \end{aligned} \tag{A.23}$$

Consequently, the speed of the shear wave for this simple material model emerges as

$$U_s = \sqrt{\frac{\mu}{\rho_R}}, \tag{A.24}$$

which coincides with the speed of acoustic shear waves in the material. The speed of the pressure shock can be established by multiplying both sides of Equation (A.15) by the normal vector $\boldsymbol{n}$ in the deformed configuration given by Equation (A.17a,b,c)$_b$ to give

$$\begin{aligned} \rho_R U_p^2([\![\boldsymbol{F}]\!]\boldsymbol{N}) \cdot \boldsymbol{n} &= \rho_R U_p^2 \frac{1}{\Lambda_A}[\![(J\boldsymbol{F}^{-T}\boldsymbol{N}) \cdot (\boldsymbol{F}\boldsymbol{N})]\!] \\ &= \frac{\rho_R U_p^2}{\Lambda_A}[\![\boldsymbol{N} \cdot J\boldsymbol{F}^{-1}\boldsymbol{F}\boldsymbol{N}]\!] \\ &= \frac{\rho_R U_p^2}{\Lambda_A}[\![J]\!](\boldsymbol{I} : \boldsymbol{N} \otimes \boldsymbol{N}) \\ &= \frac{\rho_R U_p^2}{\Lambda_A}[\![J]\!], \end{aligned} \tag{A.25}$$

where now U has been replaced by U_p to denote the speed of the pressure or volumetric shock.

In developing the normal component of the right-hand side of Equation (A.15) it is helpful to consider the case where the jump in the first Piola–Kirchhoff stress $\boldsymbol{P}$ is dominated by the jump in the pressure component of the stress while the deviatoric component can be neglected (see NL-Statics, Section 5.5.4). This is obviously a simplification, but it is instructive as a means of deriving a simple expression for the pressure wave speed U_p. Acknowledging the simplification, the jump in the traction vector in the normal direction becomes

$$\begin{aligned} [\![\boldsymbol{PN}]\!] \cdot \boldsymbol{n} &= [\![(\boldsymbol{P}' + pJ\boldsymbol{F}^{-T})\boldsymbol{N}]\!] \cdot \boldsymbol{n} \\ &= [\![p]\!](J\boldsymbol{F}^{-T}\boldsymbol{N}) \cdot \boldsymbol{n} \\ &= [\![p]\!]\Lambda_A \boldsymbol{n} \cdot \boldsymbol{n} \\ &= \Lambda_A [\![p]\!], \end{aligned} \tag{A.26}$$

where $\boldsymbol{P}'$ is the deviatoric component of $\boldsymbol{P}$ and $pJ\boldsymbol{F}^{-T}$ is the pressure component. In addition, Equation (A.17a,b,c)$_b$ was employed for the normal vector $\boldsymbol{n}$, and neglecting the deviatoric component implied $[\![\, \boldsymbol{P}' \,]\!] : (\boldsymbol{n} \otimes \boldsymbol{N}) = 0$. Combining Equations (A.25) and (A.26) gives

$$\frac{\rho_R U_p^2}{\Lambda_A^2} = \frac{[\![p]\!]}{[\![J]\!]}. \tag{A.27}$$

For example, in the very simplistic case of a nearly incompressible material in which the pressure is given by the linear relationship (see NL-Statics, Section 6.5.3)

$$p = \kappa(J - 1) \tag{A.28}$$

Equation (A.27) would yield

$$U_p = \Lambda_A \sqrt{\frac{\kappa}{\rho_R}}. \tag{A.29}$$

Again, this coincides with the speed of the acoustic (sound) wave in the material. This is of course an oversimplification of a generally more complex expression for U_p, which will invariably be higher than the speed of the acoustic wave, as will be explained in Section A.6.

A.5 ENERGY, TEMPERATURE, AND ENTROPY BEHIND THE SHOCK WAVE

Shock waves are physical phenomena that lead to changes in temperature and entropy as they travel through the material. For simple material models it is possible and instructive to derive the increase in temperature and entropy induced by a shock wave. The evaluation is achieved through the use of Equation (A.11c) for the jump of energy per unit undeformed volume, $[\![E]\!]$. Noting that E is the sum of the internal energy density $\mathcal{E}$ plus the kinetic energy per unit underformed volume gives

$$U[\![\mathcal{E} + \frac{1}{2}\rho_R \boldsymbol{v}\cdot\boldsymbol{v}]\!] = [\![\boldsymbol{Q}]\!]\cdot\boldsymbol{N} - [\![\boldsymbol{P}^T\boldsymbol{v}]\!]\cdot\boldsymbol{N}. \tag{A.30}$$

To proceed, it is necessary to explore the algebraic expression for the jump of the product of any two arbitrary variables a and b which can be expressed as

$$[\![ab]\!] = a_{av}[\![b]\!] + [\![a]\!]b_{av}, \tag{A.31a}$$

where

$$a_{av} = (a^+ + a^-)/2\ ;\ b_{av} = (b^+ + b^-)/2. \tag{A.31b}$$

This expression for the jump of a product can easily be confirmed by direct algebraic manipulation as

$$\begin{aligned}
[\![ab]\!] &= a_{av}[\![b]\!] + [\![a]\!]b_{av} \\
&= \frac{1}{2}\Big((a^+ + a^-)(b^+ - b^-) + (a^+ - a^-)(b^+ + b^-)\Big) \\
&= \frac{1}{2}\Big(a^+b^+ + \cancel{a^-b^+} - \cancel{a^+b^-} - a^-b^- + a^+b^+ + \cancel{a^+b^-} - \cancel{a^-b^+} - a^-b^-\Big) \\
&= a^+b^+ - a^-b^- \\
&= [\![ab]\!].
\end{aligned} \tag{A.32}$$

Assuming the simple case where the heat flow term $[\![\boldsymbol{Q}]\!] = \boldsymbol{0}$, and using Equations (A.31) repeatedly together with Equations (A.11a) and (A.11b) containing the velocity jump, yields, after some algebra, the jump in internal energy as follows:

$$\begin{aligned}
[\![\mathcal{E}]\!] &= -[\![\frac{1}{2}\rho_R\boldsymbol{v}\cdot\boldsymbol{v}]\!] - \frac{1}{U}[\![\boldsymbol{P}^T\boldsymbol{v}]\!]\cdot\boldsymbol{N} \\
&= -\rho_R\boldsymbol{v}_{av}\cdot[\![\boldsymbol{v}]\!] - \frac{1}{U}[\![\boldsymbol{v}]\!]\cdot(\boldsymbol{P}_{av})\boldsymbol{N} - \frac{1}{U}\boldsymbol{v}_{av}\cdot[\![\boldsymbol{P}\boldsymbol{N}]\!] \\
&= -\rho_R\boldsymbol{v}_{av}\cdot[\![\boldsymbol{v}]\!] - \frac{1}{U}\boldsymbol{P}_{av} : \Big([\![\boldsymbol{v}]\!]\otimes\boldsymbol{N}\Big) + \rho_R\boldsymbol{v}_{av}\cdot[\![\boldsymbol{v}]\!] \\
&= -\frac{1}{U}\boldsymbol{P}_{av} : \Big([\![\boldsymbol{v}]\!]\otimes\boldsymbol{N}\Big) \\
&= \boldsymbol{P}_{av} : [\![\boldsymbol{F}]\!].
\end{aligned} \tag{A.33}$$

For common, nearly incompressible materials of the type described in Section 6.4.7, it is useful to decompose Equation (A.33) into deviatoric and volumetric components as

$$[\![\mathcal{E}]\!] = \boldsymbol{P}'_{av} : [\![\boldsymbol{F}]\!] + (pJ\boldsymbol{F}^{-T})_{av} : [\![\boldsymbol{F}]\!]. \tag{A.34}$$

The volumetric term involving the pressure p in this equation can be simplified by noting that Equation (A.11b) implies that the jump in $[\![\boldsymbol{F}]\!]$ can be expressed as

$$[\![\boldsymbol{F}]\!] = [\![\boldsymbol{FN}]\!] \otimes \boldsymbol{N}. \tag{A.35}$$

Substituting this expression into the second term on the right-hand side of Equation (A.34) gives, after some simple algebra, the volumetric component of the energy jump as

$$\begin{aligned}(pJ\boldsymbol{F}^{-T})_{av} : [\![\boldsymbol{F}]\!] &= (pJ\boldsymbol{F}^{-T})_{av} : \Big([\![\boldsymbol{FN}]\!] \otimes \boldsymbol{N}\Big) \\ &= (pJ\boldsymbol{F}^{-T})_{av} \cdot [\![\boldsymbol{FN}]\!] \\ &= p_{av}(J\boldsymbol{F}^{-T}) \cdot [\![\boldsymbol{FN}]\!] \\ &= p_{av}[\![J(\boldsymbol{F}^{-T}\boldsymbol{F}^{T}) : \boldsymbol{N} \otimes \boldsymbol{N}]\!] \\ &= p_{av}[\![J]\!]. \end{aligned} \tag{A.36}$$

This allows the jump in internal energy to be split into deviatoric and pressure components:

$$[\![\mathcal{E}]\!] = \boldsymbol{P}'_{av} : [\![\boldsymbol{F}]\!] + p_{av}[\![J]\!]. \tag{A.37}$$

Equation (A.37) can be used in conjunction with a thermoelastic constitutive model to derive the jumps in temperature and entropy that result from the passing of a shock front. This is generally complex, and often analytical expressions are difficult to find. However, from a purely illustrative point of view it is instructive to consider a simplified case in which the deviatoric component of Equation (A.37) is ignored and the material is assumed to have a constant specific heat coefficient c_v and to satisfy the Mie–Grüneisen equation of state (see Section 6.4.8) with a $q = 1$ coefficient. Under these conditions, and combining Equations (6.46) and (6.55), the internal energy $\mathcal{E}$ can be written as

$$\begin{aligned}\tilde{\mathcal{E}}(\boldsymbol{F}, \theta) &= \mathcal{E}_R(\boldsymbol{F}) + c_v(\theta - \theta_R) \\ &= \Psi_R(\boldsymbol{F}) + \theta_R\eta_R(J) + c_v(\theta - \theta_R) \\ &= \Psi_R(\boldsymbol{F}) + \theta_R c_v\Gamma_0(J-1) + c_v(\theta - \theta_R) \\ &= \hat{\Psi}_R(\boldsymbol{F}) + \Psi_{R,vol}(J) + \theta_R c_v\Gamma_0(J-1) + c_v(\theta - \theta_R), \end{aligned} \tag{A.38}$$

where Γ_0 is the Mie–Grüneisen coefficient. From the above constitutive model the jump in internal energy can be expressed as

$$[\![\mathcal{E}]\!] = [\![\hat{\Psi}_R(\boldsymbol{F})]\!] + [\![\Psi_{R,vol}(J)]\!] + \theta_R c_v \Gamma_0 [\![J]\!] + c_v [\![\theta]\!]. \tag{A.39}$$

In addition, for the simple model under consideration, the pressure is evaluated using Equation (6.82) for $q = 1$ as

$$p(J,\theta) = p_R(J) - c_v \Gamma_0 (\theta - \theta_R), \tag{A.40}$$

and therefore the average value across the shock is (see Equation (6.82ab))

$$p_{av} = p_{R,av} - c_v \Gamma_0 (\theta_{av} - \theta_R) \; ; \; p_{R,av} = p_R(J_{av}) = \left.\frac{d\Psi_{R,vol}}{dJ}\right|_{J=J_{av}} . \tag{A.41a,b}$$

Substituting Equation (A.41a,b) into (A.37) for the energy jump, and substituting Equation (A.37) into the left-hand side of Equation (A.39), gives the relationship between the jump in temperature and the jump in the value of J as

$$\begin{aligned} c_v [\![\theta]\!] + \theta_R c_v \Gamma_0 [\![J]\!] &= \left(\boldsymbol{P}'_{av} : [\![\boldsymbol{F}]\!] - [\![\hat{\Psi}_R]\!]\right) + \left(p_{R,av}[\![J]\!] - [\![\Psi_{R,vol}]\!]\right) \\ &\quad - c_v \Gamma_0 (\theta_{av} - \theta_R) [\![J]\!]. \end{aligned} \tag{A.42}$$

For simple linear constitutive models it is possible to show that the first two brackets on the right-hand side of Equation (A.42) vanish. For instance, if $\Psi_R(J) = \kappa(J-1)^2/2$, then

$$\begin{aligned} p_{R,av}[\![J]\!] - [\![\Psi_{R,vol}]\!] &= \kappa(J_{av} - 1)[\![J]\!] - \frac{1}{2}\kappa[\![(J-1)^2]\!] \\ &= \kappa(J_{av} - 1)[\![J]\!] - \kappa(J_{av} - 1[\![J]\!]) \\ &= 0. \end{aligned} \tag{A.43}$$

Similarly, the first bracket on the right-hand side of Equation (A.43) vanishes for quadratic distorsional energy expressions, or it can be neglected for more general constitutive models as a high-order contribution. In either case, the jump in temperature can be approximated as

$$[\![\theta]\!] = -\Gamma_0 \, \theta_{av} [\![J]\!]. \tag{A.44}$$

An explicit expression for the temperature θ^- behind the shock in terms of the temperature θ^+ ahead of the shock is easily determined from Equation (A.44) by substituting $\theta_{av} = (\theta^+ - \theta^-)/2$ and rearranging terms to yield

$$\theta^- = \theta^+ \left(\frac{2 + \Gamma_0 [\![J]\!]}{2 - \Gamma_0 [\![J]\!]} \right), \tag{A.45}$$

or alternatively

$$[\![\theta]\!] = -\theta^+ \left(\frac{2\Gamma_0 [\![J]\!]}{2 - \Gamma_0 [\![J]\!]} \right). \tag{A.46}$$

Finally, the jump in entropy can be evaluated from the entropy temperature relationship given by Equation (6.36a,b) for $\eta(\boldsymbol{F}, \theta)$ and noting Equation (6.90a,b), for $q = 1$, to give

$$\begin{aligned} \eta(\theta, J) &= \eta_R(J) + c_v \ln \frac{\theta}{\theta_R} \\ &= c_v \Gamma_0 (J - 1) + c_v \ln \frac{\theta}{\theta_R}. \end{aligned} \tag{A.47}$$

Combining this equation with Equation (A.45) gives the jump in entropy as

$$[\![\eta]\!] = c_v \Gamma_0 [\![J]\!] + c_v \ln \left(\frac{2 - \Gamma_0 [\![J]\!]}{2 + \Gamma_0 [\![J]\!]} \right). \tag{A.48}$$

Using a Taylor series expansion it is possible to show that for small $[\![J]\!]$ the jump in entropy can be approximated by

$$[\![\eta]\!] \approx -\frac{1}{12} c_v \Gamma_0 [\![J]\!]^3. \tag{A.49}$$

The fact that the second law of thermodynamics implies that $[\![\eta]\!] \leq 0$, meaning that the entropy behind the shock is greater than that in front (see Equation (A.13)), leads to the conclusion that for a shock to be possible $[\![J]\!] \geq 0$. In other words, the shock has to progress from a more compressed material to a less compressed material, indicating that $J^- \leq J^+$. The reverse is not possible unless the Mie–Grüneisen coefficient $\Gamma_0 = 0$ and the material has simple quadratic energy functions, which is only the case in linear elasticity.

A.6 CONSTITUTIVE MODELS DERIVED FROM SHOCK DATA

The previous sections have used the jump Equations (A.11a)–(A.11c) together with established simple constitutive models, to determine the properties of the shock, such as its speed and temperature or entropy behind the shock. However, in practice it is far simpler to proceed in reverse, that is, to measure the properties associated with the propagation of shocks and then derive constitutive models from these measurements. This is in fact often the only viable experimental way in which to determine complex nonlinear constitutive models for certain materials. In their simplest form these experiments measure the speed of propagation of a compressive shock traveling through an undisturbed one-dimensional bar initially at rest, as

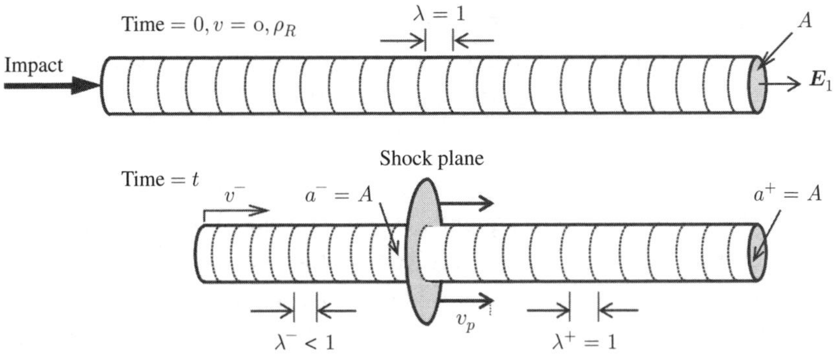

FIGURE A.3 One-dimensional rod impact test showing the transmission of a shock wave into an unstressed bar.

shown in Figure A.3. A simple uniform bar of the material being tested is impacted on the left-hand side so as to produce a sudden velocity and traction normal to the cross-sectional area. A shock wave travels along the bar until it is detected by a pressure or velocity sensor a certain distance away from the origin. In this way a measurement of the shock speed can be established for different values of the initial particle velocity applied on the left-hand side. It transpires that the shock speed grows linearly with this particle velocity and that from this simple relationship it is possible to derive useful volumetric constitutive expressions relating the pressure to the volume change J.

In the simple setting illustrated in Figure A.3, the deformation gradient $\boldsymbol{F}$ is

$$\boldsymbol{F} = \begin{bmatrix} \lambda & 0 & 0 \\ 0 & 1 & 0 \\ 0 & 0 & 1 \end{bmatrix}; \quad [\![\boldsymbol{F}]\!] = [\![\lambda]\!] \boldsymbol{E}_1 \otimes \boldsymbol{E}_1 \,; \quad [\![J]\!] = [\![\lambda]\!], \tag{A.50a,b,c}$$

where λ denotes the stretch in the direction of the bar defined by the unit vector $\boldsymbol{E}_1$. Note that at either side of the shock there is no stretch in the transverse direction (normal to the one-dimensional axis) since $a^+ = A$ and therefore $a^- = a^+ = A$. As a consequence of Equations (A.50a,b,c), the jump condition Equation (A.11c) now becomes

$$\begin{aligned} U_p [\![J]\!] &= -[\![v]\!] \\ &= -(v^+ - v^-) \\ &= v^-, \end{aligned} \tag{A.51}$$

where, recall, U_p denotes the speed of the pressure or volumetric shock and where v^- is the particle speed behind the shock in the bar. Typically, different values of

v^- can be achieved by impacting the bar at different speeds. Through experimental measurements and observations it transpires that for most metals the speed of the shock wave and v^- are related via a constant coefficient s as follows:

$$U_p = c_p + sv^-, \tag{A.52}$$

where c_p is the speed of sound. Consequently, for very weak shocks where v^- is close to zero in comparison with c_p, U_p is given by the speed of sound. This is known as the *acoustic limit*. For more severe shocks, that is, when v^- is significant with respect to the speed of sound c_p, $U_p > c_p$ as the dimensionless coefficient s is always positive.

The thermoelastic state behind the shock described in Figure A.3 is usually known as the Hugoniot state, and variables such as pressure, temperature, or entropy are typically denoted as $p_H = p^-$, $\theta_H = \theta^-$, and $\eta_H = \eta^-$ respectively. The subindex H indicates values at the Hugoniot state, that is, behind a shock propagating into undisturbed material. Note that, since the shock moves into undisturbed material, $p^+ = 0$, $\theta^+ = \theta_R$, and $\eta^+ = 0$.

In order to make use of Equation (A.52) for the purpose of developing a $p(J, \eta)$ constitutive model, it is first necessary to rewrite this equation as a relationship between the shock speed and the jump in volumetric ratio $[\![J]\!] = 1 - J^-$. To this end, we combine Equations (A.51) and (A.52) to give

$$\begin{aligned} U_p &= c_p + sv^- \\ &= c_p + sU_p[\![J]\!], \end{aligned} \tag{A.53}$$

from which U_p emerges through elimination as

$$U_p = \frac{c_p}{1 - s[\![J]\!]}. \tag{A.54}$$

Note that $s[\![J]\!] < 1$ for this equation to give physically meaningful values of the shock speed. An expression for the value of the pressure behind the shock can now be derived by recalling Equation (A.27) involving U_p, which, in the absence of cross-sectional area changes in the bar, now becomes

$$\frac{[\![p]\!]}{[\![J]\!]} = \rho_R U_p^2 = \rho_R \frac{c_p^2}{(1 - s[\![J]\!])^2}. \tag{A.55}$$

Observing that $[\![p]\!] = 0 - p_H = -p_H$ and $[\![J]\!] = 1 - J$, where J instead of J^- has been used for simplicity to denote the volume ratio behind the shock, enables the Hugoniot pressure versus J relationship to be derived as

$$p_H(J) = \rho_R c_p^2 \frac{J - 1}{\left[1 - s(1 - J)\right]^2}. \tag{A.56}$$

This relationship establishes the pressure behind a one-dimensional shock traveling into a material at rest as a function of the volumetric compression J (note that $J < 1$), where the solid mechanics convention of p positive in tension is still in use. Equation (A.56) can be related to pressure versus temperature or pressure versus entropy constitutive models by noting that $p_H(J) = p(J, \theta_H) = p(J, \eta_H)$, where θ_H and η_H are the temperature and entropy behind the shock, respectively.

Generally, the evaluation of $\theta_H(J)$ and $\eta_H(J)$ cannot be performed analytically, even after making assumptions such as ignoring the shear strength of the material. An interesting and simple alternative, however, emerges from the observation that the jump in entropy across a shock is proportional to the jump in J to the power of 3; see Equation (A.49). Hence, for moderate shocks, where the jump $[\![J]\!]$ is significantly smaller than unity, it is possible to neglect the jump in entropy and take $p_H(J)$ as the isentropic (constant entropy) equation of state $p_0(J) \approx p_H(J)$. This approximation is correct to second order in $(J - 1)$ and hence provides a useful practical model for moderate compression beyond the linear range. The resulting $p(J, \eta)$ equation of state is given by introducing $p_H(J)$ as $p_0(J)$ into Equation (6.77) to give

$$\begin{aligned} p(J, \eta) &\approx p_H(J) + c_v \theta_0'(J)\left(e^{\eta/c_v} - 1\right) \\ &= \rho_R c_p^2 \frac{J - 1}{\left[1 - s(1 - J)\right]^2} + c_v \theta_0'(J)\left(e^{\eta/c_v} - 1\right), \end{aligned} \tag{A.57}$$

where $\theta_0'(J)$ can be obtained for a simple Mie–Grüneisen thermomechanical coupling model (see Equations (6.83a,b) and (6.90a,b) as

$$\theta_0'(J) = \theta_R \Gamma_0 J^{q-1} e^{c_v \Gamma_0 \left(\frac{1 - J^q}{q}\right)}. \tag{A.58}$$

For the simple case where the Mie–Grüneisen coefficient $q = 1$, the resulting $p(J, \eta)$ equation of state becomes

$$p(J, \eta) = \rho_R c_p^2 \frac{J - 1}{\left[1 - s(1 - J)\right]^2} - c_v \theta_R \Gamma_0 e^{c_v(1-J)}\left(e^{\eta/c_v} - 1\right). \tag{A.59}$$

A.7 CONTACT-IMPACT CONDITIONS

One of the most common situations that leads to the generation of shocks takes place when solids impact against each other. Contact between two solids traveling at different speeds leads to a common speed and traction vector at the point of contact and two shock waves propagating into each solid, as illustrated in Figure A.4. The propagation of the shock in the direction normal to the contact is also shown in Figure A.5, where x_n denotes a coordinate along the direction $\boldsymbol{n}$ and p_n and t_n

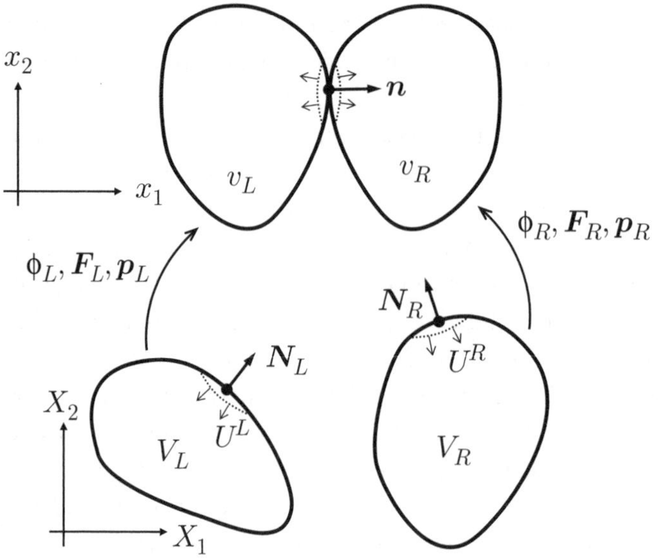

FIGURE A.4 Impact between two solids in two dimensions, showing the resulting shock waves.

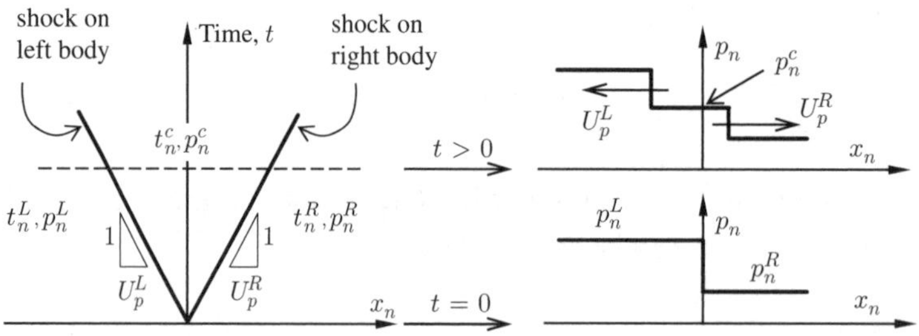

FIGURE A.5 One-dimensional representation, on the space-time plane $x_n - t$, of impact shocks generated as a result of contact between two bodies, and momentum versus x_n profiles at $t = 0$ and $t > 0$.

denote the normal components of the momentum vector $\boldsymbol{p}$ and traction vector $\boldsymbol{t}$, respectively, at the point of impact defined as

$$p_n = \boldsymbol{p} \cdot \boldsymbol{n}\,; \quad t_n = \boldsymbol{t} \cdot \boldsymbol{n} = (\boldsymbol{P}\boldsymbol{N}) \cdot \boldsymbol{n}. \tag{A.60a,b}$$

Both these values are likely to be different for the left and right bodies *just before* contact and will therefore be denoted as p_n^L and t_n^L for the left body and p_n^R and t_n^R for the right body. After contact, the normal velocities must be equal to prevent penetration, and the magnitude of the normal tractions will also be equal due to Newton's third law of action and reaction. The common values after contact will be denoted by p_n^c and t_n^c, where for simplicity equal densities of the left and right bodies have been assumed.

In the context of conservation laws, the evaluation of the common contact normal momentum (or velocity) and traction is known as the solution of a "Riemann problem." This is governed by the jump conditions across the two shocks, obtained by applying Equation (A.11a) twice, as follows:

$$U_p^L [\![\boldsymbol{p}]\!]_L = -[\![\boldsymbol{P}]\!]_L(-\boldsymbol{N}_L)\,; \quad U_p^R [\![\boldsymbol{p}]\!]_R = -[\![\boldsymbol{P}]\!]_L(-\boldsymbol{N}_R). \tag{A.61a,b}$$

Note that the negative signs in front of $\boldsymbol{N}_L$ and $\boldsymbol{N}_R$ are necessary as the shocks propagate into the body in directions opposite to $\boldsymbol{N}_L$ and $\boldsymbol{N}_R$. Multiplying Equations (A.61a,b) by the unique contact normal vector $\boldsymbol{n}$ shown in Figure A.4 gives

$$U_p^L (p_n^L - p_n^c) = t_n^L - t_n^c, \tag{A.62a}$$

$$U_p^R (p_n^R - p_n^c) = -(t_n^R - t_n^c). \tag{A.62b}$$

The difference in sign between Equations (A.62a) and (A.62b) is because $\boldsymbol{n}$ is normal to the surface of the left body and hence $t_n^R = -\boldsymbol{n} \cdot \boldsymbol{P}_R \boldsymbol{N}_R$ and $t_n^c = -\boldsymbol{n} \cdot \boldsymbol{P}_c \boldsymbol{N}_R$, whereas $t_n^L = \boldsymbol{n} \cdot \boldsymbol{P}_L \boldsymbol{N}_L$ and $t_n^c = \boldsymbol{n} \cdot \boldsymbol{P}_c \boldsymbol{N}_L$. Equations (A.62) represent a system of two equations for two unknowns t_n^c and p_n^c in terms of the left and right normal tractions and momenta before the impact and the speeds of the shocks after impact. Unfortunately, the shock speeds in the reference configuration in general will also be a function of the unknowns t_n^c and p_n^c, rendering the system of equations nonlinear. For instance, in the case described in Section A.6, where shock speed grows linearly with the particle velocity, Equations (A.62) become a system of quadratic equations. A much simpler case emerges when the speed of the shock is taken as constant in both bodies after impact and equal to the speed of sound c_p. In the material model presented in Section A.6, this implies $s = 0$, in which case Equations (A.62), usually referred to as an "acoustic Riemann solver," give an analytical solution for p_n^c and t_n^c as

$$p_n^c = \frac{c_p^L p_n^L + c_p^R p_n^R}{c_p^L + c_p^R} + \frac{t_n^R - t_n^L}{c_p^L + c_p^R}, \tag{A.63a}$$

$$t_n^c = \frac{c_p^L c_p^R}{c_p^L + c_p^R} \left(\frac{t_n^L}{c_p^L} + \frac{t_n^R}{c_p^R} \right) + \frac{c_p^L c_p^R}{c_p^L + c_p^R} (p_n^R - p_n^L). \tag{A.63b}$$

For the particular case where $c_p^L = c_p^R = c_p$, these expressions become

$$p_n^c = \frac{1}{2}(p_n^L + p_n^R) + \frac{1}{2c_p}(t_n^R - t_n^L), \tag{A.64a}$$

$$t_n^c = \frac{1}{2}(t_n^L + t_n^R) + \frac{1}{2}c_p(p_n^R - p_n^L). \tag{A.64b}$$

These equations take into account only the normal components of linear momentum (i.e. velocity) and traction. In the situation where friction is present to a sufficient degree to prevent relative sliding, similar common tangential components of the velocity and traction can be derived. Consequently, shear shocks are also generated. If $\boldsymbol{p}_t$ and $\boldsymbol{t}_t$ are the tangential components of the linear momentum and traction, respectively, defined as

$$\boldsymbol{p}_t = \boldsymbol{p} - (\boldsymbol{p} \cdot \boldsymbol{n})\boldsymbol{n}\,; \quad \boldsymbol{t}_t = \boldsymbol{t} - (\boldsymbol{p} \cdot \boldsymbol{n})\boldsymbol{n}, \tag{A.65a,b}$$

and the shear shock speed is assumed constant, a similar derivation for the common tangential traction and momentum gives

$$\boldsymbol{p}_t^c = \frac{c_s^L \boldsymbol{p}_t^L + c_s^R \boldsymbol{p}_t^R}{c_s^L + c_s^R} + \frac{\boldsymbol{t}_t^R - \boldsymbol{t}_t^L}{c_s^L + c_s^R}, \tag{A.66a}$$

$$\boldsymbol{t}_t^c = \frac{c_s^L c_s^R}{c_s^L + c_s^R}\left(\frac{\boldsymbol{t}_t^L}{c_s^L} + \frac{\boldsymbol{t}_t^R}{c_s^R}\right) + \frac{c_s^L c_s^R}{c_s^L + c_s^R}(\boldsymbol{p}_t^R - \boldsymbol{p}_t^L), \tag{A.66b}$$

where c_s^L and c_s^R are the left and right body shear shock speeds. Finally, the full common linear momentum and traction vectors post impact at the contact point can be evaluated as

$$\boldsymbol{p}_c = p_c \boldsymbol{n} + \boldsymbol{p}_t^c\,; \quad \boldsymbol{t}_c = t_c \boldsymbol{n} + \boldsymbol{t}_t^c. \tag{A.67a,b}$$

Exercises

1. Consider the dynamic equilibrium of a rod moving in a one-dimensional direction under the action of a force per unit length $f_L(X)$. Conservation of linear momentum in integral form is expressed as

$$\frac{d}{dt}\int_{X_A}^{X_B} \rho_L v\, dX = \int_{X_a}^{X_b} f_L\, dX + \big(T(X_B) - T(X_A)\big),$$

where ρ_L is the reference linear density (mass per unit initial length), $T(X)$ is the resultant internal traction at point X, and X_A and X_B denote the left and right boundaries of the rod in the Lagrangian reference configuration. Show that the corresponding conservation laws and jump conditions are

$$\frac{\partial(\rho_L v)}{\partial t} = \frac{\partial T}{\partial X} + f_L,$$

$$\frac{\partial \lambda}{\partial t} = \frac{\partial v}{\partial X},$$

$$U[\![\rho_L v]\!] = -[\![T]\!],$$

$$U[\![\lambda]\!] = -[\![v]\!].$$

2. Starting from the Eulerian conservation laws given in Chapter 5 (see for instance Table 5.1), show that the Eulerian jump conditions are,

$$u[\![\rho]\!] = (\rho \boldsymbol{v}) \cdot \boldsymbol{n},$$

$$u[\![\rho \boldsymbol{v}]\!] = [\![\rho \boldsymbol{v} \otimes \boldsymbol{v} - \boldsymbol{\sigma}]\!]\boldsymbol{n},$$

$$u[\![e]\!] = [\![\boldsymbol{q} + e\boldsymbol{v} - \boldsymbol{\sigma}\boldsymbol{v}]\!]\boldsymbol{n},$$

where u is the spatial speed of the shock and $\boldsymbol{n}$ is the normal direction to the shock surface. Show also that the jump in the inverse deformation gradient becomes

$$u[\![\boldsymbol{F}^{-1}]\!] = [\![\boldsymbol{F}^{-1}\boldsymbol{v}]\!] \otimes \boldsymbol{n}.$$

3. For the one-dimensional bar model derived in Exercise 1, show that the assumption that U varies linearly with the jump in velocity as $U = c + sv^-$, where v^- is the velocity behind a shock transmitting into the undisturbed material (see Section A.7), leads to an equation for the Hugoniot traction as

$$T_H = \rho_L c^2 \frac{(\lambda - 1)}{[1 + s(\lambda - 1)]^2}.$$

4. Starting from the Lagrangian jump equation for the energy given by Equation (A.37), show that, for a material in which the shock speed increases linearly with the jump in normal velocity as described in Section A.6, the energy behind a volumetric shock propagating into undisturbed material is

$$\mathcal{E}_H(J) = \frac{1}{2}\rho_R c_p^2 \frac{(J-1)^2}{[1 + s(J-1)]^2}.$$

5. Consider the frictionless impact of two solids made of the same material of Lagrangian density ρ_R. Both solids are initially unstressed: one is at rest and the other travels at constant velocity v normal to the impact surface. If the material model is defined by the fact that the shock propagation speed increases linearly with the jump in speed in the normal direction through a coefficient s, as explained in Section A.6, show that Equations (A.62) imply that the common post impact normal velocity is

$$v_n^c = \frac{1}{2}v,$$

and that the normal traction is

$$t_n = -\frac{1}{2}\rho_R(c_p v + \frac{1}{2}sv^2).$$

BIBLIOGRAPHY

[1] BATHE, K-J., *Finite Element Procedures in Engineering Analysis*, Prentice Hall, 1996.

[2] BELYTSCHKO, T., LIU, W. K., and MORAN, B., *Nonlinear Finite Elements for Continua and Structures*, John Wiley and Sons, 2000.

[3] BONET, J., GIL, A. J., LEE, C. H., AGUIRRE, M., and ORTIGOSA, R., A first order hyperbolic framework for large strain computational solid dynamics. Part I: Total Lagrangian isothermal elasticity, *Comput. Meths. Appl. Mech. Engrg.*, **283**, 689–732, 2014.

[4] BONET, J., GIL, A. J., and ORTIGOSA, R., A computational framework for polyconvex large strain elasticity, *Comput. Meths. Appl. Mech. Engrg.*, **283**, 1061–1094, 2015.

[5] BROOKS, A. N. and HUGHES, T. J. R., Streamline upwind/Petrov–Galerkin formulations for convection dominated flows with particular emphasis on the incompressible Navier–Stokes equations, *Comput. Meths. Appl. Mech. Engrg.*, **32**, 199–259, 1982.

[6] CRISFIELD, M. A., *Non-Linear Finite Element Analysis of Solids and Structures*, John Wiley & Sons, Volume 1, 1991.

[7] ERINGEN, A. C. and SUHUBI, E. S., *Elastodynamics (Volume 1, Finite Motions)*, Academic Press Inc., 1974.

[8] GIL, A. J., LEE, C. H., BONET, J., and AGUIRRE, M., A stabilised Petrov-Galerkin formulation for linear tetrahedral elements in compressible, nearly incompressible and truly incompressible fast dynamics, *Comput. Meths. Appl. Mech. Engrg.*, **276**, 659–690, 2014.

[9] GIL, A. J., LEE, C. H., BONET, J., and ORTIGOSA, R., A first order hyperbolic framework for large strain computational solid dynamics. Part II: Total Lagrangian compressible, nearly incompressible and truly incompressible elasticity, *Comput. Meths. Appl. Mech. Engrg.*, **300**, 146–181, 2016.

[10] GONZALEZ, O., Exact energy and momentum conserving algorithms for general models in nonlinear elasticity, *Comput. Meths. Appl. Mech. Engrg.*, **190**, 1763–1783, 2000.

[11] GONZALEZ, O. and STUART, A. M., *A First Course in Continuum Mechanics*, Cambridge University Press, 2008.

[12] GURTIN, M. E., *An Introduction to Continuum Mechanics*, Academic Press, 1981.

[13] GURTIN, M. E, FRIED, E., and ANAND, L., *The Mechanics and Thermodynamics of Continua*, Cambridge University Press, 2010.

[14] HILBERT, H. M., HUGHES, T. J. R., and TAYLOR, R., Improved numerical dissipation for time integration algorithms in structural dynamics, *Earthquake Engrg. Struct. Dyn.*, **5**, 283–292, 1977.

[15] HOLZAPFEL, G. A., *Nonlinear Solid Mechanics: A Continuum Approach for Engineering*, John Wiley & Sons, 2000.

[16] HUGHES, T. J. R., *The Finite Element Method*, Prentice Hall, 1987.

[17] HUGHES, T. J. R., FRANCA, L. P., and MALLET, M., A new finite element formulation for computational fluid dynamics: I. Symmetric forms of the compressible Euler and Navier–Stokes equations and the second law of thermodynamics, *Comput. Meths. Appl. Mech. Engrg.*, **54**, 223–234, 1986.

[18] HUGHES, T. J. R. and PISTER, K. S., Consistent linearization in mechanics of solids and structures, *Compt. & Struct.*, **8**, 391–397, 1978.

[19] HUGHES, T. J. R., SCOVAZZI, G., and FRANCA, L. P., Multiscale and stabilized methods, in Stein, E., de Borst, R., and Hughes, T. J. R., eds., *Encyclopedia of Computational Mechanics*, John Wiley and Sons, 2004.

[20] KANE, C., MARSDEN J. E., ORTIZ, M., and WEST, M., Variational integrators and the Newmark algorithm for conservative and dissipative mechanical systems, *Int. J. Num. Meth. Engrg.*, **49**, 1295–1325, 2000.

[21] LAHIRI S. K., BONET J., and PERAIRE, J., A variationally consistent mesh adaptation method for triangular elements in explicit Lagrangian dynamics, *Int. J. Num. Meth. Engrg.*, **82**, 1073–1113, 2010.

[22] LAHIRI S. K., BONET J., PERAIRE, J., and CASALS, L., A variationally consistent fractional time-step integration method for incompressible and nearly incompressible Lagrangian dynamics, *Int. J. Num. Meth. Engrg.*, **63**, 1371–1395, 2005.

[23] LEE, C. H., GIL, A. J., and BONET, J., Development of a stabilised Petrov-Galerkin formulation for conservation laws in Lagrangian fast solid dynamics, *Comput. Meths. Appl. Mech. Engrg.*, **268**, 40–64, 2014.

[24] MALVERN, L. E., *Introduction to the Mechanics of Continuous Medium*, Prentice Hall, 1969.

[25] MARSDEN, J. E. and HUGHES, T. J. R., *Mathematical Foundations of Elasticity*, Prentice Hall, 1983.

[26] MIEHE, C., Aspects of the formulation and finite element implementation of large strain isotropic elasticity, *Int. J. Num. Meth. Engrg.*, **37**, 1981–2004, 1994.

[27] ODEN, J. T., *Finite Elements of Nonlinear Continua*. McGraw-Hill, 1972. Also Dover Publications, 2006.

[28] OGDEN, R. W., *Non-Linear Elastic Deformations*, Ellis Horwood, 1984.

[29] ROMERO, I., An analysis of the stress formula for energy-momentum methods in nonlinear elastodynamics, *Comp. Mech.*, **50**, 603–610, 2012.

[30] SANZ SERNA, J. M., Symplectic integrators for Hamiltonian problems: an overview, *Acta Numerica*, **18**, 243–286, 1991.

[31] SIMMONDS, J. G., *A Brief on Tensor Analysis*, Springer, 2nd edition, 1994.

[32] SIMO, J. C. and TARNOW, N., The discrete energy-momentum method. Conserving algorithms for nonlinear elastodynamics, *Z. Angew. Math. Phys.*, **43**, 757–793, 1992.

[33] SIMO, J. C. and TARNOW, N., A new energy and momentum conserving algorithm for the non-linear dynamics of shells, *Int. J. Num. Meth. Engrg.*, **37**, 2527–2549, 1994.

[34] SPENCER, A. J. M., *Continuum Mechanics*, Longman, 1980.

[35] SUSSKIND, L. and HRABOVSKY, G., *Classical Mechanics – The Theoretical Minimum*, Penguin, 2013.

[36] WEBER, G. and ANAND, L., Finite deformation constitutive equations and a time integration procedure for isotropic, hyperelastic-viscoplastic solids, *Comput. Meths. Appl. Mech. Engrg.*, **79**, 173–202, 1990.

[37] ZIENKIEWICZ, O. C. and TAYLOR, R. L., *The Finite Element Method*, McGraw-Hill, 4th edition, Volumes 1 and 2, 1994.

INDEX